Gene Delivery

DRUGS AND THE PHARMACEUTICAL SCIENCES

A Series of Textbooks and Monographs

Series Editor

Anthony J. Hickey

RTI International, Research Triangle Park, USA

The Drugs and Pharmaceutical Sciences series is designed to enable the pharmaceutical scientist to stay abreast of the changing trends, advances and innovations associated with therapeutic drugs and that area of expertise and interest that has come to be known as the pharmaceutical sciences. The body of knowledge that those working in the pharmaceutical environment have to work with, and master, has been, and continues, to expand at a rapid pace as new scientific approaches, technologies, instrumentations, clinical advances, economic factors and social needs arise and influence the discovery, development, manufacture, commercialization and clinical use of new agents and devices.

Recent Titles in Series

Gene Delivery: Nanotechnology and Therapeutic Applications, Yashwant Pathak

Gene Delivery Systems, Yashwant Pathak

Percutaneous Absorption: Drugs, Cosmetics, Mechanisms, Methods, Nina Dragićević and Howard Maibach

Handbook of Pharmaceutical Granulation Technology, Dilip M. Parikh

Biotechnology: the Science, the Products, the Government the Business, Ronald P. Evens

Filtration and Purification in the Biopharmaceutical Industry, Third Edition, Maik W. Jornitz

Handbook of Drug Metabolism, Third Edition, Paul G. Pearson and Larry Wienkers

The Art and Science of Dermal Formulation Development, Marc Brown and Adrian C. Williams

Pharmaceutical Inhalation Aerosol Technology, Third Edition, Anthony J. Hickey and Sandro R. da Rocha

Good Manufacturing Practices for Pharmaceuticals, Seventh Edition, Graham P. Bunn

Pharmaceutical Extrusion Technology, Second Edition, Isaac Ghebre-Sellassie, Charles E. Martin, Feng Zhang, and James Dinunzio

Biosimilar Drug Product Development, Laszlo Endrenyi, Paul Declerck, and Shein-Chung Chow

High Throughput Screening in Drug Discovery, Amancio Carnero

Generic Drug Product Development: International Regulatory Requirements for Bioequivalence, Second Edition, Isadore Kanfer and Leon Shargel

Aqueous Polymeric Coatings for Pharmaceutical Dosage Forms, Fourth Edition, Linda A. Felton

Good Design Practices for GMP Pharmaceutical Facilities, Second Edition, Terry Jacobs and Andrew A. Signore

Handbook of Bioequivalence Testing, Second Edition, Sarfaraz K. Niazi

For more information about this series, please visit: www.crcpress.com/Drugs-and-the-Pharmaceutical-Sciences/book-series/IHCDRUPHASCI

Gene Delivery

Nanotechnology and Therapeutic Applications

Edited by
Yashwant Pathak

CRC Press
Taylor & Francis Group
Boca Raton London New York

CRC Press is an imprint of the
Taylor & Francis Group, an **informa** business

First edition published 2022
by CRC Press
6000 Broken Sound Parkway NW, Suite 300, Boca Raton, FL 33487-2742

and by CRC Press
4 Park Square, Milton Park, Abingdon, Oxon, OX14 4RN

CRC Press is an imprint of Taylor & Francis Group, LLC

Library of Congress Cataloging-in-Publication Data
Names: Pathak, Yashwant, editor.
Title: Gene delivery : nanotechnology and therapeutic applications / edited by Yashwant Pathak.
Description: First edition. | Boca Raton : Taylor and Francis, 2022. | Series: Drugs and the pharmaceutical sciences | Includes bibliographical references and index.
Identifiers: LCCN 2021059391 (print) | LCCN 2021059392 (ebook) | ISBN 9781032025544 (hardback) | ISBN 9781032029771 (paperback) | ISBN 9781003186083 (ebook)
Subjects: LCSH: Gene therapy. | Drug delivery systems. | Nanomedicine.
Classification: LCC RB155.8 .G455 2022 (print) | LCC RB155.8 (ebook) | DDC 615.8/95--dc23/eng/20211206
LC record available at https://lccn.loc.gov/2021059391
LC ebook record available at https://lccn.loc.gov/2021059392

ISBN: 978-1-032-02554-4 (hbk)
ISBN: 978-1-032-02977-1 (pbk)
ISBN: 978-1-003-18608-3 (ebk)

DOI: 10.1201/9781003186083

Typeset in Times
by SPi Technologies India Pvt Ltd (Straive)

To the loving memories of my parents and Dr Keshav Baliram Hedgewar, who gave proper direction to my life; to my beloved wife Seema who gave positive meaning and my son Sarvadaman who gave a golden lining to my life.

I would like to dedicate this book to the loving memories of Ma Chamanlaljee, Ma Lakshmanraojee Bhide and Ma Madhujee Limaye who mentored me selflessly and helped me to become a good and socially useful human being.

Yashwant Pathak

Contents

Preface

Gene delivery is a transport of genes of therapeutic values into the chromosomes of the cells or tissues that can be targeted to replace faulty genes. In the last two decades, many research efforts are dedicated to gene delivery for therapeutic applications. Today, gene therapy is a promising approach in the treatment of genetic diseases, including mitochondrial related diseases like blindness, muscular dystrophy, cystic fibrosis, and some cancers. Since the first clinical trial for a treatment of combined immune deficiencies in the 1990s; over 3000 clinical trials have been approved.

The delivery of genes is very complex. It entails traversing many biological barriers, and it will take years to resolve the scientific and technical issues to exploit the full potential of gene therapy. People have tried to repair genetic defects through the injection of naked plamid DNA and found it not very effective due to poor transfection efficiency. Viral vectors have shown a promising role in providing high efficiency for DNA delivery.

In recent years, nanotechnology is offering the potential for the successful delivery of genes for therapeutic applications. An extensive set of nanoparticle-based delivery systems have been reported, including polymeric nanoparticles, silica-based hybrids, gold nanoparticle-based hybrids, two-dimensional nanomaterials, and lipid-based nanoparticles. Some of the interesting nano systems that are under clinical trials include DOTAP cholesterol, polyethylenimine nanoparticles, poly (ethylene glycol) polyethylenimine–cholesterol (PEG PEI Cholesterol) PEI mannose dextrose nanoparticles, and so on.

With new developments in gene therapy using messenger RNA (mRNA), which is a new method of providing therapeutic proteins for cells, it can realize the expression of therapeutic genes in non-dividing cells without the need for nuclear entry. mRNA delivery is a potential option to solve the risk of mutagenesis, which would provide an alternative for DNA based gene therapy. The significant potential of gene therapy technologies got an enormous boost with the use of different kinds of nano particles. The scientific community needs to put more effort toward overcoming existing barriers to realize the clinical applications of gene therapy for treating diseases and disorders.

In the recent pandemic, two companies used the gene-based mRNA technology to develop vaccine for Covid-19 and successfully administered millions of doses around the world. Using mRNA as a medicine is a fundamentally different approach than treating disease with other drug classes. The mRNA plays a fundamental role in human biology. It is the set of instructions by which cells make all proteins and send them to various parts of the body. The mRNA medicines take advantage of normal biological processes to express proteins and create a desired therapeutic effect. This enables the potential treatment of a broad spectrum of diseases, many of which cannot be addressed with current technologies. Vaccines that are mRNA based are delivered using lipid-based nanoparticles. Lipid-based nanoparticles are one of the most extensively explored conduits for gene delivery owing to their optimal properties, including high biocompatibility and close resemblance to the lipidic membranes, which facilitate their penetration into the cells.

Seven scientists in the United States and Britain, who have come up with a revolutionary gene therapy cure for a rare genetic form of childhood blindness, won a 1 million euro ($1.15 million) prize in 2018 from Portugal's Champalimaud Foundation. Their gene augmentation therapy involved the delivery of healthy genes by using engineered, harmless viruses, described by the foundation as "an elegant solution."

A significant boost to gene therapy research was given in 2020 with the Noble prize in Chemistry given to Emmanuelle Charpentier of the Max Planck Unit for the Science of Pathogens and Jennifer Doudna of the University of California, Berkeley, for their discovery of the CRISPR/Cas9 genetic scissors that have revolutionized genome editing.

It is believed that mRNA has the potential to transform how medicines are discovered, developed, and manufactured – at a breadth, speed, and scale not common in our industry.

This book is dedicated to exploration of nanotechnology for gene therapy and gene delivery. This book has 15 chapters, written by prominent authors, covering various aspects, including challenges in delivering gene therapy, advances in genome editing, RNA-based gene therapy, green nanoparticles for oligonucleotide delivery, development of mRNA vaccines for Covid -19 in challenging pandemic scenario, clinical applications of gene therapy for immunodeficiencies, and so on.

I am extremely indebted to all the authors who expended a great deal of effort to complete the chapters on time and made sure that all the aspects related to gene delivery systems using nanotechnology delivery systems, are covered.

I am aware that gene delivery is a vast area, and it will be hard to cover everything related to this area of research, but our lead authors of the chapters have done phenomenal work and deserve kudos from the readers.

I am extremely thankful to Ms Hillary Lafoe and Tony Hickey of Taylor & Francis for encouraging me to edit this book. Several people from Taylor & Francis have helped me to bring this book to market, including Ms. Danielle Zarfati, Ms. Laura Piedrahita, and many others from the production team.

I would like to express my sincere thanks to my university authorities and the administration of the Taneja College of pharmacy, who always supported my endeavors in editing books. I will be failing in my duties if I do not mention my sincere thanks to my family who always must compromise their time for such efforts.

Thanks go to our readers. They are the ultimate judge to evaluate our efforts and provide feedback to improve the book. I am sure it will be a very useful reference book for the people who are working in this field. If there are any suggestions to improve, please do not hesitate to e-mail me the corrections so they can be implemented in the second edition.

Yashwant Pathak, M Pharm, Executive MBA, MSCM, PhD
Professor and Associate Dean for Faculty Affairs, USF Health Taneja College of Pharmacy, University of South Florida, 12901 Bruce B Downs Blvd, MDC 030, Tampa FL 33612, USA and Adjunct Professor, Faculty of Pharmacy, University of Airlangga, Surabaya Indonesia
E mail: ypathak1@usf.edu

Editor

Dr. Yashwant Pathak completed his Ph.D. in Pharmaceutical Technology from Nagpur University, India and his EMBA MS in Conflict Management from Sullivan University. He is Professor and Associate Dean for Faculty Affairs at the College of Pharmacy, University of South Florida. Tampa, Florida. With extensive experience in academia as well as industry, he has more than 100 publications, two patents and two patent applications, twenty edited books published, including seven books in Nanotechnology and six in Nutraceuticals and drug delivery systems. He has several books in cultural studies and conflict management. He has received several National and International awards. His recent assignment was Fulbright Senior Fellow in Indonesia for six months, where he traveled in seven countries and lectured in more than twenty universities. He was a key note speaker at the World Hindu Wisdom meeting in Bali, Indonesia, in June 2017 and co-chaired with the Governor of Bali. Dr Yashwant Pathak is also a Professor of Global Health with the College of Public Health at USF. Recently, his student completed a PhD in Global Health on homeopathy medicine and its efficacy, a survey of 1274 patients in Pune in India. Dr Pathak is also an adjunct faculty member at AirLangga University College of Public Health at Surabaya Indonesia.

List of Contributors

Ashish Kumar Agrawal
Department of Pharmaceutical Engineering & Technology
Indian Institute of Technology (Banaras Hindu University)
Varnasi, Uttar Pradesh, India

Md Meraj Anjum
Department of Pharmaceutical Engineering & Technology
Indian Institute of Technology (Banaras Hindu University)
Varnasi, Uttar Pradesh, India

Mansi N. Athalye
L. M. College of Pharmacy
Ahmedabad, India

Sonia Barua
Chung Ang University
Seoul, Republic of Korea

Praful D. Bharadia
L. M. College of Pharmacy
Ahmedabad, India

Kunjan B. Bodiwala
L. M. College of Pharmacy
Ahmedabad, India

Aiswarya Chaudhuri
Department of Pharmaceutical Engineering & Technology
Indian Institute of Technology (Banaras Hindu University)
Varnasi, Uttar Pradesh, India

Uma G. Daryai
L M College of Pharmacy
Ahmedabad, India

Khushboo Faldu
Department of Pharmacology
Institute of Pharmacy
Nirma University
Ahmedabad, Gujarat, India

Disha Jain
OPS Biological Scientist IV I Florida Department of Health
College of Public Health I University of South Florida
Tampa, Florida, USA

Devika S. Joshi
Department of Ophthalmology
Deenanath Mangeshkar Hospital & Research Center
Pune, India

Shrikant D. Joshi
Department of Ophthalmology
Deenanath Mangeshkar Hospital & Research Center
Pune, India.

Sakshi Gurbani
Department of Pharmacology
Institute of Pharmacy
Nirma University
Ahmedabad, Gujarat, India

Gaurav M. Karve
Department of Molecular Genetics
Deenanath Mangeshkar Hospital & Research Center
Pune, India.

Samarth Kumar
Formulation- R & D
Sun Pharmaceutical Industries Ltd.
Vadodara, Gujarat, India

Vineet Mahajan
Research Faculty, Department of Pathology
Pittsburgh Liver Research Center
University of Pittsburgh
Pittsburgh, Pennsylvania, USA

Mustakim M. Mansuri
L. M. College of Pharmacy
Ahmedabad, India

Dulla Naveen Kumar
Department of Pharmaceutical Engineering & Technology
Indian Institute of Technology (Banaras Hindu University)
Varnasi, Uttar Pradesh, India

Prachi Pandey
Babaria Institute of Pharmacy
BITS Edu campus
Vadodara, Gujarat, India

Komal Parmar
ROFEL
Shri G.M. Bilakhia College of Pharmacy
Vapi, Gujarat, India

Jayvadan Patel
Nootan Pharmacy College, Dean
Faculty of Pharmacy and R&D and Innovation
Sankalchand Patel University
Visnagar, Gujarat, India

Manish P. Patel
L. M. College of Pharmacy
Ahmedabad, India

Yashwant Pathak
Taneja College of Pharmacy
University of South Florida
Tampa, Florida, USA
Adjunct Professor
Faculty of Pharmacy
Airlangga University
Surabaya, Indonesia

Shruti Saptarshi
Taneja College of Pharmacy
University of South Florida
Tampa, Florida, USA

Jigna Shah
Department of Pharmacology
Institute of Pharmacy
Nirma University
Ahmedabad, Gujarat, India

Vandit Shah
Department of Pharmacology
L. M. College of Pharmacy
Ahmedabad, Gujarat, India

Divya Sheth
Department of Biomedical Engineering
University of South Florida
Tampa, Florida, USA

Sanjay Singh
Department of Pharmaceutical Engineering & Technology
Indian Institute of Technology (Banaras Hindu University)
Varnasi, Uttar Pradesh, India

Dhwani Thakkar
Department of Pharmacology
Institute of Pharmacy
Nirma University
Ahmedabad, Gujarat, India

1 Challenges in Delivering Gene Therapy

Divya Sheth
University of South Florida, Tampa, USA

Yashwant Pathak
University of South Florida, Tampa, USA
Airlangga University, Surabaya, Indonesia

CONTENTS

1.1 INTRODUCTION

Gene therapy at the simplest level is still a diverse and highly researched topic in the field of medicine. With years of research and many more years to come, unpacking the unknown within gene therapy holds wonders for the future. With such praise of gene therapy, the topic must be unpacking and explaining for conventional use. A simple overview of gene therapy encompasses the experimental nature of the technique coupled with the use of genes for possible treatment or prevention of disease. This method is non-surgical and drug free, as it allows the doctor to insert genes into patient's cells, ultimately treating or prevent sickness [1]. While this technique is still experimental, researchers have taken a plethora of complex and multidimensional approaches in hopes of further advancing the knowledge of gene therapy.

1.2 EXPERIMENTAL CONCEPTS OF GENE THERAPY

With all the research behind gene therapy, scientists continue to progress. New concepts, experimental methods, and clinical trials of gene therapy show many advancements and data collection

DOI: 10.1201/9781003186083-1

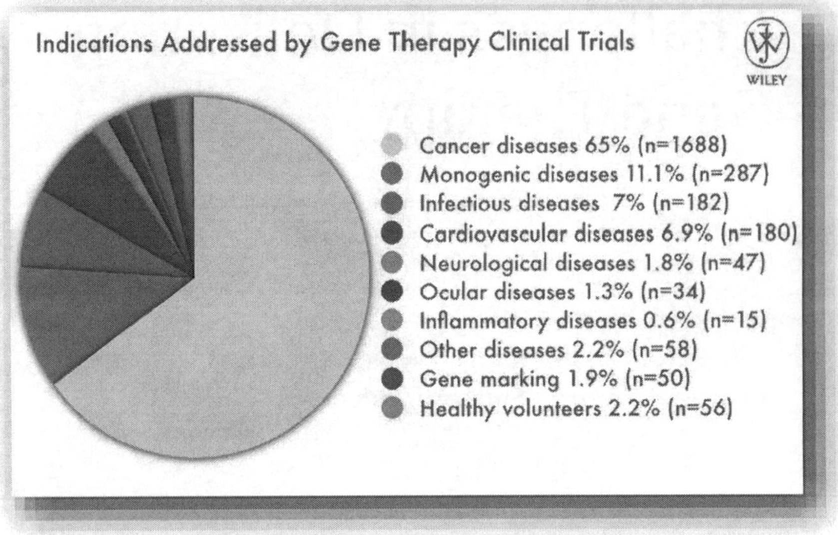

FIGURE 1.1 Indications addressed by gene therapy.

(Adapted from http://www.wiley.co.k/genmed/clinical.)

for future reference. One experimental method of using gene therapy constitutes the use of a healthy copy of a gene to replace a mutated gene that may currently or in the future cause harm to the body. This method would consist of essentially replacing or substituting the mutated gene for a new, healthy one. Other experimental approaches introduce the concept of the addition or removal of genes. One of the approaches takes a mutated gene, which is functioning improperly, and inactivates it, thus eliminating the function completely. This technique follows the basis of suppression, which can be used for cancer scenarios. Another approach introduces a new gene into the environment of the body to counteract the mutated gene's function, thus contributing to the prevention and active immunity against disease [1]. The use of another gene to make the disease more relevant to the body is such that the gene stimulates the body. All the methods stated above are experimental methods and concepts of gene therapy that can target multiple locations. Below, Figure 1.1 shows the percentage of diseases that are addressed by gene therapy clinical trials.

From Figure 1.1, it is shown that gene therapy is mostly used for cancer diseases, while a small percent is for experimental purposes with the use of healthy volunteers.

1.3 AIMS OF GENE THERAPY

Gene therapy has an overall aim that depends on the approach used. The main goal of gene therapy is to use genes to correct mutations that arise from mechanisms involving DNA within the human body. It also counts for the mutations that occur from pathogenic substances, such as virions and bacterium [2]. From the approaches stated in the previous sections, gene therapy can be divided into three distinct categories based on technique. Each technique has the main purpose of correcting a mutation, but the method of correction yields a variety of mechanisms. In one of the techniques, the mutated gene is inactivated or suppressed. One major cancer treatment is cancer suppressor gene therapy, which refers to when genes function to inhibit cell proliferation and have regulation of development, growth, and differentiation of cells [3]. Suppressing this gene, for this instance, might be better suited to being suppressed rather than replaced by a healthy copy. Just like this technique, other techniques work in different mechanisms. Another example would be substituting mutated genes for healthy genes. This would cause the body to slowly correct the expression of the abnormal gene to expression normal function.

1.4 DELIVERY SYSTEMS OF GENE THERAPY

With different treatments incorporating gene therapy, multiple mechanisms have been tested for a variety of diseases the human body may experience. The most notable use of gene therapy is to use a delivery system that delivers gene therapy to a targeted cell. Delivery systems are mechanisms of carriers for a controlled release of a specific therapy that is targeted for a specific cell. As more research is conducted on gene therapy, biomedical engineers are making progress on different delivery systems for different types of gene therapy. Delivery systems essentially control the rate at which the therapy is released and where it is released in the body. Two major categories make up the components of delivery systems. These categories can be classified by the route of delivery, which refers to the mode of medication delivery, and the vehicle with its cargo, which refers to the carrier and the treatment in simple terms [4]. Modes of medication delivery are not complicated. The vehicle is sent into the body by several modes of transport. Many people can relate to swallowing, inhaling, injecting, or rubbing medication on. Currently, there are several vehicles or vectors of delivery systems used in gene therapy. Some of the delivery systems used by gene therapy include retroviral, lentiviral, adenoviral, and non-viral vectors [3].

1.5 VECTORS OF GENE THERAPY

The vectors of gene therapy are sorted into two categories: viral and nonviral. Both vectors are used in gene therapy, but the each has its own unique purpose. Viral vectors are comprised of altered viruses that have been specifically modified to become carrying vessels. These vectors are modified to carry treatment and release a specific amount of treatment at a certain location. While viral vectors use DNA or RNA as their genetic makeup, modification are generally made to the genetic makeup, due to the possibility of the virus infecting the host cell. Helper DNA from the virus contains genes crucial for viral replication of the DNA that is delivered by a plasmid into the host chromosomal DNA. From there, the DNA is replicated and translated into the final vector for use in gene therapy [5]. Below is a visual representation of how the virus DNA is modified and how the new vector is created.

Figure 1.2 refers to a generic method of engineering a virus into a carrying vector. As more specific viral vectors are used, differences are often presented. In the previous section, the different viral vectors of gene therapy were mentioned. These vectors can now be covered in more depth after discussing the overall concept of viral vectors.

Gene therapy usually has two main ways of delivery: in vivo and ex vivo. Another name for these delivery systems would be "direct delivery" for in vivo and "cell-based delivery" for ex-vivo delivery. In in vivo, the work performed, in this instant the delivery of gene therapy, is performed within the natural condition of the organism or quite literally within the organism. Ex-vivo refers to the opposite, which would be outside the living organism [6]. Ex-vivo can almost be compared to that of invitro. Invitro, refers to the work within a set environment, such as a test tube. Ex-vivo gene therapy would use the concept of invitro, as ex-vivo refers to cells being taken out of the body and then transduced with the gene in a test tube, invitro, and then simply readministered to the body, where the gene expression can start to occur. Below in the figure is a summary of direct and cell-based delivery (Figure 1.3).

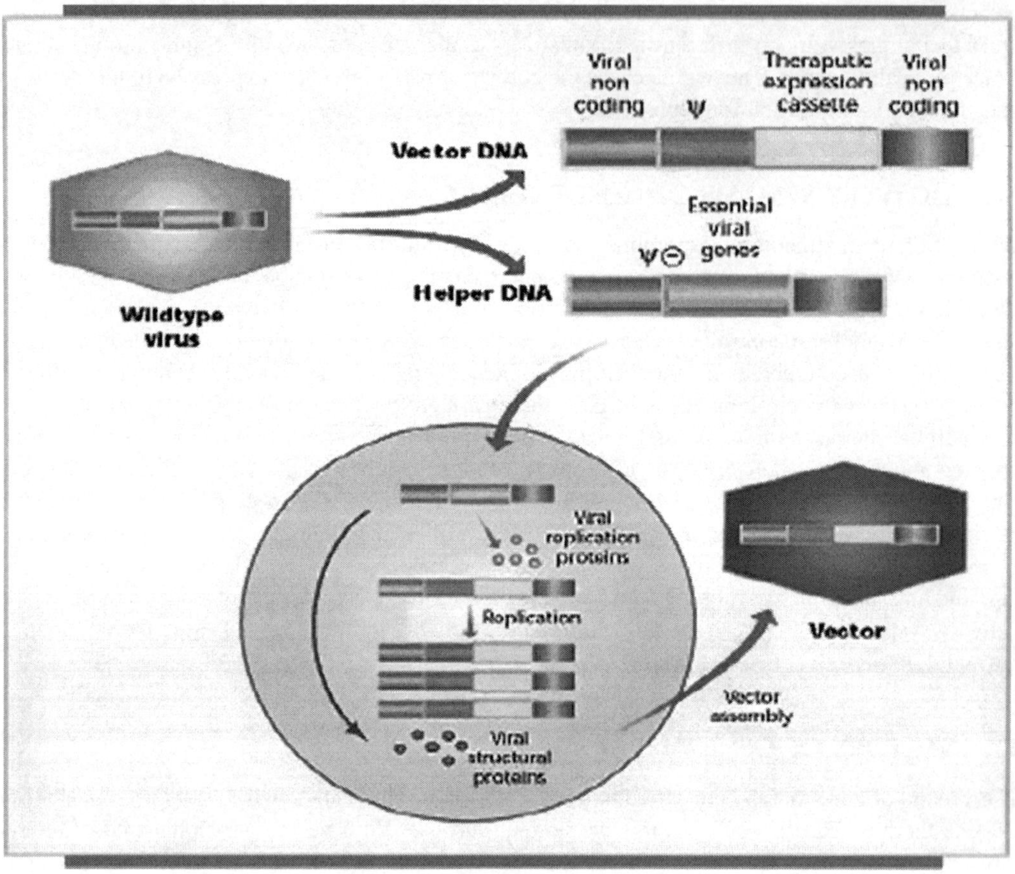

FIGURE 1.2 Engineering strategy for the modification of a virus into a vector.

(Adapted from "Viral vectors for gene therapy: the art of turning infectious agents into vehicles of therapeutics".)

1.6 RETROVIRAL VECTORS

Retroviruses are a subsection of viruses, which can place a copy of its own single stranded RNA makeup into the DNA of the virion's host cell. These viruses can retro transcribe their single-stranded RNA by retro-transcribing it into linear double-stranded DNA, which causes a seamless insertion of the virus and its contents with little immunological repercussions, if any. An advantage of retroviral vectors in gene therapy is the ability to retro-transcribe their RNA into DNA, which when integrated into the host cell's DNA, is a permanent modification [7]. With this permeant modification, it leads to an effective way to maintain self-renewing genetic material. However, the integrated DNA does not warrant stable expression of those transduced genes. In situations where a permeant outcome is the most feasible option, a retroviral vector could be coupled with targeted gene therapy for a possible treatment option.

1.7 LENTIVIRAL VECTORS

Taking a step deeper into retroviral vectors, Lentiviral vectors are a subsection of retroviruses. Essentially, lentiviruses are a specific type of retrovirus that rely on active transport of the preinitiation complex through the nuclear body of the targeted cell [5]. This method is especially useful

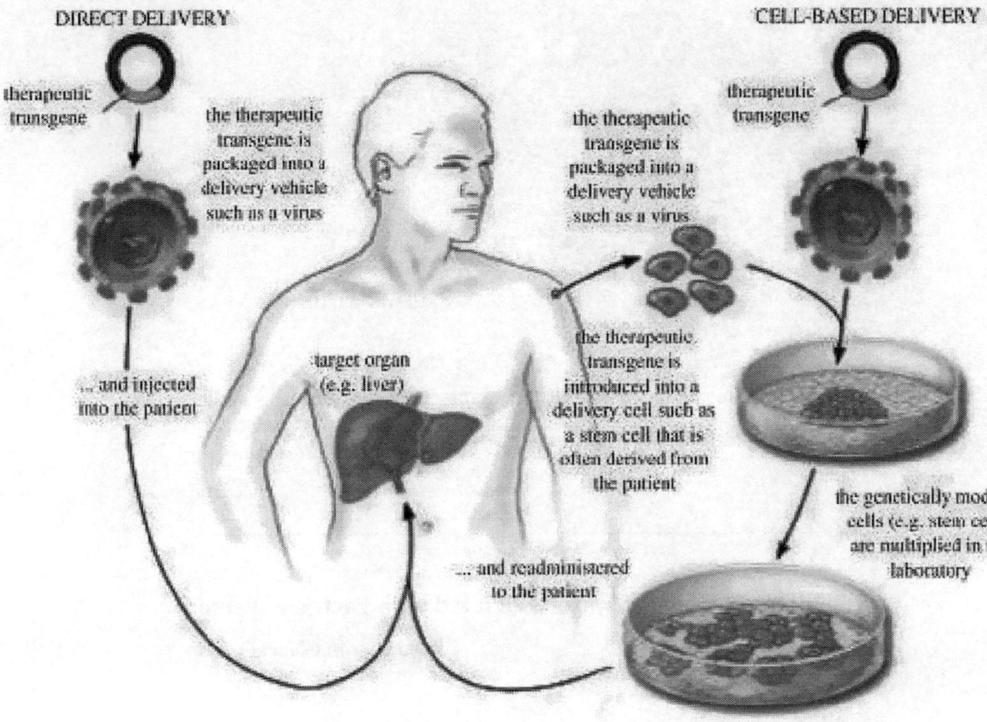

FIGURE 1.3 Summary of direct and cell-based delivery.

(Adapted from "Use of genetically modified stem cells in experimental gene therapies".)

in gene therapy, due to the nuclear facilitation of infection in non-dividing cells, as well as mitotic cells. HIV-1 is one of the most notable and studied lentiviruses. Since lentiviral vectors and retroviral vectors are similar in many concepts, it would be generalized that they have a similar system of delivery. First, the lentivirus has an interaction with the target cell and binds to the viral envelope glycoprotein. Once everything is bonded, reverse transcriptase takes the single strand RNA and retro-transcribes to a new double stranded DNA. As the newly synthesized DNA is completed, the DNA integrates to the host DNA, which is essential for gene expression. With the use of a lentiviral vector, it can integrate its newly synthesized DNA into the DNA of the genome of non-dividing cells, not capable of regular generic retroviral vectors [8, 9]. Lastly, the structural proteins and viral RNA are condensed and packaged into small particles, which are released into the plasma membrane [10]. From here, the lentiviral vector can be used in gene therapy, and the affects can be tested.

1.8 ADENOVIRAL VECTORS

The last subfamily of retroviruses are adenoviruses, another vector used in gene therapy. Adenoviruses are non-enveloped and contain an icosahedral nucleocapsid that contains a double stranded DNA genome [11]. A unique thing about adenoviruses is that their DNA is flaked on both ends, which are called inverted terminal repeats or ITRs. This helps act as a self-primer, which promotes an enzyme called promote Primase- independent DNA synthesis and makes it important in DNA multiplication. Besides that, the ITRs help to facilitate integration of its genetic makeup into the genome of the host cell [11, 12]. Due to adenoviral vectors and their low pathogenicity and mild symptoms, these vectors can be highly favorable for gene therapy. These viruses also come in use with their ease of manipulation with recombinant DNA, as well as their ability to transduce foreign genes into proliferating cells [12–14]. The most notable use of adenoviral vectors in gene therapy was its early

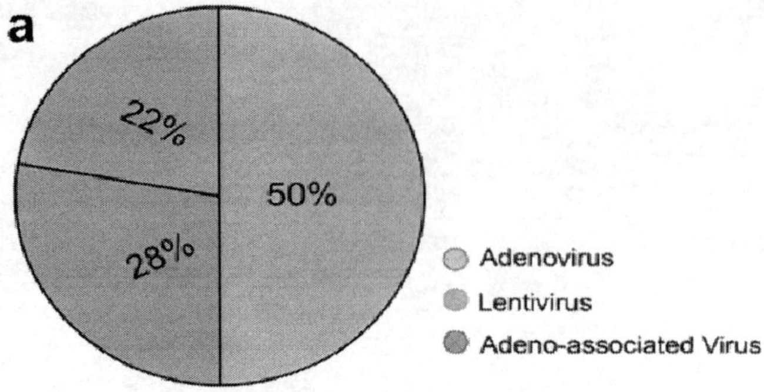

FIGURE 1.4 Percentage of viral vectors used in gene therapy clinical trials.

(Adapted from Viral vector platforms within the gene therapy landscape.)

TABLE 1.1
Number of Clinical Trials Associated with Each Vial Vector

Vectors	Number of Clinical Trials
Adenovirus	575
Adeno-associated virus	250
Lentivirus	315
Total	1,140

Source: adapted from Viral vector platforms within the gene therapy landscape

use in clinical trials for cystic fibrosis, which is an inherited disease that causes the cells in the lungs to produce mucus and fluids, causing severe damage to the lungs and other organs located in the general area. A subsection in adenoviral vectors is the adeno-associated virus (AAV) use as a vector. An easy explanation of AAV vectors compared to regular adenoviral vectors is that AAV is a small, slower version of adenoviral vectors. AAVs have a lower packaging capacity, lower protein levels, and take longer for genes to express while gene expression lasts longer. With a low packing size, AAVs must limit their treatment under a specific size, which is a disadvantage of using these viruses as vectors. However, depending on the situation, AAVs trigger low levels of immune response, thus allowing fewer of the vectors to be killed off by T-lymphocytes. Last, adenoviral vectors have much larger packing sizes, allowing more variability in adding a specific gene, while causing an onset of expression relatively fast at 16–24 hours [15].

In different clinical trials, different vectors are used, depending on the situation. Below is a figure which took three strategies for viral vectors used in gene therapy. Based on the results, adenovirus vectors where used the most, with 575 seen in the table below [16]. That was 50 percent of all viral vectors used in gene therapy (Figure 1.4 and Table 1.1).

1.9 NONVIRAL VECTORS

Nearly 70 percent of gene therapy clinical trials have used modified viruses as viral vectors. That means the other 30 percent of gene therapy clinical trials have used non-viral vectors [17]. Non-viral vectors have naked DNA and are usually particle and chemical based [17, 18]. This means that the DNA contained in the vector is not harmful, and the genetic makeup is usually made up of oligonucleotides and their analogues. Using non-viral vectors creates an alternative to viral vectors where viral vectors might not be the best choice of delivery system. The drawback to viral vectors that nonviral vectors offer is the low effect of its immunogenicity and cytotoxicity. In other words, non-viral vectors create a lower immune response and are less likely to be toxic to the contents of the human body [19]. Due to this reduced immunotoxicity, the use of non-viral vectors in clinical trials has seen an upward trend throughout the years. Other limitations the nonviral vectors have the potential to address are the limited capacity to pack DNA, difficulty of vector production, and the variability within the vector and its environment [20–22]. However, one drawback to using non-viral vectors is the lack of data present. Non-viral vectors have been shown to produce excellent results in invitro experiments, however when expressed in vivo, no significant results can be concluded [23]. For this reason, further research is being conducted with nonviral vectors, and new modifications are made every day.

1.10 CHALLENGES IN DELIVERY SYSTEMS

With gene therapy considered as a treatment of disease using healthy genes there have been a plethora of challenges in establishing a consistent and reliable method of treatment. Many treatments vary on many different factors, such as delivery method, location of target cells, size of treatment, and the type of vector. All these factors are involved in the combined effort in establishing gene therapy treatment to the harmful disease. With so many factors, and each being variable and independent of each other, there are bound to be challenges associated with treatment using gene therapy.

The first factor is dependent on the type of vector chosen. Previously mentioned, there were multiple types of viruses used to create vectors to hold the treatment to the target cell. Each virus has its advantages and disadvantages, which must be explored and researched to find the vector with the most efficiency for the target cell. In Tables 1.2 and 1.3 (the advantages and disadvantages of viral, Table 1.2, and non-viral, Table 1.3), vectors are laid out for ease of identification [24]. In Table 1.2, the viral vectors are differentiated by the type of viral virus used, coupled with the advantages and disadvantages associated with it. In Table 1.3, only the advantages and disadvantages are listed, as they are only discussed for the non-viral vectors.

While gene therapy holds promising results, the safety of the patients is taken as the primary concern. Vectors are chosen to maximize security and safety of the patient. While some vectors are safer, they cannot get the task done for multiple reasons, such as the capacity of DNA packing and the manufacturing aspect of it. Vectors chosen, once inserted into the patient, should not induce an allergic reaction or inflammatory process. This process could lead to the destruction of the vector and its treatment for the target cell. Retroviral vectors have an increased chance of mutation, which causes concern to some. Retrovirus-mediated gene transfer sets up the potential risk of insertional oncogenesis [24]. Essentially, insertional oncogenesis is the mutation of the genetic makeup of a genome by the addition of nucleotide base pairs.

Adenoviral vectors can infect and integrate into dividing and nondividing cells, which allows them to have more versatility in how the vector could be used. This causes a higher level of expression, but the duration is dissipated relatively quickly. This biggest limitation in this vector would be the strength of the immune and inflammatory response that could cause death. Adeno-associated viral vectors show the same disadvantages, in terms of immunotoxicity, as adenoviral vectors, with the addition of limited space for DNA capacity. In nonviral vectors, challenges are primarily based on the effect of the treatment rather than the dangers of use. The duration of expression is short. Thus, due to the unstained expression, repeat therapy is required for continuation of replication within the host cell and gene expression [25]. As seen, non-viral vectors have limitations based on the effect of the treatment delivered. In part, this is due to the vector properties themselves, but recent research has shown nonviral vectors to be safer to use in gene therapy [19, 21, 22].

TABLE 1.2
Summary of Advantages and Disadvantages of Viral Vectors

Vector	Advantages	Disadvantages
Adenovirus	High transfection efficiency Transfects proliferating and non-proliferating cells Substantial clinical experience	Strong immune responses Insert size limit of 7.5 kb Difficult to manufacture and quality control Poor storage characteristics Short duration of experssion
Retrovirus	Fairly prolonged experssion High transfection efficiency Substantial clinical experience Low immunogenicity	Low transfection efficiency in-vivo Insert size limit of 8kb ex-vivo Transfects only proliferating cells Difficult manufacture and quality control Safety concerns (mutagenesis)
Lentivirus	Transfects proliferating and non-cells Transfects hematopoietic stem cells	Vert difficult manufacture and quality control Poor storage characteristics Insert size limit of 8kb No clinical experience Safety concerns (Origins of HIV)
Adeno-associated virus	Efficient transfectionof wide variety of cell types in-vivo Prolonged expression Low immunogenicity	Difficult manufacture and quality control Insert size limit of 4.5kb Limited clinical experience Safety concerns (mutagenesis)

Source: adapted from "The challenge of gene therapy and DNA delivery".

TABLE 1.3
Advantages and Disadvantages of Non-viral Vectors Short

Advantages	Disadvantages
Manufacture and quality control relatively simple	Short duration of expression
Good storage characteristics	Repeat therapy required as plasmid does not replicate with host cells
Low immunogenicity	Inefficient transfection in-vivo
Good safety profile	
Efficient transfection ex-vivo	
Delivery to any somatic cell	
Non-infectious	
No limit on size of plasmid	

Source: adapted from "The challenge of gene therapy and DNA delivery".

1.11 CHALLENGES IN GENE DELIVERY

For gene therapy to work, DNA much be delivered to the target tissue and then transported to the nucleus for protein expression. Within this problem there is another underlying problem. The first problem is that a system must be established to deliver DNA to the target tissues and must be preventive to degradation. The second problem stems from the establishment of another system to build a DNA construct, and then allow the target cell to express that protein at efficient therapeutic levels [25, 26]. To establish a system which delivers DNA to the tissues and nucleus, a mechanism to circumvent extra and intracellular barriers must be taken into account. Some of these barriers include stability of the biological environment, cellular uptake, lysosomal escape, cytosolic transport, and gene unpacking [27–31]. These are some of the barriers that must be overcome, and such solutions must be provided with trials and experimentation.

DNA delivery starts from a multistep process that begins with DNA condensation, introduction of DNA to systemic circulation, targeted delivery followed by the unpacking of DNA, and translation into the eukaryotic cell [25]. The key is success and to overcome the challenge of DNA delivery, the mechanism of delivery must be thoroughly considered in design such that there is a maximization of therapeutically active DNA and a minimization of side effects. Once condensation of DNA is completed, a critical challenge of navigating through the extracellular barriers becomes prominent. The infamous immune response is the most likely to occur as an extracellular barrier. In the lungs, alveolar macrophages come in response to a triggered immune response. These macrophages are phagocytic cells that consume cells that pose a threat to the system [32]. Generally, immune responses are triggered by viral vectors, but there have been instances where non-viral vectors have activated an immune response. One example is when cationic lipoplexes are intravenously injected into the patient, they can induce an inflammatory response which involves the release of TNFα and IFNγ into the serum [33].

After the vector has made it into the cellular environment, it undergoes a process called cellular uptake. Cellular uptake is basically the endocytosis of the DNA carrying vector. To reach the target cell, the vector must pass through several energy dependent pathways. The efficiency of cellular uptake is dependent on several components. Some of these components are size, surface charge, variability in the polyplexes, and much more [34–37]. The endosome picks up the vector and mediates it to the cytosol, where the vector can then enter the nucleus and transfects the genome. To effectively transfect the genome, the vector must disassemble from the polyplex in order to activate the transcription apparatus and insert the gene of interest into the nuclear genome to attain therapeutic results [25]. Below in the figure, there is a visual representation of cell mediated gene therapy [25]. It covers the steps to induce transfection in a visual form (Figure 1.5).

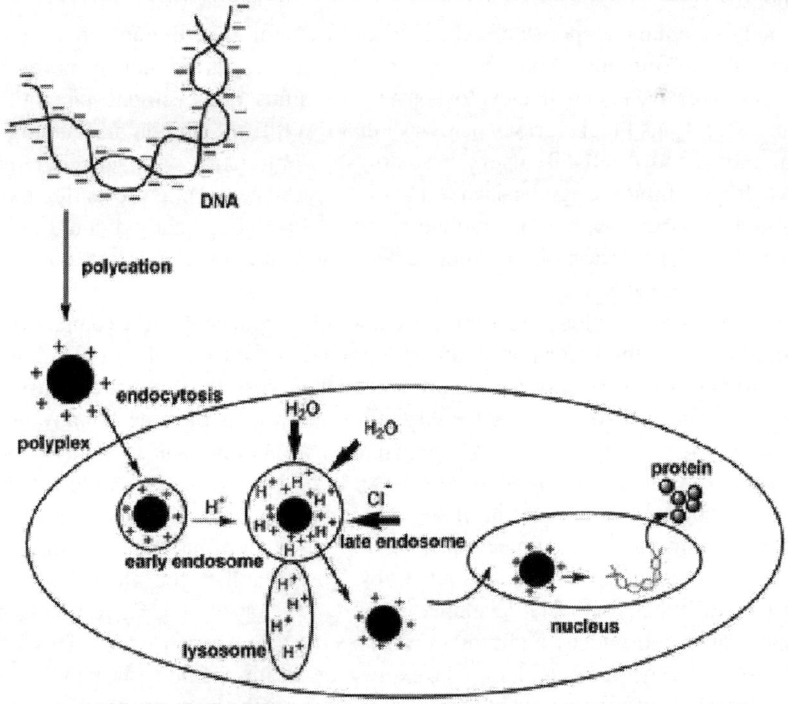

FIGURE 1.5 Cell-mediated gene delivery.

(Adapted from" non-viral gene therapy: polycation-mediated DNA delivery".)

Another common problem associated with gene therapy is the sustainability of gene expression. A major problem in the use of nonviral vectors is the duration that the therapy lasts for. Compared to viral vectors, nonviral vector induced gene expression isn't sustained and usually becomes ineffective after a period of time. For situations where there is a lack of gene expression, treatment can be applied again, causing the gene to express once more. This constant application of treatment serves as inefficient, due to the minor therapeutic activity.

1.12 IMMUNE RESPONSE CHALLENGES IN GENE DELIVERY

Immune responses caused by gene therapy are one of the most common challenges circumvented by researchers. The immune response is the body's ability to recognize and defend itself again foreign bodies that may pose a threat or be harmful. The use of gene therapy is almost a conflict of interest with the immune system, as the use of a foreign vector is used in conjunction with new therapeutic genes. Once the vector enters the body, depending on the vector, an immune response could be triggered, which can be generally known as immunostimulants. If the cellular immune response is induced, cytotoxic T cells will be generated, coupled with a possible humoral immune response, by creating specific antibodies [25]. For clinical trials to utilize vectors for gene therapy, they must be proven safe in multiple conditions that mimic the human body, and be efficient in gene expression while inducing minimal immune responses. It is apparent that when immune responses are induced, the level and duration of transgene expression is reduced, due to the ability of CD8 and T cell recognizing and eliminating the foreign body [38, 39]. In non-specific immune responses, gene expression is inhibited. IFNγ and TNFα, which are examples of non-specific cytokines, can decrease mRNA stability and cause inhibition in gene transcription [40]. Furthermore, cytokines can inhibit promoter genes that aid in the control of transgene expression [41]. With all being said, the only way for a vector and specific treatment to meet guidelines to be used in clinical trials is that these vectors, once deployed into the body, must be deemed safe and should not conflict with flare ups in immune responses.

With the body's immune response affecting the delivery for gene therapy, there must be strategies to circumvent this immune response. One way of regulating an immune response has been the use of immunosuppressive agents like cyclosporine, tacrolimus, and cyclophosphamide [42]. These immunosuppressive drugs block various pathways that result from antigen presentation all the way to the activation of B and T cells. By using these drugs in conjunction with gene therapy, the immunosuppressive drugs inhibit the synthesis and release of cytokines, while preventing the differentiation of CD4 cells, hence blocking an immune response [43]. This method could hold promising results, such that the suppression of the immune response will allow the vector to unpack the DNA and allow it to integrate the host cell.

The next strategy takes another approach to the immune response. Unlike using immunosuppressive drugs to inhibit an immune response, this strategy takes the route of administration of vectors and how it contributes to the induction of an immune response [44]. Administration of vectors using different methods can lead to different results and effectiveness of gene therapy vectors. There was a study conducted where mice were injected with a recombinant adeno-associated virus to measure the strength of a specific antibody response. In this study, mice were injected with rAAV-OVA, an adeno-associated virus, and its therapeutic gene, either by intraperitoneal, intravenous, subcutaneously, or intramuscularly injection. The strength level of an OVA-specific CTL response was measured after the injection of the gene therapy. The mice that got injected either intraperitoneal, intravenous, or subcutaneously developed a strong OVA-specific CTL response, but the mice injected intramuscularly produced a minimal OVA-specific CTL response [45]. This shows that the method of delivery affects the strength of gene expression within the body. Depending on where the gene therapy is needed, a specific type of injection could yield the best result, which would be a low response of antibodies.

Lastly, another strategy to circumvent an immune response while using gene therapy is the modifications to the promoter to drive transgene expression. When specific promoters are used for gene therapy, the expression level of certain genes vary depending on the promoter used. Regulated promoters in inflammatory conditions can usually lead to an effective approach for circumventing immune responses and gene therapy. Researchers have found that the murine acute phase protein can express under inflammatory conditions and avoid transient expression of targeted gene induction [46].

1.13 FUTURE PERSPECTIVES

With many challenges to the delivery of gene therapy, there is potential to optimize the system. For example, creating a cocktail of immunosuppressors, promoter modifications, and alternative delivery routes combined with the vector and therapy, the ideal scenario would proceed: the immune response will fail to trigger, and the vector can unpack the treatment to the cell. However, the ideal scenario will take time and research to achieve. Delivery vectors have much improvement that can still be made. As previously stated, some vectors have complications that affect the immune system of the body. Future scientists should allocate more of their efforts to suppressing an immune response induced by some vectors. These efforts can lead to further modifications and improvements within viral and non-viral vectors. The most critical aspect of these modifications must be the reduction of their immunotoxicity and cytotoxicity. Modification of vectors to inhibit immune responses is essentially the main challenge to be overcome. Beside this, alteration of DNA delivery can be researched and modified for the most efficient outcome. Extracellular and intracellular barriers are the main inhibitors in DNA delivery within the cell.

1.14 CONCLUSION

The science behind gene therapy has long been researched and continues to be a novelty in the field of molecular medicine. From the contribution of the complex systems of the body, they challenge researchers in their approach toward overcoming the hardships induced by the systems of the human body. The development of DNA vectors with a specific emphasis on efficiency of gene expression and minimal immunotoxicity can eventually contribute to the successes of gene therapy. With the multitude of vectors used in gene therapy, viral and non-viral, there hold a plethora of combinations that can maximize the efficiency of gene expression while minimizing the potential for an induced immune response. With the correct combination of vector, therapeutic genes, and immunosuppressive factors, the success of gene therapy can be within reach. Ultimately, the success of gene therapy could hold the key for treating and preventing many of the common diseases surrounded by medicine.

REFERENCES

1. What is gene therapy? https://medlineplus.gov/genetics/understanding/therapy/genetherapy/
2. Gene therapy. https://www.ndsu.edu/pubweb/~mcclean/plsc431/students98/schilling.htm
3. Sun, W., Shi, Q., Zhang, H., et al. Advances in the techniques and methodologies of cancer gene therapy. *JAMA* 2020, 27(146), 45–55.
4. Drug delivery systems. https://www.nibib.nih.gov/science-education/science-topics/drug-delivery-systems-getting-drugs-their-targets-controlled-manner
5. Kay, M.A., Glorioso, J.C., & Naldini, L. Viral vectors for gene therapy: The art of turning infectious agents into vehicles of therapeutics. *Nat Med* 2001, 7(1), 33–40.
6. Zwaka, T.P. Use of genetically modified stem cells in experimental gene therapies. 2009. See http://stemcells.nih.gov/info/Regenerative_medicine/pages/2006chapter4.aspx.
7. Anson, D.S. The use of retroviral vectors for gene therapy-what are the risks? A review of retroviral pathogenesis and its relevance to retroviral vector-mediated gene delivery. Genetic vaccines and therapy. *BioMed Central* 2004. https://www.ncbi.nlm.nih.gov/pmc/articles/PMC515179/?report=reader#__ffn_sectitle

8. Sakai, H., Kawamura, M., Sakuragi, J., Sakuragi, S., Shibata, R., Ishimoto, A., Ono, N., Ueda, S., & Adachi, A. Integration is essential for efficient gene expression of human immunodeficiency virus type 1. *J Virol* 1993, 67, 1169–1174.
9. Masuda, T., Planelles, V., Krogstad, P., & Chen, I. Genetic analysis of human immunodeficiency virus type 1 integrase and the U3 att site: Unusual phenotype of mutants in the zinc finger-like domain. *J Virol* 1995, 69, 6687–6696.
10. Klimatcheva, E., Rosenblatt, J.D., & Planelles, V. Lentiviral vectors and gene therapy. *Front Biosci* 1999, 4, D481–D496.
11. Lee, C.S., Bishop, E.S., Zhang, R., et al. Adenovirus-mediated gene delivery: Potential applications for gene and cell-based therapies in the new era of personalized medicine. *Genes & Diseases* 2017, 4(2), 43–63.
12. Ma, S., & Guan, J. Adenovirus-mediated gene therapy for allergy. *Iran J Allergy Asthma Immunol* 2018 Dec 1, 17(6), 502–516. PMID: 30644694.
13. Vorburger, S.A., & Hunt, K.K. Adenoviral gene therapy. *Oncologist* 2002, 7(1), 46–59.
14. Suzuki, T., Sasaki, T., Yano, K., Sakurai, F., Kawabata, K., Kondoh, M., et al. Development of a recombinant adenovirus vector production system free of replication-competent adenovirus by utilizing a packaging size limit of the viral genome. *Virus Res* 2011, 158, 154–160.
15. Vector Biolabs. Adenovirus vs. AAV, which should you use? 2017. [cited August 1, 2021]. https://www.vectorbiolabs.com/adenovirus-vs-aav/
16. Bulcha, J.T., Wang, Y., Ma, H., et al. Viral vector platforms within the gene therapy landscape. *Sig Transduct Target Ther* 2021, 6, 53. https://doi.org/10.1038/s41392-021-00487-6
17. Ramamoorth, M., & Narvekar, A. Non viral vectors in gene therapy: An overview. *J Clin Diagn Res* 2015, 9(1), GE01.
18. Pahle, J., & Walther, W. Vectors and strategies for nonviral cancer gene therapy. *Expert Opin Biol Ther* 2016, 16(4), 443–461. https://doi.org/10.1517/14712598.2016.1134480
19. Gloves, D.J., & Lipps, H.J. Towards safe non-viral therapeutic gene expression in humans. *Nat Rev Genet* 2005, 6, 299–310.
20. Baum, C., Kustikova, O., Modlich, U., Li, Z., & Fehse, B. Mutagenesis and oncogenesis by chromosomal insertion of gene transfer vectors. *Hum Gene Ther* 2006, 17, 253–263.
21. Bessis, N., GarciaCozar, F.J., & Boissier, M.C. Immune responses to gene therapy vectors: Influence on vector function and effector mechanisms. *Gene Ther* 2004, 11(Suppl. 1), S10–S17.
22. Thomas, C.E., Ehrhardt, A., & Kay, M.A. Progress and problems with the use of viral vectors for gene therapy. *Nat Rev Genet* 2003, 4, 346–358.
23. Jin, L., Zeng, X., Liu, M., Deng, Y., & He, N. Current progress in gene delivery technology based on chemical methods and nano carriers. *Theranostics* 2014, 4(3), 240–255.
24. Arjmand, B., Larijani, B., Sheikh Hosseini, M., Payab, M., Gilany, K., Goodarzi, P., Parhizkar Roudsari, P., Amanollahi Baharvand, M., & Hoseini Mohammadi, N.S. The horizon of gene therapy in modern medicine: Advances and challenges. *Adv Exp Med Biol* 2020, 1247, 33–64.
25. Shi, B., Zheng, M., Tao, W., Chung, R., Jin, D., Ghaffari, D., & Farokhzad, O.C. Challenges in DNA delivery and recent advances in multifunctional polymeric DNA delivery systems. *Biomacromolecules* 2017, 18(8), 2231–2246.
26. Phillips, A.J. The challenge of gene therapy and DNA delivery. *J Pharm Pharmacol* 2001, 53(9), 1169–1174.
27. Yin, H., Kanasty, R.L., Eltoukhy, A.A., Vegas, A.J., Dorkin, J.R., & Anderson, D.G. Non-viral vectors for gene-based therapy. *Nat Rev Genet* 2014, 15, 541–555. https://doi.org/10.1038/nrg3763
28. Lachelt, U., & Wagner, E. Nucleic acid therapeutics using polyplexes: A journey of 50 years (and beyond). *Chem Rev* 2015, 115, 11043–11078. https://doi.org/10.1021/cr5006793
29. Mintzer, M.A., & Simanek, E.E. Nonviral vectors for gene delivery. *Chem Rev* 2009, 109, 259–302. https://doi.org/10.1021/cr800409e
30. Liu, Z.Z., Zheng, M., Meng, F.H., & Zhong, Z.Y. Non-viral gene transfection in vitro using endosomal pH-sensitive reversibly hydrophobilized polyethylenimine *Biomaterials* 2011, 32, 9109–9119. https://doi.org/10.1016/j.biomaterials.2011.08.017
31. Miyata, K., Kakizawa, Y., Nishiyama, N., Harada, A., Yamasaki, Y., Koyama, H., & Kataoka, K. Block catiomer polyplexes with regulated densities of charge and disulfide cross-linking directed to enhance gene expression. *J Am Chem Soc* 2004, 126, 2355–2361. https://doi.org/10.1021/ja0379666

32. McCray P.B. Jr., Wang, G., Kline, J.N., Zabner, J., Chada, S., Jolly, D.J., et al. Alveolar macrophages inhibit retrovirus-mediated gene transfer to airway epithelia. *Hum Gene Ther* 1997, 8, 1087–1093.
33. Dow, S.W., Fradkin, L., GLiggitt, D.H., Willson, A.P., Heath, T.D., & Potter, T.A. Lipid-DNA complexes induce potent activation of innate immune responses and antitumor activity when administered intravenously. *J Immunol* 1999, 163, 1552–1561.
34. LaManna, C.M., Lusic, H., Camplo, M., McIntosh, T.J., Barthelemy, P., & Grinstaff, M.W. Charge-reversal lipids, peptidebased lipids, and nucleoside-based lipids for gene delivery. *Acc Chem Res* 2012, 45, 1026–1038.
35. Kamaly, N., Xiao, Z.Y., Valencia, P.M., Radovic-Moreno, A.F., & Farokhzad, O.C. Targeted polymeric therapeutic nanoparticles: Design, development and clinical translation. *Chem Soc Rev* 2012, 41, 2971–3010.
36. Liu, Z.H., Zhang, Z.Y., Zhou, C.R., & Jiao, Y.P. Hydrophobic modifications of cationic polymers for gene delivery. *Prog Polym Sci* 2010, 35, 1144–1162.
37. Thomas, M., & Klibanov, A.M. Non-viral gene therapy: Polycation-mediated DNA delivery. *Appl Microbiol Biotechnol* 2003, 62(1), 27–34.
38. Jooss, K., Ertl, H.C., & Wilson, J.M. Cytotoxic T-lymphocyte target proteins and their major histocompatibility complex class I restriction in response to adenovirus vectors delivered to mouse liver. *J Virol* 1998, 72, 2945–2954.
39. Gagandeep, S., Ott, M., Sokhi, R.P., & Gupta, S. Rapid clearance of syngeneic transplanted hepatocytes following transduction with E-1-deleted adenovirus indicates early host immune responses and offers novel ways for studying viral vector, target cell and host interactions. *Gene Ther* 1999, 6, 729–736.
40. Ghazizadeh, S., Carroll, J.M., & Taichman, L.B. Repression of retrovirus-mediated transgene expression by interferons: Implications for gene therapy. *J Virol* 1997, 71, 9163–9169.
41. Qin, L., Ding, Y., Pahud, D.R., Chang, E., Imperiale, M.J., & Bromberg, J.S. Promoter attenuation in gene therapy: Interferon-gamma and tumor necrosis factor-alpha inhibit transgene expression. *Hum Gene Ther* 1997, 8, 2019–2029.
42. Ritter, T., Lehmam, M., & Volk, H.-D. Improvements in gene therapy: Averting the immune response to adenoviral vectors. *BioDrugs* 2002, 16, 3–10.
43. Brown, B.D., & Lillicrap, D. Dangerous liaisons: The role of "danger" signals in the immune response to gene therapy. *Blood* 2002, 100, 1133–1140.
44. Ge, Y., Powell, S., Van Roey, M., & McArthur, J.G. Factors influencing the development of an anti-factor IX (FIX) immune response following administration of adeno-associated virus-FIX. *Blood* 2001, 97, 3733–3737.
45. Brockstedt, D.G., Podsakoff, G.M., Fong, L., Kurtzman, G., Mueller-Ruchholtz, W., & Engleman, E.G. Induction of immunity to antigens expressed by recombinant adeno-associated virus depends on the route of administration. *Clin Immunol* 1999, 92, 67–75.

2 Advances in Genome Editing

The Technology of Choice for Precise and Efficient Disease Treatment, with Special Focus on Nano Delivery Systems

Komal Parmar
Shri G.M. Bilakhia College of Pharmacy, Vapi, India

Jayvadan Patel
Sankalchand Patel University, Visnagar, India

CONTENTS

2.1 INTRODUCTION

Genetics is in a fantastic place right now, thanks to improvements in genetic analysis and manipulation. Gene therapy is the process of changing or introducing new genes to a person's genome in order to treat or increase their ability to tackle disease. One facet of gene therapy is genome editing. Existing gene therapy procedures are founded on the findings of longer apparent laboratory investigations on individual cells and nonhuman species, demonstrating the ability to complement, eliminate, or alter genes in living beings. The advent of very adaptable genome-editing technology has given researchers the ability to rapidly and cost-effectively incorporate sequence-specific alterations into the genomes of a wide spectrum of cell types and organisms over the past few years (Metje-Sprink et al. 2019, Tsanova et al. 2021). The invention of genetic engineering in the late 20th century ushered in a new era in genome editing (Scherer and Davis, 1979; Rothstein, 1983; Smithies et al., 1985; Thomas et al., 1986). The true origins of this technique can be traced back to forerunners in genome engineering (Ishino et al., 1987; Nakata et al., 1989). Basically, genome editing refers to

DOI: 10.1201/9781003186083-2

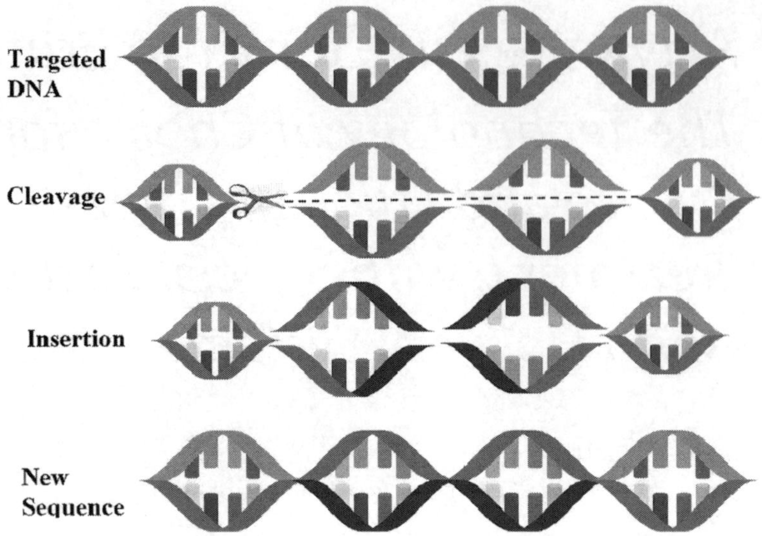

FIGURE 2.1 Basics of genome editing.

the employment of molecular tools, wherein genomic DNA is sliced in certain sites to facilitate tailored changes in the DNA pattern. These DNA splits then initiate cellular DNA repair processes, allowing site-specific epigenetic amendments to be introduced more easily (Bibikova et al., 2002; Urnov et al., 2005) (Figure 2.1). Initially, investigators found that when a length of DNA with similar arms on both ends is put into a cell, it may be assimilated into the host genome by homologous recombination and can control desired modifications in the cell (Capecchi 1989). However, the technique was limited to only dividing cells (Saleh-Gohari and Helleday, 2004). Later, more robust and effective gene targeting was achieved by the introduction of the double stranded break at a definite genomic target. The activation of cellular DNA repair mechanisms is then triggered, allowing for the insertion of site-specific genomic alterations. (Rouet et al., 1994). Several gene editing strategies have concentrated on the innovation and application of various endonuclease-based mechanisms to produce these breaks with high accuracy (Jacinto et al., 2020). Specifically, zinc finger nucleases (ZFNs), transcription activator-like effector nuclease (TALENs), meganucleases or homing endonucleases (MegNs), and clustered regularly interspaced short palindromic repeats (CRISPR)/CRISPR-associated protein 9 (Cas9) are four key sites-specific genome editing strategies that have cleared the path for potential agricultural and medical advancements. In recent years, the genome editing technique is used fruitfully in varied clinical applications (Foss et al., 2019; Memi et al., 2018; Mahmoudian-sani et al., 2018). The chapter focuses on genome editing tools and its applications in treatment of various diseases employing nano delivery systems.

2.2 THREE MAJOR GENOME EDITING METHODS

2.2.1 Zinc-Finger Nucleases (ZFNs)

ZFNs are made by connecting a non-sequence specific cleavage domain to a zinc finger-loaded site-specific DNA-binding domain (Urnov et al., 2010). In Xenopus oocytes, the zinc-finger protein with site-specific DNA binding characteristics was identified in 1985 as part of the interaction of Xenopus protein transcription factor IIIa with 5S RNA (Klug 2010). The developed zinc-finger domain's functional specificity is obtained from an assembly of Cys_2His_2 zinc fingers, which are produced from highly conserved couplings of their zinc-finger domains with homologous DNA sequences. The most prevalent form of DNA-binding domain in eukaryotic transcription factors is

Cys_2-His_2-ZF, which is a flexible DNA recognition domain (Mueller et al., 2020). The DNA-binding and catalytic domains can be tuned separately, making redirection and platform optimization efforts easier. The Zinc Finger Protein region allows a ZFN to bind a distinct base sequence (Wolfe et al., 2000). The FokI type II restriction endonuclease creates a DNA-cleaving domain that can be used in a dimer to target specific sequences within the genome for successful gene editing (Smith et al., 2000). Because the FokI nuclease must dimerize in order to cleave DNA, two ZFN molecules are normally required to attach to the target site in the correct orientation, resulting in a twofold increase in the number of specifically identified base pairs. After ZFNs cleave DNA in eukaryotic cells, double strand breaks are created at a specific site of the genome, causing the desired changes in the endogenous NHEJ or HDR repair processes.

2.2.2 Transcription Activator-Like Effector Nucleases (TALEN)

The TALEN (Transcription Activator-Like Effector Nucleases) tool was introduced in 2011 to improve the efficiency, reliability, and accessibility of genome editing. The transcription activator-like effectors (TALES) generated by the phytopathogenic bacteria Xanthomonas genus gave rise to the TALEN system (Moore et al., 2014). The activator proteins are members of the DNA binding protein family and, like transcription factors in eukaryotic genomes, can be utilized to promote the expression of their target genes. The TALE proteins' DNA binding domain is constituted of monomers of 34 amino acid residues, each of which binds one nucleotide in the target nucleotide sequence (Christian et al., 2012). Apart from two hypervariable amino acids (the repeat variable di-residues) at positions 12 and 13, the amino acid sequence of each repeat is fundamentally similar and in charge of identifying a specific nucleotide. (Mak et al., 2013). To produce TALEN-induced targeted genomic alteration, the functional endonuclease FokI is artificially coupled to DNA binding domains to form site-specific DSBs and encourage DNA recombination. The FokI cleavage domain must be dimerized in order to cleave the two strands of the targeted DNA. As a result, TALEN modules are built in pairs to engage opposing DNA target loci, with sufficient separation (12–20 bp) between the two binding sites (Li et al., 2011).

2.3 CLUSTERED REGULARLY INTERSPACED SHORT PALINDROMIC REPEATS (CRISPR/CAS)

CRISPR–Cas is a bacterial adaptive immune system that cleaves invading nucleic acids. CRISPRs were first discovered in E. coli in 1987 during an examination of genes involved in phosphate metabolism, and later in a variety of other bacterium species (Ishino et al., 2018). CRISPR-Cas systems are classified into two groups (class I and II), based on the structural variation of the Cas genes and the way they are organized. The classes are further subdivided into six types (type I–VI). Class I includes type I, III, and IV, and class II includes type II, V, and VI (Makarova and Koonin, 2015). Class 1 CRISPR–Cas systems have multiprotein effector complexes, whereas class II systems only have a single effector protein. Presently, six subtypes of the Type I system (Type I-A through Type I-F) have been found, each with a different number of Cas genes. All Type I systems, with the exception of cas1, cas2, and cas3, encode a Cascade-like complex. Cascade is the name given to the effector complex of type I systems (CRISPR-associated complex for antiviral defense) (Brouns et al., 2008). Cascade engages crRNA and locates the target, and the majority of variations are also in charge of crRNA processing. In rare circumstances, cascade can also help with spacer acquisition. Cas3 is a component of the Cascade complex in the Type I-A system, a protein with both helicase and DNase domains responsible for degrading the target (Huo et al., 2014; Gong et al., 2014). Cas1 and Cas2, the Cas9 hallmark protein, and occasionally a fourth protein are encoded by Type II CRISPR-Cas systems (Csn2 or Cas4) (Chylinski et al., 2014). Cas9 helps with adaptation, engages in crRNA processing, and cleaves the target DNA with the help of crRNA and tracrRNA (Karvelis et al., 2013). Type II systems are split into subtypes II-A, II-B, and II-C. In Type II-A and Type II-B, respectively, the csn2 and cas4 genes encoding adaptation proteins are present, whereas Type II-C lacks a fourth gene (Rath et al., 2015). The Type III CRISPR-Cas systems contain the hallmark protein Cas10 with unclear function (Dorsey et al., 2019).

The type II CRISPR/Cas9 system, which relies on a single Cas protein from Streptococcus pyogenes (SpCas9) targeting specific DNA sequences and is hence an appealing gene editing tool, is the most commonly utilized subtype of CRISPR systems (Hsu et al., 2014; Mir et al., 2018; Tang and Fu, 2018; Wang et al., 2020). The CRISPR/Cas9 system is made up of two parts: a single-stranded guide RNA (sgRNA) and a Cas9 endonuclease. The sgRNA often provides a specific 20-base-pair (bp) sequence befitting the target DNA site in a sequential manner, and this must be accompanied by a short DNA sequence upstream integral for Cas9 protein suitability, known as the "protospacer adjacent motif" (PAM) of a "NGG" or "NAG" protein. Cas9 accurately cleaves the DNA to induce a DSB after the sgRNA attaches to the target sequence via Watson–Crick base pairing. DNA-DSB repair mechanisms begin genome repair after the DSB. Targeted genomic alterations, including the inclusion of tiny insertions and deletions (indels), can be generated with the CRISPR/Cas9 system via NHEJ or high-fidelity HDR pathways.

With the CRISPR/Cas9 platform, three common genome editing techniques have been developed: 1) *In vitro*, the plasmid-based CRISPR/Cas9 method, which uses a plasmid to encode Cas9 protein and sgRNA21,22, assembles Cas9 gene and sgRNA into the same plasmid. This method produces longer-lasting Cas9 and sgRNA expression and eliminates the need for multiple transfections; 2) major disadvantage of direct intracellular delivery of Cas9 messenger RNA (mRNA) and sgRNA is the poor stability of mRNA, which results in transitory mRNA production and a limited duration of gene modification; 3) Cas9 protein and sgRNA56 are delivered directly, which has a number of advantages, including quick action, high stability, and low antigenicity (Liu et al., 2017).

2.3.1 Applications of Genome Editing in Therapy

The use of chimeric gene editing tools to target gene adjustment is a potent way of assessing gene function and thereby manipulating cellular behaviour and functioning. Thanks to these genome editing techniques, the investigators have been able to employ genetically altered animals to better understand the epidemiology of many diseases and to elucidate underlying pathways that can be explored for better therapeutic approaches. The execution for disease treatment can be done in two ways: first, cells can be withdrawn from a patient or donor and edited outside the body before being reintroduced into the patient, or the genome editor can be introduced directly into the in vivo, using nano delivery systems (Li et al., 2020a; Mills et al., 2020). It's vital to note that gene editing's therapeutic effectiveness is dependent on the number of parameters, including editing efficacy, which varies greatly relying on the type of cell, ageing condition, and cell cycle stage of the recipient (Rodríguez-Rodríguez et al., 2019). Other variables that impact therapeutic efficacy include cell proficiency, which alludes to the viability of achieving a therapeutic reconfiguration baseline, and the effective transmission of programmable nuclease structure to the target tissue. Along with these factors, precision of the editing process is considered the important facet of the technique (Wang et al., 2017). In cell lines, disease models, and humans, genome editing procedures have been used to treat a variety of genetic ailments (Table 2.1).

2.3.2 Nano Delivery Approaches in Genome Editing for Disease Treatment

2.3.2.1 Lipid Nanoparticles

Lipid nanoparticles is one of the traditional delivery methods for nucleic acid transfer and has been widely investigated (Noh et al., 2011; Buck et al., 2019; Mohammadinejad et al., 2020). Negatively charged nucleic acids and positively charged lipids form a compound through mediated activation

TABLE 2.1
Genome Editing Method in Treatment of Various Diseases

Genome Editing Method	Treatment of Disease	Reference
CRISPR	Hemophilia	Chen et al. (2019)
NHEJ knock-in	Hemophilia	Zhang et al. (2019a)
CRISPR	Mucopolysaccharidosis	Ou et al. (2020),
Zinc finger nucleases	Murine Hurler Syndrome	Ou et al. (2019)
CRISPR/Cas9	Mucopolysaccharidosis	de Carvalho et al. (2018), Schuh et al. (2020)
TALEN	Sickle cell disease	Sun and Zhao (2014)
CRISPR/Cas9	Wilson's disease	Pöhler et al. (2020)
CRISPR/Cas9	Colon cancer	Li et al. (2020b)
CRISPR/Cas9	Leukemia	Valletta et al. (2015)
Zinc finger nucleases	Usher Syndrome	Overlack et al. (2012)
CRISPR/Cas9	Alzheimer disease	Ortiz-Virumbrales et al. (2017), György et al. (2018)
Zinc finger nucleases and TALEN	Cancer	Piganeau et al. (2013)
CRISPR/Cas9	Angelman syndrome	Schmid et al. (2021)
CRISPR/Cas9	Ovarian cancer	Walton et al. (2016)
Zinc finger nucleases	Sickle cell disease	Porteus (2006)
CRISPR/Cas9	Acute Myeloid Leukemia	Brabetz et al. (2017)
CRISPR/Cas9	Chronic Myeloid Leukemia	Valletta et al. (2015)
CRISPR/Cas9	Targeting of PCSK9 in Human Hepatocytes to prevent cardiovascular disease	Wang et al. (2016), Ding et al. (2014)
CRISPR/Cas9	Dyslipidemia	Chadwick and Musunuru (2017)
CRISPR/Cas9	Hearing loss	Gao et al. (2018)
TALEN	Cystic fibrosis	Xia et al. (2019)
Meganuclease and TALEN	Xeroderma pigmentosum group C	Dupuy et al. (2013)
CRISPR/Cas9	Cervical cancer	Zhen et al. (2014), Kennedy et al. (2014)
CRISPR/Cas13a	Cervical cancer	Chen et al. (2020)
CRISPR/Cas9, TALEN, ZFN	β-Thalassemia	Antony et al. (2018)
CRISPR/Cas9, TALEN	β-Thalassemia	Xu et al. (2015)
CRISPR/Cas9	Dystrophic cardiomyopathy	El Refaey et al. (2017), Xu et al. (2019)
TALEN	β-Thalassemia	Ma et al. (2013)
CRISPR/Cas9	B-Lymphoma	Bai et al. (2020)

and electrostatic forces, which is then later taken via the endocytosis process. Nanoparticles can transmit plasmids or mRNAs safely and effectively to the target cells while also preventing them from being degraded by nucleases. Lipid nanoparticles have been shown to transfer siRNA and mRNA in investigational studies.

Schuh and co-workers studied targeting of CRISPR/Cas9 and a donor oligonucleotide aiming at Mucopolysaccharidosis type I gene editing in vitro. The DNA were complexed with nanoemulsions, which showed effective introduction into MPS I p.Trp402∗ patient's fibroblasts (Schuh et al. 2018a). CRISPR/Cas9 plasmids were delivered effectively into a triple-negative breast malignancy, using an antibody-conjugated tumor-targeted nanolipogel. The Lipocalin 2 gene knocking efficiency of 81 percent resulted in a 77 percent reduction in cancer growth. These investigations showed that tumor-targeted nanolipogel are a secure, precise, and efficient delivery vehicle for CRISPR-mediated genome editing with target specificity (Guo et al., 2019).

A novel delivery vehicle in the form of thread-like DNA nanoclews was fabricated. In this approach, the Cas9 protein and sgRNA were integrated with DNA as a core and encased in polyetherimide. Endosomal escape is triggered by the positive charge on the polyetherimide surface, which enhances transfection efficiency (Sun et al., 2015). Besides, in another study, researchers described a novel dual-function gene carrier viz. α-helical polypeptide poly (PPABLG). With its high positive charge, PPABLG can bind Cas9 proteins and sgRNAs to form nanoparticles with the help of polyethylene glycol while maintaining its helical structure, allowing it to penetrate membranes more effectively. In vitro performance of the nanoparticles in gene editing was up to the level of 47.3 percent. Further in vivo, the nanoparticles can target and suppress the tumor growth by more than 71 percent by eliminating the Plk1 gene 66.7 percent in HeLa tumor tissues and thereby extending survival rate by 60 percent (Wang et al., 2018). Self-assembled long-circulating pH-sensitive cationic nano-liposomes were prepared for delivery of CRISPR-associated protein 9 for gene therapy in cervical cancer. CRISPR/Cas9 system efficiently repressed proliferation of human papillomavirus (HPV) 16-positive cervical tumor cells and induced programmed cell death by disabling the HR-HPV16E6/E7 oncogene (Zhen et al., 2020). In vivo genome editing of mucopolysaccharidosis I (MPS I) mice using the CRISPR/Cas9 system was investigated employing cationic liposomes as a carrier. The liposomal system was fabricated by microfluidization process by the addition of DNA at +4/−1 charge ratio. The outcome exhibited complexes of about 110 nm, with positive zeta potential of +30 mV. The incubation of the complexes with fibroblasts from MPS I patients led to a striking increase in alpha-L-iduronidase activity and decrease in lysosomal abnormalities (Schuh et al., 2018b). Lipid nanoparticles were employed for efficient genome editing with Cas9-sgRNA in vivo. Using these lipid nanoparticle formulations of these enhanced sgRNAs (e-sgRNA) and mRNA encoding Cas9, investigators proposed that a single intravenous injection into mice brings >80 percent editing of Pcsk9 in the liver. Thus, serum Pcsk9 is reduced to imperceptible levels, and cholesterol levels are suggestively dropped about 35 percent to 40 percent in animals (Yin et al., 2017). Miller and co-workers developed zwitterionic amino lipids to (co)deliver long RNAs, including Cas9 mRNA and sgRNAs. Low sgRNA doses (15 nm) delivered through nanoparticles suppress protein expression in cells by over 90 percent. In vitro (600 pM) and in vivo (1 mg/kg), mRNA delivery by zwitterionic amino lipids leads to high protein expression at modest dosages. In the liver, kidneys, and lungs of transgenic mice, intravenous co-delivery of Cas9 mRNA and sgLoxP produced floxed tdTomato expression (Miller et al., 2017). In one work, authors illustrated that by targeting the TTR or PCSK9 genes with mRNA encoding zinc finger nucleases packaged into lipid nanoparticles, the result was >90 percent knockout of gene expression in mice at mRNA concentrations 10-fold lower than previously reported (Conway et al., 2019).

Self-assembled micellar delivery using F127/PPO-NMe_3 of plasmids was developed for human papillomavirus (HPV) E7 oncogene and *Natronobacterium gregoryi* Argonaute (NgAgo) manipulation. In vitro and in vivo, Cas9-mediated E7 deletion diminishes HPV-induced carcinogenic activity, whereas NgAgo has little effect on E7 suppression in the xenograft mouse model (Lao et al., 2018).

2.3.2.2 Polymeric Nanoparticles

Polymers are among the most commonly investigated materials for drug and gene delivery as nanocarriers. This is attributed to a variety of inherent features, including as structural flexibility, biocompatibility, and simplicity of synthesis, all of which have aided in polymeric nanoparticle design in gene editing. These polymers can be electrostatically coupled with genetic material at physiological pH, thereby making gene transport easier (Thomas et al., 2019). Cationic polymer carriers have higher chemical diversity and functional potential than cationic lipid carriers, allowing for more versatile structural designs (Barua et al., 2011). Polymer nanoparticle carriers, like lipid carriers, can pass through the membrane via endocytosis and shield loaded contents from immune reaction and nuclease destruction (Kou et al., 2013).

Recently, biodegradable poly(beta-amino ester) (PBAE) nanoparticles were investigated for the efficient therapy of pediatric CNS malignancies. The nanoparticles were tested for plasmid delivery of a suicide gene therapy to pediatric brain cancer models, specifically the herpes simplex virus type I thymidine kinase (HSVtk), which causes transfected cells to undergo regulated apoptosis. *In vivo*, PBAE-HSVtk treated groups had a greater survival rate in mice implanted with atypical teratoid/rhabdoid tumors ($P = 0.0083$ vs. control) and medulloblastoma ($P < 0.0001$ vs. control) (Choi et al., 2020). In another study, authors demonstrated supramolecular nanoparticle vectors by mixing three molecular building blocks, i.e., β-cyclodextrin-grafted branched polyethyleneimine, adamantane-grafted polyamidoamine dendrimer, and Ad-grafted poly(ethylene glycol) employed for introduction of Cas9/sgRNA-plasmid and Donor-RS1/GFP-plasmid in BALB/c-strain mice. Results suggested CRISPR/Cas9-mediated knockin of 3.0-kb Retinoschisin 1/green fluorescent protein gene into the Rosa26 site in mice retinas for X-Linked Juvenile Retinoschisis (Chou et al., 2020). Yang and co-workers exhibited supramolecular nanosubstrate (Ad-grafted silicon nanowire substrates)–mediated delivery to facilitate CRISPR-Cas9 knockin of the hemoglobin beta gene into the adeno-associated virus integration site 1 safe-harbor site of an engineered K562 3.21 cell line harboring the sickle cell disease mutation. Stepwise treatments of the two supramolecular nanoparticle vectors encapsulating a Cas9.single-guide RNA complex and a hemoglobin beta/green fluorescent protein–encoding plasmid, CRISPR-Cas9 knockin was efficaciously accomplished through endogenous homology-directed repair pathway (Yang et al., 2020). Zhang and co-workers reported co-delivery of Sorafenib and CRISPR/Cas9 via polyamidoamine-aptamer-coated hollow mesoporous silica nanoparticle. In all nine identical sites, the nanocomplex showed >60 percent EGFR-editing efficacy without off-target effects, inhibiting angiogenesis through controlling the EGFR-PI3K-Akt pathway, and exhibiting a synergistic effect on cell proliferation. This gene-chemo-combination therapy provided 85 percent tumor suppression in a mouse model with no apparent negative effects on normal tissues (Zhang et al., 2020).

2.3.2.3 Extra-Cellular Vesicles

Extracellular vesicles are an emerging new type of biological RNA delivery mechanism that is progressively being studied. Extracellular vesicles are nanosized particles discharged into the extracellular environment by generating cells to interact with and act on remote and nearby cells. Extracellular vesicles can transport a variety of cellular material, including DNA, messenger RNA, microRNA, and proteins to specific cells (Jiang et al., 2017; Massaro et al., 2020).

Recently, epithelial cell and cancer cell -derived micro-vesicles were investigated for delivery of CRISPR/Cas9 plasmids in tumors of HepG2 xenografts in vivo. qPCR results exhibited that miR-21 and miR-181a expression were upregulated in HepG2 cells treated with cancer cell -derived micro-vesicles. Furthermore, results suggested CRISPR/Cas9 conferring synergistic anti-tumor effect with sorafenib (He et al., 2020). In one study, authors report cancer derived exosomes for potential delivery of CRISPR/Cas9 plasmids in ovarian cancer tumors of SKOV3 xenograft mice. Results exhibited that CRISPR/Cas9-loaded exosomes can suppress expression of poly (ADP-ribose) polymerase-1 (PARP-1), resulting in the initiation of programmed cell death in ovarian cancer. Consequently, PARP-1 suppression via CRISPR/Cas9-mediated genome editing improved cisplatin chemosensitivity, indicating synergistic cytotoxicity (Kim et al., 2017). In another study, extracellular vesicles decorated via cholesterol to valency-controlled tetrahedral DNA nanostructures conjugated with DNA aptamer; were studied for tumor cell targeting. The intracellular transport of RNP by TDN1-extra-cellular vesicles efficiently achieved RNP's subsequent genome editing, resulting in GFP or WNT10B downregulation in certain cells. This approach was eventually used to lower WNT10B protein expression, which resulted in significant tumor growth inhibition *in vitro, ex vivo*, and *in vivo*

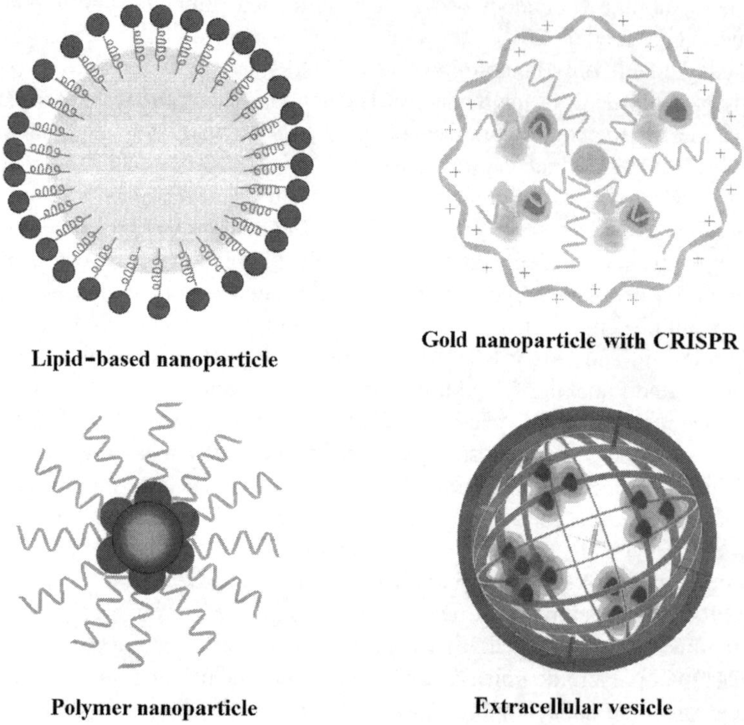

FIGURE 2.2 Nanoparticle delivery for gene editing tools.

(Zhuang et al., 2020). An extracellular nanovesicle-based ribonucleoprotein delivery system named NanoMEDIC was developed for CRISPR-Cas9 nuclease and gRNA. Authors demonstrate efficient genome editing in various hard to-transfect cell types, including human induced pluripotent stem cells, neurons, and myoblasts. NanoMEDIC also achieved over 90 percent exon skipping efficiencies in skeletal muscle cells derived from Duchenne muscular dystrophy patient induced pluripotent stem cells (Gee et al., 2020). In another study, restoration of dystrophin protein expression by exon skipping was achieved by CRISPR-Cas9 system in myoblasts from induced pluripotent stem cells (Ifuku et al., 2018). Figure 2.2 illustrates different nano-deliveries for gene editing tools. Table 2.2 tabulates various nano-tools for genome editing.

2.3.2.4 Miscellaneous Nanoparticles

Different approaches in nanostructures are investigated by various researchers for genome editing for varied purpose. Recently, Bi-functionalized aminoguanidine-PEGylated periodic mesoporous organosilica nanoparticles were investigated for intracellular delivery of the Cas9-sgRNA ribonucleoprotein complex. Gene-editing was observed with an efficiency of about 40 percent as measured by GFP gene knockdown of HT1080-GFP cells, with no notable change in the morphology of the cells (Salekdeh et al., 2021). In another study, biomimetic cancer cell coated zeolitic imidazolate frameworks were reported for genome editing carrying CRISPR-Cas9. Incubation of C3-ZIFMCF with MCF-7, HeLa, HDFn, and aTC cell lines showed the highest uptake by MCF-7 cells and negligible uptake by the healthy cells. A three-fold repression in the enhanced green fluorescent protein expression was observed (Alyami et al., 2020). CRISPR-Cas12a based nucleic acid amplification-free fluorescent biosensor was developed to detect cfDNA by a metal-enhanced fluorescence using DNA-functionalized Au nanoparticle to detect breast cancer gene-1 with very high sensitivity in 30 minutes (Choi et al., 2021).

TABLE 2.2
Nano Delivery in Genome Editing

Type of Nano Delivery	Genome Editing	Knockout Protein	Reference
Lipid nanoparticle	Cas9 messenger RNA and sgRNA	Proprotein convertase subtilisin/ kexin type 9	Liu et al. (2019)
Lipid nanoparticle	sgRNA	Mouse transthyretin (Ttr) gene	Finn et al. (2018)
Cationic polymer polyethyleneimine-β-cyclodextrin	Cas9 and sgRNA	Hemoglobin subunit beta and rhomboid 5 homolog 1	Zhang et al. (2019b)
Cationic liposomes	Cas9 and sgRNA plasmids	GFP gene	Hosseini et al. (2019)
Supramolecular nanoparticles	Cas9•sgRNA ribonucleoprotein	dystrophin gene	Ban et al. (2021)
Poly(lactic-co-glycolic acid) nanoparticle	CRISPR–Cas9 plasmid	Induce expression of bacterial Cas9 in murine bone marrow	Jo et al. (2020)
Polyethylenimine based magnetic nanoparticles	CRISPR–Cas9 plasmid	Blue florescent protein gene and green florescent protein gene	Rohiwal et al. (2020)
Extracellular vesicles	CRISPR–Cas9 plasmid	Myc oncogene	Xu et al. (2020)
Exosome	CRISPR–Cas9 plasmid	Green florescent protein gene	Ye et al. (2020)
Exosome-liposome hybrid nanoparticles	CRISPR–Cas9 plasmid	Mesenchymal stem cells	Lin et al. (2018)
Liposome-Templated Hydrogel Nanoparticles	CRISPR–Cas9 plasmid	Polo-like kinase 1	Chen et al. (2017)
Aptamer-cationic liposome	CRISPR–Cas9 plasmid	Polo-like kinase 1	Zhen et al. (2017)
Aptamer-functionalized lipopolymer	CRISPR–Cas9 plasmid	Vascular endothelial growth factor A	Liang et al. (2017)
Extracellular vesicles	CRISPR–Cas9 plasmid	CHME-5 cells and HIV LTR (long terminal repeat regions)	Campbell et al. (2019)
Lipid nanoparticles	CRISPR-Cas9 mRNA	Angptl3 gene	Qiu et al. (2021)
Lipid nanoparticles	CRISPR-Cas9	PLK1	Rosenblum et al. (2020)
PLGA-Nanoparticles	CRISPR-Cas9	CD34+ hematopoietic stem and progenitor cells	Cruz et al. (2021)

2.4 FUTURE PERSPECTIVES

The exploratory research to date has asserted important contributions of genome editing systems to utilize treatment modalities for diverse human ailments, with the CRISPR/Cas9 system being incredibly beneficial by interacting directly with target gene position or providing multipurpose tools. Hereafter, a blend of consolidated CRISPR screening and current data on the genetic and epigenetic properties of deadlier diseases will be able to find synthetic fatal interactions in the genome and speed up the emergence of new therapeutic targets. Overall, the technology has potential to elucidate the disease mechanism and progression, thereby providing novel therapies. Nanotechnology provides a toolbox for the development of genome editing tools. Amalgamation of nanotechnology with gene editing tools will surface the way for genome editing in clinical disease treatment in the coming years.

ABBREVIATIONS

DNA deoxyribonucleic acid
RNA ribonucleic acid

NHEJ non-homologous end joining
HDR homology-directed repair
SpCas9 Cas protein from Streptococcus pyogenes
PAM protospacer adjacent motif
DSB Double-strand break
MPS I Mucopolysaccharidosis I
EGFR Estimated glomerular filtration rate
GFP Green fluorescent protein
qPCR quantitative polymerase chain reaction
PLGA poly D,L-lactic-co-glycolic acid

REFERENCES

Alyami MZ, Alsaiari SK, Li Y, Qutub SS, Aleisa FA, Sougrat R, Merzaban JS, & Khashab NM. Cell-type-specific CRISPR/Cas9 delivery by biomimetic metal organic frameworks. *J Am Chem Soc*. 2020; 142(4): 1715–1720. Doi: 10.1021/jacs.9b11638.

Antony JS, Latifi N, Ashiqul Haque AKM, Lamsfus-Calle A, Daniel-Moreno A, Graeter S, Baskaran P, et al. Gene correction of HBB mutations in CD34+ hematopoietic stem cells using Cas9 mRNA and ssODN donors. *Mol Cell Pediatr*. 2018; 5(1): 9. Doi: 10.1186/s40348-018-0086-1.

Bai B, Myklebust JH, & Wälchli S. Gene editing in B-lymphoma cell lines using CRISPR/Cas9 technology. *Methods Mol Biol*. 2020; 2115: 445–454. Doi: 10.1007/978-1-0716-0290-4_25.

Ban Q, Yang P, Chou SJ, Qiao L, Xia H, Xue J, Wang F, et al. Supramolecular nanosubstrate-mediated delivery for CRISPR/Cas9 gene disruption and deletion. *Small*. 2021: e2100546. Doi: 10.1002/smll.202100546.

Barua S, Ramos J, Potta T, Taylor D, Huang HC, Montanez G, & Rege K. Discovery of cationic polymers for non-viral gene delivery using combinatorial approaches. *Comb Chem High Throughput Screen*. 2011; 14(10): 908–924. Doi: 10.2174/138620711797537076.

Bibikova M, Golic M, Golic KG, & Carroll D. Targeted chromosomal cleavage and mutagenesis in Drosophila using zinc-finger nucleases. *Genetics*. 2002; 161(3): 1169–1175.

Brabetz O, Alla V, Angenendt L, Schliemann C, Berdel WE, Arteaga MF, & Mikesch JH. RNA-guided CRISPR-Cas9 system-mediated engineering of acute myeloid leukemia mutations. *Mol Ther Nucleic Acids*. 2017; 6: 243–248. Doi: 10.1016/j.omtn.2016.12.012.

Brouns SJ, Jore MM, Lundgren M, Westra ER, Slijkhuis RJ, Snijders AP, Dickman MJ, Makarova KS, Koonin EV, & van der Oost J. Small CRISPR RNAs guide antiviral defense in prokaryotes. *Science*. 2008; 321(5891): 960–964. Doi: 10.1126/science.1159689.

Buck J, Grossen P, Cullis PR, Huwyler J, & Witzigmann D. Lipid-based DNA therapeutics: Hallmarks of non-viral gene delivery. *ACS Nano*. 2019; 13: 3754–3782. Doi: 10.1021/acsnano.8b07858.

Campbell LA, Coke LM, Richie CT, Fortuno LV, Park AY, & Harvey BK. Gesicle-mediated delivery of CRISPR/Cas9 ribonucleoprotein complex for inactivating the HIV provirus. *Mol Ther*. 2019; 27(1):151–163. Doi: 10.1016/j.ymthe.2018.10.002.

Capecchi MR. Altering the genome by homologous recombination. *Science*. 1989; 244(4910):1288–1292. Doi: 10.1126/science.2660260.

Chadwick AC, & Musunuru K. Treatment of dyslipidemia using CRISPR/Cas9 genome editing. *Curr Atheroscler Rep*. 2017; 19(7): 32. Doi: 10.1007/s11883-017-0668-8.

Chen H, Shi M, Gilam A, Zheng Q, Zhang Y, Afrikanova I, Li J, Gluzman Z, Jiang R, Kong LJ, & Chen-Tsai RY. Hemophilia A ameliorated in mice by CRISPR-based in vivo genome editing of human factor VIII. *Sci Rep*. 2019; 9(1): 16838. Doi: 10.1038/s41598-019-53198-y.

Chen Y, Jiang H, Wang T, He D, Tian R, Cui Z, Tian X, et al. In vitro and in vivo growth inhibition of human cervical cancer cells via human papillomavirus E6/E7 mRNAs' cleavage by CRISPR/Cas13a system. *Antiviral Res*. 2020; 178: 104794. Doi: 10.1016/j.antiviral.2020.104794.

Chen Z, Liu F, Chen Y, Liu J, Wang X, Chen AT, Deng G, et al. Targeted delivery of CRISPR/Cas9-mediated cancer gene therapy via liposome-templated hydrogel nanoparticles. *Adv Funct Mater*. 2017; 27(46): 1703036. Doi: 10.1002/adfm.201703036.

Choi J, Rui Y, Kim J, Gorelick N, Wilson DR, Kozielski K, Mangraviti A, et al. Nonviral polymeric nanoparticles for gene therapy in pediatric CNS malignancies. *Nanomedicine*. 2020; 23: 102115. Doi: 10.1016/j.nano.2019.102115.

Choi JH, Lim J, Shin M, Paek SH, & Choi JW. CRISPR-Cas12a-based nucleic acid amplification-free DNA biosensor via Au nanoparticle-assisted metal-enhanced fluorescence and colorimetric analysis. *Nano Lett.* 2021; 21(1): 693–699. Doi: 10.1021/acs.nanolett.0c04303

Chou SJ, Yang P, Ban Q, Yang YP, Wang ML, Chien CS, Chen SJ, et al. Dual supramolecular nanoparticle vectors enable CRISPR/Cas9-mediated knockin of retinoschisin 1 gene: A potential nonviral therapeutic solution for X-linked juvenile retinoschisis. *Adv Sci (Weinh)*. 2020; 7(10): 1903432. Doi: 10.1002/advs.201903432.

Christian ML, Demorest ZL, Starker CG, Osborn MJ, Nyquist MD, Zhang Y, Carlson DF, et al. Targeting G with TAL effectors: A comparison of activities of TALENs constructed with NN and NK repeat variable di-residues. *PLoS One*. 2012; 7(9): e45383. Doi: 10.1371/journal.pone.0045383.

Chylinski K, Makarova KS, Charpentier E, & Koonin EV. Classification and evolution of type II CRISPR-Cas systems. *Nucleic Acids Res*. 2014; 42(10): 6091–6105. Doi: 10.1093/nar/gku241.

Conway A, Mendel M, Kim K, McGovern K, Boyko A, Zhang L, Miller JC, et al. Non-viral delivery of zinc finger nuclease mRNA enables highly efficient in vivo genome editing of multiple therapeutic gene targets. *Mol Ther*. 2019; 27(4): 866–877. Doi: 10.1016/j.ymthe.2019.03.003.

Cruz LJ, van Dijk T, Vepris O, Li Tarcy MWY, Schomann T, Baldazzi F, Kurita R, et al. PLGA-nanoparticles for intracellular delivery of the CRISPR-complex to elevate fetal globin expression in erythroid cells. *Biomaterials*. 2021; 268: 120580. Doi: 10.1016/j.biomaterials.2020.120580.

de Carvalho TG, Schuh R, Pasqualim G, Pellenz FM, Filippi-Chiela EC, Giugliani R, Baldo G, & Matte U. CRISPR-Cas9-mediated gene editing in human MPS I fibroblasts. *Gene*. 2018; 678: 33–37. Doi: 10.1016/j.gene.2018.08.004.

Ding Q, Strong A, Patel KM, Ng SL, Gosis BS, Regan SN, Cowan CA, Rader DJ, & Musunuru K. Permanent alteration of PCSK9 with in vivo CRISPR-Cas9 genome editing. *Circ Res*. 2014; 115(5): 488–492. Doi: 10.1161/CIRCRESAHA.115.304351.

Dorsey BW, Huang L, & Mondragon A. Structural organization of a Type III-A CRISPR effector subcomplex determined by X-ray crystallography and cryo-EM. *Nucleic Acids Res*. 2019; 47(7):3765–3783. Doi: 10.1093/nar/gkz079.

Dupuy A, Valton J, Leduc S, Armier J, Galetto R, Gouble A, Lebuhotel C, et al. Targeted gene therapy of xeroderma pigmentosum cells using meganuclease and TALEN™. *PLoS One*. 2013; 8(11): e78678. Doi: 10.1371/journal.pone.0078678.

El Refaey M, Xu L, Gao Y, Canan BD, Adesanya TMA, Warner SC, Akagi K, et al. In vivo genome editing restores dystrophin expression and cardiac function in dystrophic mice. *Circ Res*. 2017; 121(8): 923–929. Doi: 10.1161/CIRCRESAHA.117.310996.

Finn JD, Smith AR, Patel MC, Shaw L, Youniss MR, van Heteren J, Dirstine T, et al. A single administration of CRISPR/Cas9 lipid nanoparticles achieves robust and persistent in vivo genome editing. *Cell Rep*. 2018; 22(9): 2227–2235. Doi: 10.1016/j.celrep.2018.02.014.

Foss DV, Hochstrasser ML, & Wilson RC. Clinical applications of CRISPR-based genome editing and diagnostics. *Transfusion*. 2019; 59(4): 1389–1399. Doi: 10.1111/trf.15126.

Gao X, Tao Y, Lamas V, Huang M, Yeh WH, Pan B, Hu YJ, et al. Treatment of autosomal dominant hearing loss by in vivo delivery of genome editing agents. *Nature*. 2018; 553(7687): 217–221. Doi: 10.1038/nature25164.

Gee P, Lung MSY, Okuzaki Y, Sasakawa N, Iguchi T, Makita Y, Hozumi H, et al. Extracellular nanovesicles for packaging of CRISPR-Cas9 protein and sgRNA to induce therapeutic exon skipping. *Nat Commun*. 2020; 11(1): 1334. Doi: 10.1038/s41467-020-14957-y.

Gong B, Shin M, Sun J, Jung CH, Bolt EL, van der Oost J, & Kim JS. Molecular insights into DNA interference by CRISPR-associated nuclease-helicase Cas3. *Proc Natl Acad Sci U S A* 2014; 111(46): 16359–16364. Doi: 10.1073/pnas.1410806111.

Guo P, Yang J, Huang J, Auguste DT, & Moses MA. Therapeutic genome editing of triple-negative breast tumors using a noncationic and deformable nanolipogel. *Proc Natl Acad Sci U S A*. 2019; 116(37): 18295–18303. Doi: 10.1073/pnas.1904697116.

György B, Lööv C, Zaborowski MP, Takeda S, Kleinstiver BP, Commins C, Kastanenka K et al. CRISPR/Cas9 mediated disruption of the Swedish APP allele as a therapeutic approach for early-onset Alzheimer's disease. *Mol Ther Nucleic Acids*. 2018; 11: 429–440. Doi: 10.1016/j.omtn.2018.03.007.

He C, Jaffar Ali D, Xu H, Kumaravel S, Si K, Li Y, Sun B, Ma J, & Xiao Z. Epithelial cell -derived microvesicles: A safe delivery platform of CRISPR/Cas9 conferring synergistic anti-tumor effect with sorafenib. *Exp Cell Res*. 2020; 392(2): 112040. Doi: 10.1016/j.yexcr.2020.112040.

Hosseini ES, Nikkhah M, & Hosseinkhani S. Cholesterol-rich lipid-mediated nanoparticles boost of transfection efficiency, utilized for gene editing by CRISPR-Cas9. *Int J Nanomedicine*. 2019; 14: 4353–4366.

Hsu PD, Lander ES, & Zhang F. Development and applications of CRISPR-Cas9 for genome engineering. *Cell*. 2014; 157(6): 1262–1278. Doi: 10.1016/j.cell.2014.05.010.

Huo Y, Nam KH, Ding F, Lee H, Wu L, Xiao Y, Farchione MD Jr et al. Structures of CRISPR Cas3 offer mechanistic insights into Cascade-activated DNA unwinding and degradation. *Nat Struct Mol Biol*. 2014; 21(9): 771–777. Doi: 10.1038/nsmb.2875.

Ifuku M, Iwabuchi KA, Tanaka M, Lung MSY, & Hotta A. Restoration of dystrophin protein expression by exon skipping utilizing CRISPR-Cas9 in myoblasts derived from DMD patient iPS cells. *Methods Mol Biol*. 2018; 1828: 191–217. Doi: 10.1007/978-1-4939-8651-4_12.

Ishino Y, Krupovic M, & Forterre P. History of CRISPR-Cas from encounter with a mysterious repeated sequence to genome editing technology. *J Bacteriol*. 2018; 200(7): e00580. Doi: 10.1128/JB.00580-17.

Ishino Y, Shinagawa H, Makino K, Amemura M, & Nakata A. Nucleotide sequence of the iap gene, responsible for alkaline phosphatase isozyme conversion in Escherichia coli, and identification of the gene product. *J Bacteriol*. 1987; 169(12): 5429–5433. Doi: 10.1128/jb.169.12.5429-5433.1987.

Jacinto FV, Link W, & Ferreira BI. CRISPR/Cas9-mediated genome editing: From basic research to translational medicine. *J Cell Mol Med*. 2020; 24(7): 3766–3778. Doi: 10.1111/jcmm.14916.

Jiang L, Vader P, & Schiffelers R. Extracellular vesicles for nucleic acid delivery: Progress and prospects for safe RNA-based gene therapy. *Gene Ther*. 2017; 24: 157–166. Doi: 10.1038/gt.2017.8

Jo A, Ringel-Scaia VM, McDaniel DK, Thomas CA, Zhang R, Riffle JS, Allen IC, & Davis RM. Fabrication and characterization of PLGA nanoparticles encapsulating large CRISPR-Cas9 plasmid. *J Nanobiotechnology*. 2020; 18(1): 16. Doi: 10.1186/s12951-019-0564-1.

Karvelis T, Gasiunas G, Miksys A, Barrangou R, Horvath P, & Siksnys V. crRNA and tracrRNA guide Cas9-mediated DNA interference in streptococcus thermophilus. *RNA Biol*. 2013; 10(5): 841–851. Doi: 10.4161/rna.24203.

Kennedy EM, Kornepati AV, Goldstein M, Bogerd HP, Poling BC, Whisnant AW, Kastan MB, & Cullen BR. Inactivation of the human papillomavirus E6 or E7 gene in cervical carcinoma cells by using a bacterial CRISPR/Cas RNA-guided endonuclease. *J Virol*. 2014; 88(20): 11965–11972. Doi: 10.1128/JVI.01879-14.

Kim SM, Yang Y, Oh SJ, Hong Y, Seo M, & Jang M. Cancer-derived exosomes as a delivery platform of CRISPR/Cas9 confer cancer cell tropism-dependent targeting. *J Control Release*. 2017; 266: 8–16. Doi: 10.1016/j.jconrel.2017.09.013.

Klug A. The discovery of zinc fingers and their development for practical applications in gene regulation and genome manipulation. *Q Rev Biophys*. 2010; 43(1): 1–21. Doi: 10.1017/S0033583510000089.

Kou L, Sun J, Zhai Y, & He Z. The endocytosis and intracellular fate of nanomedicines: Implication for rational design. *Asian J Pharm Sci*. 2013; 8(1): 1–10. Doi: 10.1016/j.ajps.2013.07.001.

Lao YH, Li M, Gao MA, Shao D, Chi CW, Huang D, Chakraborty S et al. HPV oncogene manipulation using nonvirally delivered CRISPR/Cas9 or natronobacterium gregoryi argonaute. *Adv Sci (Weinh)*. 2018; 5(7): 1700540. Doi: 10.1002/advs.201700540.

Li T, Huang S, Jiang WZ, Wright D, Spalding MH, Weeks DP, & Yang B. TAL nucleases (TALNs): Hybrid proteins composed of TAL effectors and FokI DNA-cleavage domain. *Nucleic Acids Res*. 2011; 39(1): 359–372. Doi: 10.1093/nar/gkq704.

Li Y, Glass Z, Huang M, Chen ZY, & Xu Q. Ex vivo cell-based CRISPR/Cas9 genome editing for therapeutic applications. *Biomaterials*. 2020a; 234: 119711. Doi: 10.1016/j.biomaterials.2019.119711.

Li Y, Li X, Qu J, Luo D, & Hu Z. Cas9 mediated correction of β-catenin mutation and restoring the expression of protein phosphorylation in colon cancer HCT-116 cells decrease cell proliferation in vitro and hamper tumor growth in mice in vivo. *Onco Targets Ther*. 2020b; 13: 17–29. Doi: 10.2147/OTT.S225556.

Liang C, Li F, Wang L, Zhang ZK, Wang C, He B, Li J et al. Tumor cell-targeted delivery of CRISPR/Cas9 by aptamer-functionalized lipopolymer for therapeutic genome editing of VEGFA in osteosarcoma. *Biomaterials*. 2017; 147: 68–85. Doi: 10.1016/j.biomaterials.2017.09.015.

Lin Y, Wu J, Gu W, Huang Y, Tong Z, Huang L, & Tan J. Exosome-liposome hybrid nanoparticles deliver CRISPR/Cas9 system in MSCs. *Adv Sci (Weinh)*. 2018; 5(4): 1700611. Doi: 10.1002/advs.201700611.

Liu C, Zhang L, Liu H, & Cheng K. Delivery strategies of the CRISPR-Cas9 gene-editing system for therapeutic applications. *J Control Release*. 2017; 266: 17–26. Doi: 10.1016/j.jconrel.2017.09.012.

Liu J, Chang J, Jiang Y, Meng X, Sun T, Mao L, Xu Q, & Wang M. Fast and efficient CRISPR/Cas9 genome editing in vivo enabled by bioreducible lipid and messenger RNA nanoparticles. *Adv Mater*. 2019; 31(33): e1902575. Doi: 10.1002/adma.201902575

Ma N, Liao B, Zhang H, Wang L, Shan Y, Xue Y, Huang K et al. Transcription activator-like effector nuclease (TALEN)-mediated gene correction in integration-free β-thalassemia induced pluripotent stem cells. *J Biol Chem*. 2013; 288(48): 34671–34679. Doi: 10.1074/jbc.M113.496174.

Mahmoudian-Sani MR, Farnoosh G, Mahdavinezhad A, & Saidijam M. CRISPR genome editing and its medical applications. *Biotechnol Biotechnol Equip*. 2018; 32(2): 286–292. Doi: 10.1080/13102818.2017.1406823.

Mak AN, Bradley P, Bogdanove AJ, & Stoddard BL. TAL effectors: Function, structure, engineering and applications. *Curr Opin Struct Biol*. 2013; 23(1): 93–99. Doi: 10.1016/j.sbi.2012.11.001.

Makarova KS, & Koonin EV. Annotation and classification of CRISPR-Cas systems. *Methods Mol Biol*. 2015; 1311: 47–75. Doi: 10.1007/978-1-4939-2687-9_4.

Massaro C, Sgueglia G, Frattolillo V, Baglio SR, Altucci L, & Dell'Aversana C. Extracellular vesicle-based nucleic acid delivery: Current advances and future perspectives in cancer therapeutic strategies. *Pharmaceutics*. 2020; 12(10): 980. Doi: 10.3390/pharmaceutics12100980.

Memi F, Ntokou A, & Papangeli I. CRISPR/Cas9 gene-editing: Research technologies, clinical applications and ethical considerations. *Semin. Perinatol*. 2018; 42(8): 487–500. Doi: 10.1053/j.semperi.2018.09.003.

Metje-Sprink J, Menz J, Modrzejewski D, & Sprink T. DNA-free genome editing: Past, present and future. *Front Plant Sci*. 2019; 9: 1957. Doi: 10.3389/fpls.2018.01957.

Miller JB, Zhang S, Kos P, Xiong H, Zhou K, Perelman SS, Zhu H, & Siegwart DJ. Non-viral CRISPR/Cas gene editing in vitro and in vivo enabled by synthetic nanoparticle co-delivery of Cas9 mRNA and sgRNA. *Angew Chem Int Ed Engl*. 2017; 56(4): 1059–1063. Doi: 10.1002/anie.201610209.

Mills EM, Barlow VL, Luk LYP, & Tsai YH. Applying switchable Cas9 variants to in vivo gene editing for therapeutic applications. *Cell Biol Toxicol*. 2020; 36(1): 17–29. Doi: 10.1007/s10565-019-09488-2.

Mir A, Edraki A, Lee J, & Sontheimer EJ. Type II-C CRISPR-Cas9 biology, mechanism, and application. *ACS Chem Biol*. 2018; 13(2): 357–365. Doi: 10.1021/acschembio.7b00855.

Mohammadinejad R, Dehshahri A, Sagar Madamsetty V, Zahmatkeshan M, Tavakol S, Makvandi P, Khorsandi D et al. In vivo gene delivery mediated by non-viral vectors for cancer therapy. *J Control Release*. 2020; 325: 249–275. Doi: 10.1016/j.jconrel.2020.06.038.

Moore R, Chandrahas A, & Bleris L. Transcription activator-like effectors: A toolkit for synthetic biology. *ACS Synth Biol*. 2014; 3(10): 708–716. Doi: 10.1021/sb400137b.

Mueller AL, Corbi-Verge C, Giganti DO, Ichikawa DM, Spencer JM, MacRae M, Garton M, Kim PM, & Noyes MB. The geometric influence on the Cys2His2 zinc finger domain and functional plasticity. *Nucleic Acids Res*. 2020; 48(11): 6382–6402. Doi: 10.1093/nar/gkaa291.

Nakata A, Amemura M, & Makino K. Unusual nucleotide arrangement with repeated sequences in the Escherichia coli K-12 chromosome. *J Bacteriol*. 1989; 171(6): 3553–3556. Doi: 10.1128/jb.171.6.3553-3556.1989.

Noh SM, Han SE, Shim G, Lee KE, Kim CW, Han SS, Choi Y, Kim YK, Kim WK, & Oh YK. Tocopheryl oligochitosan-based self-assembling oligomersomes for siRNA delivery. *Biomaterials*. 2011; 32: 849–857. Doi: 10.1016/j.biomaterials.2010.09.027.

Ortiz-Virumbrales M, Moreno CL, Kruglikov I, Marazuela P, Sproul A, Jacob S, Zimmer M et al. CRISPR/Cas9-correctable mutation-related molecular and physiological phenotypes in iPSC-derived Alzheimer's PSEN2 N141I neurons. *Acta Neuropathol Commun*. 2017; 5(1): 77. Doi: 10.1186/s40478-017-0475-z.

Ou L, DeKelver RC, Rohde M, Tom S, Radeke R, St Martin SJ, Santiago Y et al. ZFN-mediated in vivo genome editing corrects murine hurler syndrome. *Mol Ther*. 2019; 27(1): 178–187. Doi: 10.1016/j.ymthe.2018.10.018.

Ou L, Przybilla MJ, Ahlat O, Kim S, Overn P, Jarnes J, O'Sullivan MG, & Whitley CB. A highly efficacious PS gene editing system corrects metabolic and neurological complications of mucopolysaccharidosis type I. *Mol Ther*. 2020; 28(6): 1442–1454. Doi: 10.1016/j.ymthe.2020.03.018

Overlack N, Goldmann T, Wolfrum U, & Nagel-Wolfrum K. Gene repair of an Usher syndrome causing mutation by zinc-finger nuclease mediated homologous recombination. *Invest Ophthalmol Vis Sci*. 2012; 53(7): 4140–4146. Doi: 10.1167/iovs.12-9812.

Piganeau M, Ghezraoui H, De Cian A, Guittat L, Tomishima M, Perrouault L, René O et al. Cancer translocations in human cells induced by zinc finger and TALE nucleases. *Genome Res*. 2013; 23(7): 1182–1193. Doi: 10.1101/gr.147314.112.

Pöhler M, Guttmann S, Nadzemova O, Lenders M, Brand E, Zibert A, Schmidt HH, & Sandfort V. CRISPR/Cas9-mediated correction of mutated copper transporter ATP7B. *PLoS One*. 2020; 15(9): e0239411. Doi: 10.1371/journal.pone.0239411.

Porteus MH. Mammalian gene targeting with designed zinc finger nucleases. *Mol Ther*. 2006; 13(2): 438–446. Doi: 10.1016/j.ymthe.2005.08.003.

Qiu M, Glass Z, Chen J, Haas M, Jin X, Zhao X, Rui X et al. Lipid nanoparticle-mediated codelivery of Cas9 mRNA and single-guide RNA achieves liver-specific in vivo genome editing of Angptl3. *Proc Natl Acad Sci U S A*. 2021; 118(10): e2020401118. Doi: 10.1073/pnas.2020401118.

Rath D, Amlinger L, Rath A, & Lundgren M. The CRISPR-Cas immune system: Biology, mechanisms and applications. *Biochimie*. 2015; 117: 119–128. Doi: 10.1016/j.biochi.2015.03.025.

Rodríguez-Rodríguez DR, Ramírez-Solís R, Garza-Elizondo MA, Garza-Rodríguez ML, & Barrera-Saldaña HA. Genome editing: A perspective on the application of CRISPR/Cas9 to study human diseases (Review). *Int J Mol Med*. 2019; 43(4): 1559–1574. Doi: 10.3892/ijmm.2019.4112.

Rohiwal SS, Dvorakova N, Klima J, Vaskovicova M, Senigl F, Slouf M, Pavlova E et al. Polyethylenimine based magnetic nanoparticles mediated non-viral CRISPR/Cas9 system for genome editing. *Sci Rep*. 2020; 10(1): 4619. Doi: 10.1038/s41598-020-61465-6.

Rosenblum D, Gutkin A, Kedmi R, Ramishetti S, Veiga N, Jacobi AM, Schubert MS et al. CRISPR-Cas9 genome editing using targeted lipid nanoparticles for cancer therapy. *Sci Adv*. 2020; 6(47): eabc9450. Doi: 10.1126/sciadv.abc9450.

Rothstein RJ. One-step gene disruption in yeast. *Methods Enzymol*. 1983; 101: 202–211. Doi: 10.1016/0076-6879(83)01015-0.

Rouet P, Smih F, & Jasin M. Introduction of double-strand breaks into the genome of mouse cells by expression of a rare-cutting endonuclease. *Mol Cell Biol*. 1994; 14(12): 8096–8106. Doi: 10.1128/mcb.14.12.8096-8106.1994.

Saleh-Gohari N, & Helleday T. Conservative homologous recombination preferentially repairs DNA double-strand breaks in the S phase of the cell cycle in human cells. *Nucleic Acids Res*. 2004; 32(12): 3683–3688. Doi: 10.1093/nar/gkh703.

Salekdeh PR, Ma'mani L, Tavakkoly-Bazzaz J, Mousavi H, Modarressi MH, & Salekdeh GH. Bi-functionalized aminoguanidine-PEGylated periodic mesoporous organosilica nanoparticles: A promising nanocarrier for delivery of Cas9-sgRNA ribonucleoproteine. *J Nanobiotechnology*. 2021; 19(1): 95. Doi: 10.1186/s12951-021-00838-z.

Scherer S, & Davis RW. Replacement of chromosome segments with altered DNA sequences constructed in vitro. *Proc Natl Acad Sci U S A*. 1979; 76: 4951–4955. Doi: 10.1073/pnas.76.10.4951.

Schmid RS, Deng X, Panikker P, Msackyi M, Breton C, & Wilson JM. CRISPR/Cas9 directed to the Ube3a antisense transcript improves Angelman syndrome phenotype in mice. *J Clin Invest*. 2021; 131(5): e142574. Doi: 10.1172/JCI142574.

Schuh RS, de Carvalho TG, Giugliani R, Matte U, Baldo G, & Teixeira HF. Gene editing of MPS I human fibroblasts by co-delivery of a CRISPR/Cas9 plasmid and a donor oligonucleotide using nanoemulsions as nonviral carriers. *Eur J Pharm Biopharm*. 2018a; 122: 158–166. Doi: 10.1016/j.ejpb.2017.10.017.

Schuh RS, Gonzalez EA, Tavares AMV, Seolin BG, Elias LS, Vera LNP, Kubaski F et al. Neonatal nonviral gene editing with the CRISPR/Cas9 system improves some cardiovascular, respiratory, and bone disease features of the mucopolysaccharidosis I phenotype in mice. *Gene Ther*. 2020; 27(1–2): 74–84. Doi: 10.1038/s41434-019-0113-4.

Schuh RS, Poletto É, Pasqualim G, Tavares AMV, Meyer FS, Gonzalez EA, Giugliani R et al. In vivo genome editing of mucopolysaccharidosis I mice using the CRISPR/Cas9 system. *J Control Release*. 2018b; 288: 23–33. Doi: 10.1016/j.jconrel.2018.08.031.

Smith J, Bibikova M, Whitby FG, Reddy AR, Chandrasegaran S, & Carroll D. Requirements for double-strand cleavage by chimeric restriction enzymes with zinc finger DNA-recognition domains. *Nucleic Acids Res*. 2000; 28(17): 3361–3369. Doi: 10.1093/nar/28.17.3361.

Smithies O, Gregg RG, Boggs SS, Koralewski MA, & Kucherlapati RS. Insertion of DNA sequences into the human chromosomal beta-globin locus by homologous recombination. *Nature*. 1985; 317(6034): 230–234. Doi: 10.1038/317230a0.

Sun N, & Zhao H. Seamless correction of the sickle cell disease mutation of the HBB gene in human induced pluripotent stem cells using TALENs. *Biotechnol Bioeng*. 2014; 111(5): 1048–1053. Doi: 10.1002/bit.25018.

Sun W, Ji W, Hall JM, Hu Q, Wang C, Beisel CL, & Gu Z. Self-assembled DNA nanoclews for the efficient delivery of CRISPR-Cas9 for genome editing. *Angew Chem Int Ed Engl*. 2015; 54(41): 12029–12033. Doi: 10.1002/anie.201506030.

Tang Y, & Fu Y Class 2 CRISPR/Cas: An expanding biotechnology toolbox for and beyond genome editing. *Cell Biosci*. 2018; 8: 59. Doi: 10.1186/s13578-018-0255-x.

Thomas KR, Folger KR, & Capecchi MR. High frequency targeting of genes to specific sites in the mammalian genome. *Cell*. 1986; 44: 419–428. Doi: 10.1016/0092-8674(86)90463-0.

Thomas TJ, Tajmir-Riahi HA, & Pillai CKS. Biodegradable polymers for gene delivery. *Molecules*. 2019; 24(20): 3744. Doi: 10.3390/molecules24203744.

Tsanova T, Stefanova L, Topalova L, Atanasov A, & Pantchev I. DNA-free gene editing in plants: A brief overview. *Biotechnol Biotechnol Equip*. 2021; 35(1): 131–138. Doi: 10.1080/13102818.2020.1858159.

Urnov FD, Miller JC, Lee YL, Beausejour CM, Rock JM, Augustus S, Jamieson AC et al. Highly efficient endogenous human gene correction using designed zinc-finger nucleases. *Nature*. 2005; 435(7042): 646–651. Doi: 10.1038/nature03556.

Urnov FD, Rebar EJ, Holmes MC, Zhang HS, & Gregory PD. Genome editing with engineered zinc finger nucleases. *Nat Rev Genet*. 2010; 11(9): 636–646. Doi: 10.1038/nrg2842.

Valletta S, Dolatshad H, Bartenstein M, Yip BH, Bello E, Gordon S, Yu Y, Shaw J et al. ASXL1 mutation correction by CRISPR/Cas9 restores gene function in leukemia cells and increases survival in mouse xenografts. *Oncotarget*. 2015; 6(42): 44061–44071. Doi: 10.18632/oncotarget.6392.

Walton J, Blagih J, Ennis D, Leung E, Dowson S, Farquharson M, Tookman LA et al. CRISPR/Cas9-mediated Trp53 and Brca2 knockout to generate improved murine models of ovarian high-grade serous carcinoma. *Cancer Res*. 2016; 76(20): 6118–6129. Doi: 10.1158/0008-5472.CAN-16-1272.

Wang HX, Song Z, Lao YH, Xu X, Gong J, Cheng D, Chakraborty S et al. Nonviral gene editing via CRISPR/Cas9 delivery by membrane-disruptive and endosomolytic helical polypeptide. *Proc Natl Acad Sci U S A*. 2018; 115(19): 4903–4908. Doi: 10.1073/pnas.1712963115.

Wang J, Zhang C, & Feng B. The rapidly advancing Class 2 CRISPR-Cas technologies: A customizable toolbox for molecular manipulations. *J Cell Mol Med*. 2020; 24(6): 3256–3270. Doi: 10.1111/jcmm.15039.

Wang M, Glass ZA, & Xu Q. Non-viral delivery of genome-editing nucleases for gene therapy. *Gene Ther*. 2017; 24(3):144–150. Doi: 10.1038/gt.2016.72.

Wang X, Raghavan A, Chen T, Qiao L, Zhang Y, Ding Q, & Musunuru K. CRISPR-Cas9 targeting of PCSK9 in human hepatocytes in vivo-brief report. *Arterioscler Thromb Vasc Biol*. 2016; 36(5): 783–786. Doi: 10.1161/ATVBAHA.116.307227.

Wolfe SA, Nekludova L, & Pabo CO. DNA recognition by Cys2His2 zinc finger proteins. *Annu Rev Biophys Biomol Struct*. 2000; 29: 183–212. Doi: 10.1146/annurev.biophys.29.1.183.

Xia E, Zhang Y, Cao H, Li J, Duan R, & Hu J. TALEN-mediated gene targeting for cystic fibrosis-gene therapy. *Genes (Basel)*. 2019; 10(1): 39. Doi: 10.3390/genes10010039.

Xu L, Lau YS, Gao Y, Li H, & Han R. Life-long AAV-mediated CRISPR genome editing in dystrophic heart improves cardiomyopathy without causing serious lesions in mdx mice. *Mol Ther*. 2019; 27(8): 1407–1414. Doi: 10.1016/j.ymthe.2019.05.001.

Xu P, Tong Y, Liu XZ, Wang TT, Cheng L, Wang BY, Lv X, Huang Y, & Liu DP. Both TALENs and CRISPR/Cas9 directly target the HBB IVS2-654 (C>T) mutation in β-thalassemia-derived iPSCs. *Sci Rep*. 2015; 5: 12065. Doi: 10.1038/srep12065.

Xu Q, Zhang Z, Zhao L, Qin Y, Cai H, Geng Z, Zhu X et al. Tropism-facilitated delivery of CRISPR/Cas9 system with chimeric antigen receptor-extracellular vesicles against B-cell malignancies. *J Control Rel*. 2020; 326: 455–467. Doi: 10.1016/j.jconrel.2020.07.033.

Yang P, Chou SJ, Li J, Hui W, Liu W, Sun N, Zhang RY et al. Supramolecular nanosubstrate-mediated delivery system enables CRISPR-Cas9 knockin of hemoglobin beta gene for hemoglobinopathies. *Sci Adv*. 2020; 6(43): eabb7107. Doi: 10.1126/sciadv.abb7107.

Ye Y, Zhang X, Xie F, Xu B, Xie P, Yang T, Shi Q et al. An engineered exosome for delivering sgRNA:Cas9 ribonucleoprotein complex and genome editing in recipient cells. *Biomater Sci*. 2020; 8(10): 2966–2976. Doi: 10.1039/d0bm00427h.

Yin H, Song CQ, Suresh S, Wu Q, Walsh S, Rhym LH, Mintzer E et al. Structure-guided chemical modification of guide RNA enables potent non-viral in vivo genome editing. *Nat Biotechnol*. 2017; 35(12): 1179–1187. Doi: 10.1038/nbt.4005.

Zhang BC, Luo BY, Zou JJ, Wu PY, Jiang JL, Le JQ, Zhao RR, Chen L, & Shao JW. Co-delivery of sorafenib and CRISPR/Cas9 based on targeted core-shell hollow mesoporous organosilica nanoparticles for synergistic HCC therapy. *ACS Appl Mater Interfaces*. 2020; 12(51): 57362–57372. Doi: 10.1021/acsami.0c17660.

Zhang JP, Cheng XX, Zhao M, Li GH, Xu J, Zhang F, Yin MD et al. Curing hemophilia A by NHEJ-mediated ectopic F8 insertion in the mouse. *Genome Biol*. 2019a; 20(1): 276. Doi: 10.1186/s13059-019-1907-9.

Zhang Z, Wan T, Chen Y, Chen Y, Sun H, Cao T, Songyang Z et al. Cationic polymer-mediated CRISPR/Cas9 plasmid delivery for genome editing. *Macromol Rapid Commun*. 2019b; 40(5): e1800068.

Zhen S, Hua L, Takahashi Y, Narita S, Liu YH, & Li Y. In vitro and in vivo growth suppression of human papillomavirus 16-positive cervical cancer cells by CRISPR/Cas9. *Biochem Biophys Res Commun*. 2014; 450(4): 1422–1426. Doi: 10.1016/j.bbrc.2014.07.014.

Zhen S, Liu Y, Lu J, Tuo X, Yang X, Chen H, Chen W, & Li X. Human papillomavirus oncogene manipulation using clustered regularly interspersed short palindromic repeats/Cas9 delivered by pH-sensitive cationic liposomes. *Hum Gene Ther*. 2020; 31(5–6): 309–324. Doi: 10.1089/hum.2019.312.

Zhen S, Takahashi Y, Narita S, Yang YC, & Li X. Targeted delivery of CRISPR/Cas9 to prostate cancer by modified gRNA using a flexible aptamer-cationic liposome. *Oncotarget*. 2017; 8(6): 9375–9387. Doi: 10.18632/oncotarget.14072.

Zhuang J, Tan J, Wu C, Zhang J, Liu T, Fan C, Li J, & Zhang Y. Extracellular vesicles engineered with valency-controlled DNA nanostructures deliver CRISPR/Cas9 system for gene therapy. *Nucleic Acids Res*. 2020; 48: 8870–8882. Doi: 10.1093/nar/gkaa683

3 Extracellular Vesicles for Nucleic Acid Delivery

Progress and Prospects for Safe RNA-Based Gene Therapy

Md Meraj Anjum, Dulla Naveen Kumar, Aiswarya Chaudhuri, Sanjay Singh and Ashish Kumar Agrawal
Indian Institute of Technology (Banaras Hindu University), Varnasi, India

CONTENTS

3.1 INTRODUCTION

Small RNA molecules are a category of RNAs comprising less than 200 nucleotides that are capable of regulating gene expression in both prokaryote and eukaryote [1]. These small RNAs, like siRNA and miRNA, demonstrate potential for being a remarkable tool to counter several pathological conditions through regulation of expression of a specific gene via another small RNA called RNA interference (RNAi) [2]. The RNAi is a strategy in which gene silencing occurs after the synthesis of protein (post transcriptional) that is stimulated by miRNA of genome origin, germ line-specific Piwi-interacting RNA (piRNA), and small interfering RNA (siRNA) [3]. siRNAs are exogenous molecules of synthetic origin extensively implemented in biological research [4]. siRNAs can be distinguished from microRNA in that siRNA prohibits the expression of definite targeted mRNA, whereas microRNA regulates the mRNA expression [5]. Such convenient properties of siRNA have

DOI: 10.1201/9781003186083-3

drawn much attention in the past to employ the delivery of siRNA in the RNAi-based drug delivery approaches. The siRNAs are composed of ds RNA molecules with typical 21 to 23 nucleotides in either strand, along with having two nucleotides overhanging at both 3' ends. In cytoplasm, siRNAs develop enzyme-based systems comprising of multiple units of protein, which are expressed as RNA induced silencing complex (RISC) [6]. These units of protein regulate deterioration and elimination of the sense strand from RISC. Therefore, the antisense strand that is remaining is attached to the complementary mRNA to commence the deterioration through RISC [7]. Taking these facts into account, siRNA can be remodeled to downregulate the target gene; therefore, its therapeutic potential treats a wide array of diseases varying from infections to genetic diseases, as well as cancers [8]. In cancer treatment particularly, the target gene is broadly regulated via oncogenes, angiogenesis, proliferation, cell cycle, etc [9]. Several approaches, including clinical studies, have been taken into consideration to have successful gene therapy. The massive research numbers are motivated towards overcoming the setbacks and challenges linked with traditional cancer therapy strategies, and these include transfer of both viral and non-viral genes, offering promising therapeutic options for cancer patients [10]. Moreover, there are certain obvious issues related to gene therapy, out of which the delivery associated drawbacks are more common. The naked RNA molecule cannot invade the target cells due to their dimensions, hydrophilic nature, and negative charge. So, in order to have a solution to these problems, various approaches, like application of viruses, polymers, lipids, nucleic acid etc, in drug carrier systems, are under scrutiny [11]. Such approaches are not often prudent due to absence of efficacy, non-targetability, toxicity, or immunogenicity. For instance, it was observed that the cationic lipids employed in liposomal formulations often invoke unwanted immune responses, while cationic polymers employed demonstrate higher cytotoxicity [12]. Such detrimental immune responses and cytotoxicity have also been observed in the delivery of viral gene vectors. Colloidal nanocarriers are being identified by the immune system as foreign entities, leading to allergic reaction, or infusion reaction, which are explained as complement (C) activation-related pseudoallergy (CARPA) [13]. A novel category of biological RNA delivery systems is emerging now that are being extensively investigated. Such a novel biological RNA delivery system is denoted as "Extracellular Vesicles" (EVs) [14]. EVs are nano-vesicles that have an outer phospholipid bilayer and that are released by all the cells existing in living organisms of all domains, like eukaryotes, bacteria, and archaea. EVs are also involved in cell-cell interaction. Since EVs are endogenously expelled, they may provide a safe alternative for gene delivery [15]. These are biocompatible drug delivery carriers which have emerged as a potential tool over the last few years; increased demand is due to their unique characteristics, for instance they are capable of protecting and transporting the therapeutic molecule throughout the body [16].

3.2 EXTRACELLULAR VESICLES (EVS)

Therapeutic efficacy of any drug is not only dependent upon the inherent efficacy of the drug molecule but also dependent upon its availability at target site. To make the drug molecules available at a target site, different nanoformulations based approaches came into the picture as a result of decades of scientific endeavors. Although a variety of natural and synthetic material based nanoformulations *viz.* lipid [17–33], polymers [34–50], protein [51–53], metals [54], etc., have been stated to improve the therapeutic efficacy of the drugs, there is a desperate need for biocompatible carrier systems which can deliver as well as safeguard the therapeutic carriers throughout the entire body. In past years, extracellular vesicles have gained remarkable attention as drug delivery systems and cell replacement therapeutic approaches. Extracellular vesicles are particles of a non-replicating nature that are composed of phospholipid bilayers, released by almost all types of cells found in the human body, and in several plants and animals [55]. EVs are also found in various biological fluids like blood, saliva, milk, cerebrospinal fluids and malignant ascites, as well as in tissue culture supernatants, making up particles of heterogeneous population that are categorized into three specific types depending on their production: exosomes, microvesicles, and apoptotic bodies (Table 3.1) [56].

TABLE 3.1
Different Types of Extracellular Vesicles

Type	Size (nm)	Density (g/ml)	Shape	Markers	Origin	Biogenesis Process	Composition	References
Exosomes	30–150	1.13–1.19	Spherical	CD9, CD63, CD80, TSG 101, Alix.	Late endosomes	Inward budding	mRNA, miRNA, DNA, lipids heat shock proteins.	[57–61]
Micro-vesicles	50–1000	1.04–1.07	Ovel	Integrins, selectins, CD40, Flotiline-2.	Plasma membrane	Outward budding	mRNA, miRNA, DNA, Cytoplasmic proteins.	
Apoptotic bodies	50–5000	1.16–1.28	Hetero-geneous	Annexin V positivity, Phosphatidyl serine	After cell death	Cell shrinkage	Cell organelles, DNA, RNA, histone.	

3.2.1 Origin and Biogenesis

Exosomes deriving from precursors are termed as intraluminal vesicles (ILVs). They are formed in the lumen of multivesicular bodies (MVBs) of the endocytic pathway. In more profound way, it could be described that the ILVs are formed via inward budding of endosome membrane followed by fusion with the plasma membrane [62]. Further, when the cells released ILVs, they are referred to as exosomes. The mechanism of biogenesis of MVBs are widely categorized into two classes; (i) endosomal sorting complex required for transport (ESCRT)-dependent, and (ii) ESCRT-independent. ESCRT-dependent pathways comprises of five complexes that include ESCRT-0, ESCRT-I, ESCRT-II, ESCRT-III and AAA ATPase vacuolar protein sorting-associated protein 4 (Vps4) complex [63, 64]. ESCRT-0 is largely liable for arranging transmembrane proteins that are labeled with ubiquitin followed by invaginations in the endosomal membrane. This function is generally consummated through a phosphatidylinositol 3-phosphate binding protein called hepatocyte growth factor-regulated tyrosine kinase substrate (HRS), which, along with a signal transducing adaptor molecule (STAM), may identify the protein signals that are ubiquitinated. Similarly, ESCRT-I and ESCRT-II complexes also identify the cargos that are ubiquitinylated and further govern the membrane bud formation [65]. Furthermore, activation of ESCRT-III results in the growth of the transient filament of the complex that drives vesicle scission and recruits de-ubiquitinases that recycle ubiquitin. Ultimately, the AAA ATPase Vps4 undergo dissociation and recycles the ESCRT-dependent complexes [66]. The ESCRT independent pathway is considered another pathway of MVB biogenesis. In this pathway for instance, lipids like phosphatidic acid can be used that induce a negative membrane curvature, or ceramide can be used which triggers budding process of vesicles, converting them into MVBs [67]. Moreover, tetraspanins, like CD63, and heat shock proteins, like chaperone HSC70, may also be employed to induce the formation of MVBs. After formation, MVBs are either fused with lysosomes or with plasma membrane. The former lead to the deterioration of its content, while the latter results in the discharge of exosomes [68]. It was further observed that certain proteins and complexes plays an important role in the secretion of exosomes, that includes Ras-associated binding (RAB) proteins and soluble N-ethylmaleimide-sensitive fusion (NSF) attachment protein receptor (SNARE) complexes [69]. The RAB proteins comprises of over more than 60 GTPases, where some of these GTPase regulate various steps of intracellular trafficking like vesicle formation, vesicle mobility, and docking to target compartments that further results in the fusion of the MVB with the target membranes [70]. After vesicle docking, the further facilitation of MVB fusion with plasma membrane is assisted by SNARE proteins. On the other hand, microvesicles (MVs) are formed from the cell membrane via direct shedding and are supposed to have a more heterogeneous population, but their biogenesis is less defined and the process seems to become active when the cells are under any stress factor [71]. The MVs are also formed via outward budding that seems to be facilitated by the changes in the local membrane microdomains and curvature-mediated changes in the composition of lipid and protein, that is supplemented via calcium-dependent enzymes like flippase and floppase, leading to redistribution of phospholipids [65]. The crucial steps in the EVs biogenesis are schematically shown in Figure 3.1.

3.2.2 RNA-Sorting Mechanisms

The availability of a distinctive RNA within the cell is a critical aspect for their loading into the EVs. It was observed that the RNA species do not exhibit preferential sorting into the EVs by their own. Instead, they normally are distributed to both the compartments as per statistical analysis. However, there exist certain sequence motifs that promote the preferential sorting of EVs in some cell types [72]. For example, specific sequence motifs like 'ACCAGCCU', 'CAGUGAGC' or 'UAAUCCCA' have been found to act as cis-acting elements that help in sorting mRNA into the EVs. Similarly, a sequence motif like 'CTGCC' is more prone to package mRNA that is comprised of zipcode, like 25-nt sequence and miR-1289, binding site in their 3'-untranslated region into EVs that are extracted from melanoma cells as well as human primary glioblastoma cells [73]. Moreover, distinct targets have been recognized in miRNAs highlighting 'GGAG' or 'CCCU' sequences, which promote sorting of mRNAs from T-lymphoblasts into EVs. This in turn seems to be regulated by binding to A2B1, a heterogeneous nuclear ribonucleoprotein. Such sequence motifs used for sorting miRNA into EVs have also been corroborated in gamma-herpesvirus-infected lymphoma cells. The sorting sequences also aid in sorting miRNAs obtained from B cells having 3′-end uridylated post transcriptional modifications into EVs [74]. Such observation was confirmed in human urine derived EVs. Growing curiosity on sequence motifs that facilitate RNA loading into EVs may be utilized to modify EVs with RNAs of interest for gene therapy purposes. In the case of OMVs also, the loading of RNA appears to be loading, but also appears to be inflected, while their sorting process has not been described clearly.

3.2.3 Composition of Extracellular Vesicles (EVs)

3.2.3.1 Proteins and Lipids

The molecular make up of EVs depends on the type of EV, its precursors, and physiological conditions like stress or hypoxia. It was observed that there are certain subtypes of EVs that are enriched with specific proteins, like the EVs derived from mammalian cells exhibit an enriched degree of tetraspanins on their surface that includes CD9, CD63 and CD81 and have been associated in the modulation of a mechanism involved in the loading of drug-cargo. There also exist certain transmembrane receptors like Integrins and selectins which facilitate intracellular interaction as well as cell to matrix interactions [75]. Such surface proteins play an important role in the cellular uptake of EVs. In addition to the surface proteins, other proteins associated with the biogenesis process *viz.* Alix, Tsg101, clathrin, and ubiquitin, and involved in membrane trafficking may also be identified in EVs [76]. These proteins are comprised of the family of Rab proteins and annexins. EVs derived from the Antigen-presenting cell can also reveal the effective implications of MHC II in immune regulatory activity. Like lipids, EVs are also comprised of ceramides, sphingomyelin, phosphatidylserine (PS) and cholesterol as compared to the parental cell membranes [77]. The molecular configuration of OMVs derived from gram-negative bacteria demonstrates that the outer membrane is composed of typical outer membrane proteins (OMPs), like outer membrane protein A (OmpA),

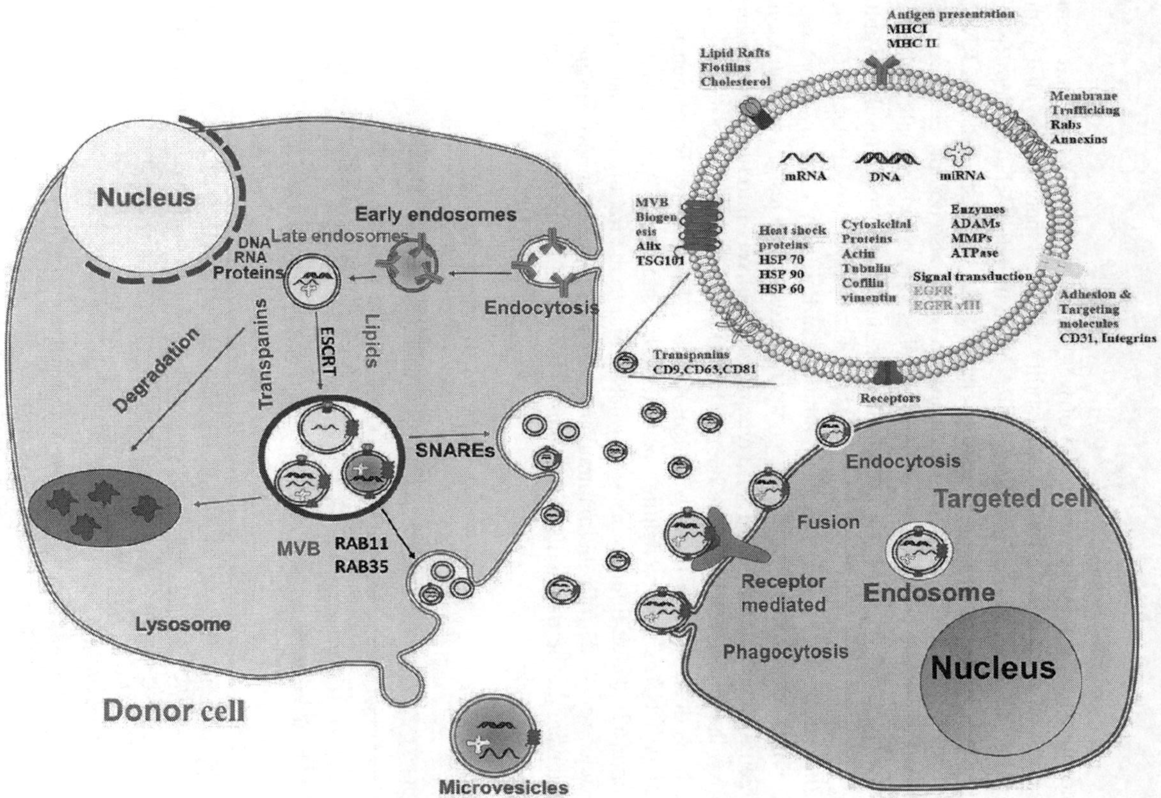

FIGURE 3.1 Intracellular mechanism of biosynthesis of EVs along with their production from the different types of cells: At the start, exosomes are made in the form of ILVs, followed by the invagination of the plasma membrane and budding into endosomes and MVBs. ESCRT, lipids and tetraspanins are some of the moieties involved in the biosynthesis of ILVs. But there still remains a question of whether the moieties act on the same target or different targets is still unclear. These MVBs can be further processed through the formation of lysosome or amalgamation with plasma membrane which releases them in the extracellular matrix. Various proteins from RAB family *viz.* (RAB11, RAB27, and RAB35) have been found to play an important role in the transit of MVBs to plasma membrane, which further leads to the production of exosomes. Apart from the RAB proteins, another protein has been identified that plays an important role in the fusion of MVBs to the plasma membrane, namely, SNARE proteins. Other type of vesicles, like Microvesicles, are also supposed to be formed directly from the plasma membrane via budding. The vesicles released are further taken up by another cells through a number of different mechanisms *viz.* endocytosis, fusion with membrane, receptor mediated endocytosis, and phagocytosis to establish cell to cell communication.

OmpC, OmpF, Cytolysin A (ClyA), and lipoproteins. OMVs are enriched with less proteins, as compared to the cellular outer membrane; however, toxins are found to be at enhanced levels in OMVs [78]. It was observed that lipopolysaccharides (LPS) form the major lipid class available on the outer surface of OMVs. For example, in Pseudomonas aeruginosa, OMVs were exclusively comprised of B-band LPS, as compared to the A-band LPS, however, this lipid was found in abundance at much lesser concentrations in the outer membrane. Proteins arising from the cellular periplasmic space are principally recovered in the OMV lumen level; however, the levels of cytoplasmic proteins are low [79].

3.2.3.2 RNAs

Apart from proteins and lipids, EVs are also comprised of RNA species. The EVs derived from mammalian cells are extensively scrutinized for the availability of RNA species, provided with latest methods of sequencing. The RNA includes messenger RNAs, non-coding RNAs, microRNA, transfer RNA, ribosomal RNA, small nuclear RNA, small nucleolar RNA, small cytoplasmic RNA, Y-RNA, and vault RNA [55]. Apart from RNAs, the mammalian EVs are also comprised of certain degraded products. However, the relative availability of various RNAs vary between donor cells and extravesicular cells. For instance, EVs derived from dendritic cells showed enhanced levels of vault RNA, SRPRNA, and Y-RNA as compared to the parent cells [73]. Overexpressed cellular miRNAs, like miR92a-1 and let-7b, were found to be in low concentrations in EVs, while highly expressed EV derived miRNAs, like miR-223, miR-142, and miR-93, were expressed at a lesser concentration in the parent cells. Such an instance implies that RNAs are sorted selectively into the extracellular vesicles during their biosynthesis. Nevertheless, the bacterial OMVs showed a smaller amount of expressed RNAs. It has been observed that the huge range of RNAs, like transfer RNA, transfer-messenger RNAs, ribosomal RNA, signal recognition particle-RNA, and 6S RNA, were found in EVs released from bacteria [62]. These findings are somewhat similar to the finding obtained from that of extracellular vesicles released from the mammalian cell, where the distinctive sorting of distinct RNAs has been unveiled. The detailed summary of protein, lipid components, and RNAs found in exosomes is given in Table 3.2.

TABLE 3.2
Composition of Extracellular Vesicles

Composition	Example
Proteins, lipids, and nucleic acids	
Transport protein	Tubulin, Actin, Radixin, Vimentin, Cofilin-1, Moesin, and Acting binding molecule
Transpanins	CD9, CD63, CD81, CD82, CD37, CD53
Proteins involved in biogenesis	Alix, Tsg101, Vps, Rab proteins.
Anti-gen presentation proteins	MHC-II
Membrane trafficking proteins	Rab proteins, Annexin-I, II, IV, V, Dynamin, and Syntaxin-3.
Heat shock proteins	HSP70, HSP90, HSP27, and HSP60.
Cell adhesion proteins	Integrins, Lactadherin, Intracellular adhesion molecule-1.
Growth factors and Cytokine	TNF-α, TGF-β.
RNAs	microRNA, messenger RNA, silencing RNA, transfer RNA.
References	[55, 68, 75–77, 80, 81]

3.2.4 Inherent Capacity of Extracellular Vesicles in Crossing Physical Barriers

The main difficulty that the traditional drug delivery systems suffer is the limited ability in potentially bypassing the biological barriers, comprised of tissue, cellular, and intracellular hindrances. Multiple studies have demonstrated the ability of EVs to potentially bypass these biological barriers that resulted in functional changes within the target sites [82]. For example, on the issue level, it was observed that the EVs have bypassed one of the most challenging barriers of drug delivery, i.e., the blood brain barrier (BBB). The BBB is considered the major barrier, blocking the delivery of the drug carrier system inside the brain for the treatment of central nervous system disorders. It was reported that the BBB hinders the passage of almost 98% of small molecules or drugs [83]. Several research works revealed the physiological role of EVs in intercellular communication between neuronal cells. It was observed that the EVs can preserve neuronal integrity while maintaining synaptic plasticity and brain integrity. Furthermore, increasing evidences highlight the capability of EVs to transfer therapeutic cargo across the BBB [84]. Although various studies proved the fact regarding the efficient crossing of EVs through BBB, its specific mechanism still remains to be unveiled. Moreover, it was observed that the EVs reach the brain parenchyma across BBB by crossing the blood-cerebrospinal fluid barrier, followed by crossing the choroid plexus. At cellular level, it was found that due to the presence of various ligand/receptors on the surface of EVs, they could interact with the plasma membrane [85]. Such phenomena resulted in more efficient internalization in comparison to synthetic nano-systems. To support such findings, a study was performed where the synthetic nanocarrier showed more accumulation on the surface of the cell of interest with minimum internalization; but, the EVs showed enhanced accumulation in less time with no accumulation over the EVs surface [86]. Hence, it seemed that for cellular uptake, the EVs employed endogenous mechanisms that appeared beneficial for the intracellular delivery of therapeutics. As it is well documented that the EVs derived from the same cell can interact distinctly with a specific recipient cell, therefore, the intercellular communication and trafficking of the EV solely depends on the distinct features of the deriving and recipient cells. It could further be stated that a clear understanding of the intercellular transfer of EV can enhance the concept of an EV-facilitated drug delivery system [87]. At the cellular level, the internalization of EVs takes place via endocytosis, although different cell types display different types of endocytotic pathways. It was further observed that EVs deliver the RNAs to the target sites via an active pathway, which implies that they procure an endogenous pathway for the RNA delivery. Thus, it has been assumed that although the particular mechanisms revolving around the delivery involved are still yet to be discovered, the EVs may fuse with endosomal membranes via endocytotic pathways [88]. Recently, it has been observed that the EVs fuse with the proteins expressed on the surface of the membrane at a lower pH [89]. However, the challenge still remains to develop more intricate technology, which could provide much elaborated justification of the mechanisms responsible for the fusion of EVs with the membrane and the release of the drugs from the EVs. Conclusively, it is inferred that as EVs are able to bypass various levels of physiological barriers, and specifically deliver the drugs at the target site, the EVs are considered an efficient tool in the field of drug delivery [90]. However, despite the promising venture of EVs in drug delivery, it might create certain limitations taking into consideration the concept of the rapid clearance profile of certain drugs.

3.3 ROLE OF EVS IN NUCLEIC ACID TRANSFER AND COMMUNICATION

After being released by the donor cells, the EVs are taken up by the recipient cells through various routes. Several studies have indicated that either EVs can directly be fused with target cells or be internalized through endocytosis. Now, EVs experience endocytosis in two ways, namely, clathrin-dependent or clathrin-independent pathways [91]. The clathrin-mediated endocytosis includes phagocytosis, micropinocytosis, and lipid raft-mediated endocytosis. Bacterial cell derived EVs may also infiltrate the host mammalian cells through adhesion-receptor-mediated attachment, followed by either internalization or fusion or both. After internalization, the derived EVs may arbitrate an exchange of genetic material among cells [92].

Ratajczak *et al.* (2016) observed that the EVs derived from an embryonic stem cell transferred mRNA and protein to adjacent cells for reprogramming the hematopoietic progenitor cells, which aided in enhancing the survival of the cells [93]. In the past also, Valadi *et al.* (2007) observed that the EVs derived from a mouse mast-cell transferred the mRNA to human mast cells, which are later transcribed into specific proteins [94]. Similarly, Pegtel *et al.* (2010) stated that the infected EBV-activated B cells derived EVs mediated the transfer of EBV-miRNA to monocyte-derived DCs, resulting in the down regulation of genes targeting miRNA [95]. Since then, there are numerous examples documented that reviewed the transfer of functional EV-mediated RNA *in vitro*. Recently, various studies have shown that the evidence for the transfer of EV-mediated RNA in vivo has been provided by a technique named Cre-loxP-based tracing technique. The Cre mRNA obtained from engineered glioma tumor cells was found to be transferred to mouse stromal cells through EVs, resulting in translation into Cre protein that ultimately leads to various recombination events [96]. Additionally, by employing such method, one could also visualize the transfer of Cre mRNA from EVs to extremely metastatic mammary cancer cells like MDA-MB-231 cancer cells or less metastatic cancer cells like T47D cancer cells in vivo, indicating the metastatic behavioral changes of T47D cells. Considering the prokaryotes, the genetic transfer facilitated by OMV-mediated is considered significant for maintaining the survival of bacterial cells. Moreover, the interaction between the bacteria and host is considered crucial in the pathogenesis of different infectious diseases. In this context, the OMVs exhibits inter-species communication as the OMVs secreted by the nematode parasite may transfer miRNAs, Y-RNA, and nematode Argonaute proteins to the host, minimizing their immunity response [97]. Furthermore, human cells are capable of communicating with gut microbiota through EVs contained in the fecal miRNA, influencing the bacterial growth condition as well as the transcription of bacterial gene. The ability of the derived EVs in transferring the genetic materials makes them important moieties in intra-species, inter-species, and inter-kingdom communication, given the fact that exchanging the genetic material between donor and recipient cells may alter the phenotype of the recipient cells [92]. This novel technique of intracellular communication is progressively being identified as of principal significance in different physiological and pathological processes.

3.3.1 EVs as Nucleic Acid Delivery Tool

Systemic application of nucleic acid delivery is considered even more challenging when the applications need to be delivered into any targeted site Moreover, the nucleic acids are prone to degradation due to their enlarged dimensions and charged behavior, which necessitated the availability of a carrier system sufficient to protect and deliver them into the targeted site [98]. For the delivery of nucleic acids, it was observed from the various studies that the nucleic acids (being negatively charged species) are associated with the positively charged species, like lipids and polymers, forming an electrostatic complex, which then forms nano-ranged particles. Further, such electrostatic complexes protect the nucleic acid from degradation by serum nucleases and aid in cellular internalization [99]. In spite of the promising results, the attached positively charged moiety of the carrier system exhibits cytotoxicity [100]. It was further observed that the delivering nucleic acid outside the liver remained inefficient, which necessitates the development of a novel nucleic acid delivery system to treat various diseases. Thus, EVs became an encouraging way for the delivery of nucleic acids [101]. Now, it was further observed from various studies that the loading of nucleic acids into exosomes is a very challenging task. Due to the macro-size and charged behavior, the nucleic acids were not able to penetrate the outer phospholipid bilayer of the EVs and be retained within their lumen. So, to overcome the problem of being macro-sized, most of the nucleic acids loaded within the EVs are usually siRNA and miRNA, which are RNAs with small base pairs, usually —20 to 25. But, then an issue is raised which states that such small base-paired RNAs might interfere with the complementary sequences of the targeted mRNA, resulting in the transformation of the recipient's phenotypes [102]. Thus, loading methods can be categorized into two categories: loading before isolation of EVs and loading after isolation of EVs[103].

3.3.2 Loading After EV Isolation

Regardless of the origin, all types of EVs can exhibit loading after the isolation of EVs. Such types of loading strategy are also preferable from the pharmaceutical point of view, as the conditions required for the loading could be controlled preferably. The various methods employed for loading nucleic acids into EVs after their isolation forms respective origins are described below.

3.3.2.1 Simple Incubation

On incubation of EVs with hydrophobic molecules, the outer phospholipid bilayer of EVs interact with the hydrophobic moieties that offer spontaneous loading of hydrophobic molecules within the EVs. It was observed that drugs of less molecular weight, like curcumin, doxorubicin, paclitaxel, etc., can be easily loaded into the EVs via a simple incubation method [104–109]. For example, hydrophobically engineered siRNA was found to be loaded successfully in EVs derived from glioblastoma U87 by employing the simple incubation method for treating Huntington disorder [110]. It was observed from such studies that the hydrophobicity increases the interaction of the phospholipid bilayer of EVs with the nucleic acids, thereby increasing their loading efficacy within the EVs. One such instance was followed during loading miR-150 into the EVs derived from the T-cells [111].

However, the simple incubation technique is somewhat dissimilar to that of the spontaneous transport method because in the spontaneous transport method, the interaction of nucleic acids with the phospholipid bilayer is considered negligible, as the hydrophobicity of the bilayer prevents the charged moieties from passing through them. However, the rationale behind the relevant loading remains unexplained, as it remained unclear as how the miRNA exposed over the surface of vesicle protects itself from the degradable enzymes as well as provide stability [112].

3.3.2.2 Electroporation

The compositions of the cellular membrane and the EVs membrane are quite similar, which makes the electroporation a better-entrenched technique for cell transfection. Electroporation technique assists active delivery of nucleic acid within the lumen of EVs. In electroporation, an electric field is applied which creates small pores over the surface of EVs and allows the easy penetration of nucleic acid into the EVs [113]. There are various examples that showed that the nucleic acids could be successfully loaded into the EVs via the electroporation method. For example, in one of the studies it was observed that ~ 25% entrapment efficiency was observed for Glyceraldehyde 3-phosphate dehydrogenase (GAPDH) siRNA and (beta-secretase 1) BACE1 siRNA when loaded into EVs derived from modified dendritic cells via electroporation [114]. In a similar context, the electroporation technique was employed during loading opioid receptor Mu siRNA and mitogen-activated protein kinase 1 (MAPK1) siRNA in EVs derived from engineered HEK-293T cells and plasma respectively. Also, KSP siRNA was loaded into a mutant *E.coli* strain derived OMVs via electroporation. From the above studies, we also observed that for optimum loading efficiency by electroporation method, various process variables are taken into consideration, which includes voltage applied, concentration of EVs, amount of nucleic acids, nature and volume of electroporation media, etc. [115]. Kooijmans *et al.* (2013) unveiled that while performing their experiment, it so happened that the application of electroporation caused aggregation of siRNA, which they falsely interpreted as loading of siRNA into EVs. After finding such circumstances, they manipulated the process variables as discussed above, but no change occurred in the overall nucleic acid loading [113]. From the above study, it came into the limelight that the application of electroporation does change the integrity of nucleic acid and EVs. Hence, such outcomes necessitate the fabrication of alternate approaches for loading of nucleic acid within the EVs.

3.3.2.3 Sonication

Like electroporation, sonication also facilitates active loading of nucleic acid, and allow easy penetration of nucleic acid through the pores created by sonication, but unlike electroporation, sonication does not cause nucleic acid aggregation. Sonication method was used to load HER2 siRNA into EVs derived from MCF-7 breast cancer cells [102]. But, like electroporation technique, in sonication method also, the various process parameters are taken into consideration for maintaining the integrity of EVs and nucleic acid which includes the power of sonication, amplitude of sonication, sonication time and the temperature [116].

3.3.3 Loading before EV isolation

Another loading approach used is the loading before isolating the EVs. In this method, the nucleic acids are directly transfected into EV derived cells followed by sorting through intrinsic endogenous approach of the cell. However the entire mechanism of sorting still need to be elucidated [117]. Various studies were performed where such method was employed for loading. EVs derived from HeLa cells and ascetic fluids successfully loaded RAD 51 siRNA and RAD 52 siRNA respectively. Also, EVs derived from Hela-229 cells was found to be successfully loaded with miR-130b. In addition to these, the transfected THP-1 macrophages with modified miR-143 was found to be successfully released in EVs. In further studies, it was observed that the viruses were utilized to produce Jurkat and Raji cells expressing small RNA that were also found to be loaded successfully in EVs [118]. Despite easy loading, such approach also experiences certain drawbacks. Likewise, after loading, it would become difficult to remove the transfection agents which will later lead to EVs contamination. In addition to this, such approaches offer poor nucleic acid loading efficiency, are dependent on cell-EVs interaction and also it was found that such approaches are not suitable for all kinds of RNAs such as the miRNA that affect cellular growth or viability inhibits the growth of EVs derived cells [119]. Such observations provide proofs regarding utilization of EVs as a therapeutic delivery system for nucleic acid, however, their loading into the EVs remained as a hurdle for clinical practice.

3.4 FUNCTIONALIZED EVS FOR TARGETED DELIVERY

In addition to efficient drug loading, EVs also show an enhanced characteristic in targetability, which plays an important role in the drug delivery system. It was confirmed from the various studies that EVs do possess an intrinsic targeting ability due to the presence of lipids and protein in their contents that may offer EVs tropism to distinct organs [120]. For instance, the different categories of integrins present in the EVs offer improved pharmacokinetic characteristics that lead to increased accumulation of the EVs in the desired target site, which includes brain, lungs, liver, etc [121]. Likewise, EVs comprised of Tspan8-integrin alpha4 complex were preferably targeted to the pancreatic cells. Furthermore, it was also observed that due to the presence of phospholipids like phosphatidylserine in the outer membrane of EVs, they are preferentially taken up by the macrophages. EVs can further be modified for improved targeting by engineering the precursor cells or the cells producing EVs. Many such studies have been carried out to provide targetability of EVs. In this context, the Lamp2b expressing EVs derived from dendritic cells was combined with siRNA, followed by fusion with either neuron-specific rabies viral glycoprotein (RVG) peptide or muscle specific peptide, led to knockdown of a specific gene in neuronal cells (Neuro2A) or muscle cells (C2C12), respectively [122, 123]. Similarly, for targeting, the EVs derived from cardiosphere-derived cells (CDC) were engineered for fusing the N-terminus of Lamp2b, as expressed over the surface of EVs, with the cardiomyocyte specific peptide (CMP). Such CMP-targeted EVs showed an 18-fold increase in cellular uptake by neonatal mouse cardiomyocytes as compared to EVs without modification. To establish the targetability, such an approach was performed in different other cell lines, which revealed that there existed no such difference in targeting approach between CMP targeted and non-targeted EVs, proving the peptide specificity for the cardiomyocytes [124]. Furthermore, in the *in vivo* studies, it was observed that on intramyocardial injection, the CMP-targeted EVs showed more retention in the cardiomyocytes as compared to the EVs without any modification [125]. In another study, it was observed that the EVs derived from immature dendritic cells (imDCs) were further modified to target $\alpha v\beta 3$ integrin that is highly expressed in angiogenic vasculature. They have fused the expressed Lamp2b of the EVs surface with the iRGD peptide. In this context, EVs modified with iRGD peptide were encapsulated with Doxorubicin (Dox), which was then administered parenterally to the triple negative breast cancer induced mice [126]. As a result, the iRGD targeted EVs showed enhanced accumulation in the cancer site, along with reduced tumor growth as compared to the non-targeted Dox loaded EVs. In spite of such promising outcomes, there still remains a concern regarding the efficacy of the iRGD peptide as it is known that during the synthesis of EVs, the surface proteins may undergo degradation when exposed to endosomal protease [127]. It was observed that the N-terminus bound peptides of Lamp2b showed acid dependent degradation during the biogenesis of EVs. So, to prevent the peptide from degradation, the Lamp2b proteins are conjugated with GNSTM glycosylation motifs. Also, for overcoming the peptide degradation, the peptides being targeted might be tethered directly with the plasma membrane [128]. An example of such a technique includes fusion of the GE11 peptide targeting EGFR with the transmembrane domain of platelet derived growth factor receptor. Such a modification was performed on HEK293 cells. Such GE11 surface modified EVs showed much more accumulation efficiency into mice with EGFR positive tumors on intravenous application. In another study, it was observed that during the maturation process of reticulocyte, the glycosylphosphatidylinositol (GPI)- anchored protein decay-accelerating factor (DAF) was released selectively in EVs, which then fused with anti-EGFR nanobodies (EGa1), found abundantly over the EVs surface. Such conjugation was then cleaved by GPI transamidase enzymes to release DAF peptides, resulting in increased binding of EVs to EGFR overexpressed cells during post translational modification [129]. Such observation can be followed both in static as well as in dynamic conditions. It could be inferred that conjugating the surface of EVs with targeting motifs by employing endogenous cellular machinery provides a promising approach in the delivery of drug loaded EVs to their desired target site. Such approaches aid in improving the therapeutic efficiency, while limiting the off-target adverse effects of the loaded drug [130].

3.5 THERAPEUTIC APPLICATIONS OF EVS IN NUCLEIC ACID DELIVERY

Nucleic acid-based therapeutical strategies have gained much attention in therapies in both preclinical and clinical platforms. It was observed that the cell derived EVs are comprised of a high amount of various macromolecules, like nucleic acids and proteins, and also exhibit a high expression of specific receptors. Moreover, the EVs are considered biocompatible and biodegradable. By considering the excellent properties of EVs as described above, the researchers have considered the EVs as an appropriate carrier system for the nucleic acid delivery [131]. EVs are derived from the mesenchymal stem cells, which are enriched with nucleic acids and proteins, such as siRNAs, miRNAs, etc., that are supposed to exhibit tissue regeneration properties, anti-inflammatory activity, and tumor suppressing or tumor promoting activity. Such studies are still in pre-clinical and clinical stages. Hence, scientists are expecting that such studies will make the stem cell derived EVs a next generation delivery system for the treatment of various cancers [132]. Recent reports stated that nucleic acid loaded EVs can deliver an optimum amount of nucleic acids to the targeted site without being affected by the physiological environment. In recent times, the scientists are discovering a novel method that will help in loading a high amount of nucleic acids into the EVs without affecting the integrity of the EVs. Alvarez-Erviti *et al.* (2011) was the first to successfully deliver Lamp2b expressed EVs loaded with *GAPDH* siRNA in mice for the treatment of Alzheimer's disease. Such study further encouraged the scientist to proceed with various studies associated with the delivery of nucleic acid loaded EVs for the treatment of various disorders [133]. Recently, Li *et al.* 2019 engineered EVs to load sufficient amounts of siRNA. They transfused CD9 protein, a highly expressed protein, over the surface of EVs with HuR, an RNA binding protein, to a significant amount of miR-155. Interestingly, it was observed that a significant amount of miR-155 was loaded into engineered EVs via transfection vectors, which was further confirmed with western blotting analysis. Based on such observations, this novel method could be considered efficient in loading the desired amount of nucleic acids within the EVs [134]. Guo *et al.* (2019) performed an interesting study where the nucleotide loaded EVs are fabricated for the treatment of spinal cord injury (SCI). They loaded phosphatase and tensin homolog small interfering RNA (PTEN-siRNA) into EVs by using the extrusion method. Interestingly, the loaded PTEN-siRNA has bypassed the blood brain barrier and entered the CNS, where it silenced the PTEN and improved the axonal outgrowth and neovascularization, along with decreasing the microgliosis and astrogliosis. Overall, the PTEN-siRNA loaded EVs showed full recovery in SCI induced mice. Hence, it could be inferred from the study that nucleic acid loaded EVs could be employed for the treatment of CNS associated diseases [135]. In a recent study, it was demonstrated that for efficient loading of siRNA, miRNA, and ssDNA into the EVs, the EVs were protonated in order to develop a pH gradient across the membrane of EVs [102]. The study further reported that the protonation did not affect their cellular uptake and also did not exhibit any toxicity in the mice. Such an EVs based approach could be used efficiently in the development of RNA nanotechnology. In this context, for targeting the cell as well as to govern the cell-cell communication of siRNA and miRNA, the arrangement of arrow-shaped RNA was adjusted so that a specific interaction could be established between the ligand expressed on the membrane of the EVs and the receptor on the targeted site. The positioning of the docked membrane cholesterol at the tail of the arrow-shaped RNA exhibited the expression of RNA aptamer on the exterior of the EVs membrane [102]. Further, it was observed that arranging the cholesterol at the head of the arrow-shaped RNA offered partial loading of RNAs within the EVs. Using this concept, it was established that by employing the RNA as a ligand over the EVs membrane, one can distinctively deliver siRNA into the target cell, leading to inhibition of the growth of a tumor in the case of prostate, breast, and colorectal cancers. In addition to this, the EVs were further undergoes cellular engineering for enhanced targeting. For tissue-specific delivery of nucleic acids, the vector targeted exosomes were engineered with an exosomal protein fused peptide ligand that led to efficient transfection with targeted cells. Kuroda and co-workers showed that the GE11 surface engineered extracellular vesicles successfully delivered Let-7a miRNA into EGFR targeted xenograft mice model, which further exhibited reduced growth of the tumor. [136]. In a recent study, it was observed that EVs derived from transfected parental cells bear suicide gene mRNAs and proteins, and such EVs were found to be injected into Schwannoma tumor in an orthotropic mouse model. Collectively with systemic prodrug treatments, the vesicles demonstrated successful tumor therapy [137]. However, various analytical methods need to be evaluated that are concerned with the sorting of EVs and establishing the interaction between the protein and lipids of exosomes and recipient cells. Such evaluations will provide a much-enhanced knowledge, which could further be translated in the clinical field. The detailed summery of EVs for nucleic acid delivery has been given in Table 3.3.

TABLE 3.3
Extracellular Vesicle Based Nucleic Acid Delivery

Source of EVs	Delivered Component	Disease	Observations	Reference
HEK293T	miR-155	Liver Fibrosis	miR-155 loaded exosomes effectively reduced liver fibrosis	[134]
MSC	PTEN-si RNA	Spinal cord Injury	It Significantly silenced the PTEN and improved the axonal outgrowth and neovascularization, along with decreasing the microgliosis and astrogliosis.	[135]
MSC	Mir-113b	Stroke	The expression of miR-113b is enhanced in extracellular vesicles by employing knock-in and knock-out technologies. The expressed miR-113b further enhanced the plasticity of axon as well as remodeled neurite in the ischemic boundary zone (IBZ)	[138]
Natural killer (NK) cell	miR-186	Neuroblastoma	Significant cytotoxicity was observed against MYCN amplified neuroblastoma cells.	[139]
Milk	siRNA	Lung cancer	Successfully delivered siRNA loaded FA functionalized exosomes. Significant dose dependent anti-proliferative activity was absorbed and siKRAS loaded exosomes significantly inhibited tumor size	[140]
HEK293	let-7a miRNA	Breast cancer	GE11 surface engineered extracellular vesicles successfully delivered Let-7a miRNA into an EGFR targeted xenograft mice model, which further exhibited reduced growth of tumor.	[136]
HLSCs	mi-RNA	Hepatocellular cancer	Selected miRNAs are successfully delivered hepatocellular cancer cells, promoting the reduction of progression of cancer in SCID mice	[141]
Mouse fibroblast L929	TGF-β1 siRNA	Mouse sarcomas	The TGF-β1 siRNA loaded extracellular vesicles exhibited improved reduction of expression of TGF-β1 along with reduced progression and metastasis of cancer cells in mice	[142]
MSC	miR-379	Breast	Efficient decrease of tumor growth	[143]
MSC	miR-199a	Glioma	Inducing apoptosis promoted chemosensitivity for temozolomide	[144]
FHC	miR-128-3p	Colorectal cancer	miR-128-3p limited EMT and enhanced the accumulation of oxaliplatin in the cancer cells	[145]
Human gastric cell line SGC-7901	siRNA	Gastric cancer	siRNA loaded extracellular vesicles reduced the growth of tumors as well as decreased the rate of angiogenesis.	[146]
Human cell derived	miR497	Lung cancer	miR497 modified extracellular vesicles significantly reduced the growth of tumor, as well as the expression of the genes concerned with its progression.	[147]
MSC	miR-100	Breast	miR-100 loaded extracellular vesicles showed a dose-dependent reduction in their expression and also reduced VEGF production.	[148]
Adipose tissue derived MSCs	miR-122	Liver	Decreasing chemoresistance and tumor volume	[149]

MSC: Mesenchymal stem cells, HLSCs: human adult liver stem cells, TGF- β1: transforming growth factor β1. HEK-293T: human embryo kidney epithelial cell line.

3.6 CONCLUSION

Gene therapy offers an approachable, and efficient platform for the treatment of various diseases, like cancer, infections, and central nervous system disorders, etc. Moreover, it could be inferred that the gene therapy somewhat provides a sense of personalized therapy approach. However, gene therapy does offer certain limitations that restricted them in translating such an approach in the clinical field, which also includes the delivery of nucleic acids to the diseased site. EVs exhibited an encouraging platform for the delivery of nucleic acid from a clinical point of view, because they are much more biocompatible, biodegradable, have a better safety profile, when compared to polymeric, lipidic, and virus-based carrier systems. Their flexible molecular composition enables them to be perpetrated across the membrane of the target tissue or cell. The various limitations that prevent their successful clinical translation include recognition of suitable EVs deriving cells, identifying an effective process for loading nucleic acid into the EVs, and enhancing their targetability. Appropriate measures to answer such limitations will further promote the application of EV applications clinically in a much safer and more efficient way in the field of RNA-based gene therapy.

REFERENCES

1. Wilson, R.C., & Doudna, J.A., Molecular mechanisms of RNA interference. *Annual Review of Biophysics*, 2013. **42**: pp. 217–239.
2. Chiu, Y.-L., & Rana, T.M., siRNA function in RNAi: a chemical modification analysis. *RNA*, 2003. **9**(9): pp. 1034–1048.
3. Pecot, C.V., et al., RNA interference in the clinic: challenges and future directions. *Nature Reviews Cancer*, 2011. **11**(1): pp. 59–67.
4. Kanasty, R., et al., Delivery materials for siRNA therapeutics. *Nature Materials*, 2013. **12**(11): pp. 967–977.
5. Hu, B., et al., Therapeutic siRNA: state of the art. *Signal Transduction and Targeted Therapy*, 2020. **5**(1): pp. 1–25.
6. Pei, Y., et al., Quantitative evaluation of siRNA delivery in vivo. *RNA*, 2010. **16**(12): pp. 2553–2563.
7. Noland, C.L., Ma, E., & Doudna, J.A., siRNA repositioning for guide strand selection by human Dicer complexes. *Molecular Cell*, 2011. **43**(1): pp. 110–121.
8. Dong, Y., Siegwart, D.J., & Anderson, D.G., Strategies, design, and chemistry in siRNA delivery systems. *Advanced Drug Delivery Reviews*, 2019. **144**: pp. 133–147.
9. Tatiparti, K., et al., siRNA delivery strategies: a comprehensive review of recent developments. *Nanomaterials*, 2017. **7**(4): pp. 77.
10. Wirth, T., & Ylä-Herttuala, S., Gene therapy used in cancer treatment. *Biomedicine*, 2014. **2**(2): pp. 149–162.
11. Ji, W., Sun, B., & Su, C., Targeting microRNAs in cancer gene therapy. *Genes*, 2017. **8**(1): p. 21.
12. Semple, S.C., et al., Rational design of cationic lipids for siRNA delivery. *Nature Biotechnology*, 2010. **28**(2): pp. 172–176.
13. Elzoghby, A.O., Samy,W.M., & Elgindy, N.A., Protein-based nanocarriers as promising drug and gene delivery systems. *Journal of Controlled Release*, 2012. **161**(1): pp. 38–49.
14. Shao, H., et al., New technologies for analysis of extracellular vesicles. *Chemical Reviews*, 2018. **118**(4): pp. 1917–1950.
15. Van der Pol, E., et al., Classification, functions, and clinical relevance of extracellular vesicles. *Pharmacological Reviews*, 2012. **64**(3): pp. 676–705.
16. Andaloussi, S.E., et al., Extracellular vesicles: biology and emerging therapeutic opportunities. *Nature Reviews Drug Discovery*, 2013. **12**(5): pp. 347–357.
17. Anjum, M.M., et al., Anacardic acid encapsulated solid lipid nanoparticles for Staphylococcus aureus biofilm therapy: chitosan and DNase coating improves antimicrobial activity. *Drug Delivery and Translational Research*, 2020. **11**: pp. 1–13.
18. Patel, K.K., et al., Effect of penetration enhancers and amorphization on transdermal permeation flux of raloxifene-encapsulated solid lipid nanoparticles: an ex vivo study on human skin. *Applied Nanoscience*, 2019. **9**: pp. 1–12.

19. Gade, S., et al., An ex vivo evaluation of moxifloxacin nanostructured lipid carrier enriched in situ gel for transcorneal permeation on goat cornea. *Journal of Pharmaceutical Sciences*, 2019. **108**(9): pp. 2905–2916.
20. Singh, S., et al., Insulin- and quercetin-loaded liquid crystalline nanoparticles: implications on oral bioavailability, antidiabetic and antioxidant efficacy. *Nanomedicine (London, England)*, 2018. **13**(5): pp. 521–537.
21. Kushwah, V., et al., Improved antitumor efficacy and reduced toxicity of docetaxel using anacardic acid functionalized stealth liposomes. *Colloids and Surfaces B: Biointerfaces*, 2018. **172**: pp. 213–223.
22. Jain, S., et al., α-Tocopherol as functional excipient for resveratrol and coenzyme Q10-loaded SNEDDS for improved bioavailability and prophylaxis of breast cancer. *Journal of Drug Targeting*, 2017. **25**(6): pp. 554–565.
23. Agrawal, A.K., et al., "Liquid crystalline nanoparticles": rationally designed vehicle to improve stability and therapeutic efficacy of insulin following oral administration. *Molecular Pharmaceutics*, 2017. **14**(6): pp. 1874–1882.
24. Harde, H., et al., Development of dual toxoid-loaded layersomes for complete immunostimulatory response following peroral administration. *Nanomedicine (London, England)*, 2015. **10**(7): pp. 1077–1091.
25. Harde, H., et al., Development of a topical adapalene-solid lipid nanoparticle loaded gel with enhanced efficacy and improved skin tolerability. *RSC Advances*, 2015. **5**(55): pp. 43917–43929.
26. Harde, H., Agrawal, A.K., & Jain, S., Tetanus toxoid-loaded layer-by-layer nanoassemblies for efficient systemic, mucosal, and cellular immunostimulatory response following oral administration. *Drug Delivery and Translational Research*, 2015. **5**(5): pp. 498–510.
27. Harde, H., Agrawal, A.K., & Jain,S., Trilateral'3P'mechanics of stabilized layersomes technology for efficient oral immunization. *Journal of Biomedical Nanotechnology*, 2015. **11**(3): pp. 363–381.
28. Jain, S., et al., Oral mucosal immunization using glucomannosylated bilosomes. *Journal of Biomedical Nanotechnology*, 2014. **10**(6): pp. 932–947.
29. Jain, S., et al., Improved stability and immunological potential of tetanus toxoid containing surface engineered bilosomes following oral administration. *Nanomedicine*, 2014. **10**(2): pp. 431–440.
30. Agrawal, A.K., et al., Improved stability and antidiabetic potential of insulin containing folic acid functionalized polymer stabilized multilayered liposomes following oral administration. *Biomacromolecules*, 2014. **15**(1): pp. 350–360.
31. Jain, S., et al., Enhanced transfection efficiency and reduced cytotoxicity of novel lipid–polymer hybrid nanoplexes. *Molecular Pharmaceutics*, 2013. **10**(6): pp. 2416–2425.
32. Jain, S., et al., Oral delivery of doxorubicin using novel polyelectrolyte-stabilized liposomes (layersomes). *Molecular Pharmaceutics*, 2012. **9**(9): pp. 2626–2635.
33. Patel, K.K., et al., Antibiofilm potential of silver sulfadiazine-loaded nanoparticle formulations: a study on the effect of DNase-I on microbial biofilm and wound healing activity. *Molecular Pharmaceutics*, 2019. **16**(9): pp. 3916–3925.
34. Patel, K.K., et al., DNase-I functionalization of ciprofloxacin-loaded chitosan nanoparticles overcomes the biofilm-mediated resistance of Pseudomonas aeruginosa. *Applied Nanoscience*, 2020. **10**(2): pp. 563–575.
35. Urimi, D., et al., Polyglutamic acid functionalization of chitosan nanoparticles enhances the therapeutic efficacy of insulin following oral administration. *AAPS PharmSciTech*, 2019. **20**(3): pp. 131.
36. Patel, K.K., et al., Alginate lyase immobilized chitosan nanoparticles of ciprofloxacin for the improved antimicrobial activity against the biofilm associated mucoid P. aeruginosa infection in cystic fibrosis. *International Journal of Pharmaceutics*, 2019. **563**: pp. 30–42.
37. Anjum, M.M., et al., Development of anacardic acid/hydroxypropyl-β-cyclodextrin inclusion complex with enhanced solubility and antimicrobial activity. *Journal of Molecular Liquids*, 2019. **296**: p. 112085.
38. Kushwah, V., et al., Implication of linker length on cell cytotoxicity, pharmacokinetic and toxicity profile of gemcitabine-docetaxel combinatorial dual drug conjugate. *International Journal of Pharmaceutics*, 2018. **548**(1): pp. 357–374.
39. Kushwah, V., et al., Co-delivery of docetaxel and gemcitabine using PEGylated self-assembled stealth nanoparticles for improved breast cancer therapy. *Nanomedicine: Nanotechnology, Biology and Medicine*, 2018. **14**(5): pp. 1629–1641.
40. Jain, S., et al., Enhanced antitumor efficacy and reduced toxicity of docetaxel loaded estradiol functionalized stealth polymeric nanoparticles. *Molecular Pharmaceutics*, 2015. **12**(11): pp. 3871–3884.

41. Jain, S., et al., Hyaluronic acid–PEI–cyclodextrin polyplexes: implications for in vitro and in vivo transfection efficiency and toxicity. *RSC Advances*, 2015. **5**(51): pp. 41144–41154.
42. Harde, H., et al., Divalent toxoids loaded stable chitosan–glucomannan nanoassemblies for efficient systemic, mucosal and cellular immunostimulatory response following oral administration. *International Journal of Pharmaceutics*, 2015. **487**(1–2): pp. 292–304.
43. Harde, H., Agrawal, A.K., & Jain, S., Tetanus toxoids loaded glucomannosylated chitosan based nanohoming vaccine adjuvant with improved oral stability and immunostimulatory response. *Pharmaceutical Research*, 2015. **32**(1): pp. 122–134.
44. Agrawal, A.K., et al., Folate appended chitosan nanoparticles augment the stability, bioavailability and efficacy of insulin in diabetic rats following oral administration. *RSC Advances*, 2015. **5**(127): pp. 105179–105193.
45. Jain, S., et al., Combinatorial bio-conjugation of gemcitabine and curcumin enables dual drug delivery with synergistic anticancer efficacy and reduced toxicity. *RSC Advances*, 2014. **4**(55): pp. 29193–29201.
46. Harde, H., Agrawal, A.K., & Jain, S., Development of stabilized glucomannosylated chitosan nanoparticles using tandem crosslinking method for oral vaccine delivery. *Nanomedicine*, 2014. **9**(16): pp. 2511–2529.
47. Das, M., et al., Macromolecular bipill of gemcitabine and methotrexate facilitates tumor-specific dual drug therapy with higher benefit-to-risk ratio. *Bioconjugate Chemistry*, 2014. **25**(3): pp. 501–509.
48. Jain, S., et al., Surface stabilized efavirenz nanoparticles for oral bioavailability enhancement. *Journal of Biomedical Nanotechnology*, 2013. **9**(11): pp. 1862–1874.
49. Choudhary, H., et al., Evaluation and optimization of preparative variables for controlled-release floating microspheres of levodopa/carbidopa. *Pharmazie*, 2010. **65**(3): pp. 194–198.
50. Agrawal, A.K., et al., Development and characterization of in situ gel system for nasal insulin delivery. *Pharmazie*, 2010. **65**(3): pp. 188–193.
51. Shilpi, D., et al., Improved stability and enhanced oral bioavailability of atorvastatin loaded stearic acid modified gelatin nanoparticles. *Pharmaceutical Research*, 2017. **34**(7): pp. 1505–1516.
52. Kushwah, V., et al., Novel gemcitabine conjugated albumin nanoparticles: a potential strategy to enhance drug efficacy in pancreatic cancer treatment. *Pharmaceutical Research*, 2017. **34**(11): pp. 2295–2311.
53. Kushwah, V., et al., Co-delivery of docetaxel and gemcitabine by anacardic acid modified self-assembled albumin nanoparticles for effective breast cancer management. *Acta Biomaterialia*, 2018. **73**: pp. 424–436.
54. Kaler, A., et al., An investigation of in vivo wound healing activity of biologically synthesized silver nanoparticles. *Journal of Nanoparticle Research*, 2014. **16**(9): p. 2605.
55. Yáñez-Mó, M., et al., Biological properties of extracellular vesicles and their physiological functions. *Journal of Extracellular Vesicles*, 2015. **4**(1): p. 27066.
56. Wiklander, O.P., et al., Advances in therapeutic applications of extracellular vesicles. *Science Translational Medicine*, 2019. **11**(492).
57. Revenfeld, A.L.S., et al., Diagnostic and prognostic potential of extracellular vesicles in peripheral blood. *Clinical Therapeutics*, 2014. **36**(6): pp. 830–846.
58. Kowal, J., et al., Proteomic comparison defines novel markers to characterize heterogeneous populations of extracellular vesicle subtypes. *Proceedings of the National Academy of Sciences*, 2016. **113**(8): pp. E968–E977.
59. Kalluri, R., The biology and function of exosomes in cancer. *The Journal of Clinical Investigation*, 2016. **126**(4): pp. 1208–1215.
60. Kerr, J.F., Wyllie, A.H., & Currie, A.R., Apoptosis: a basic biological phenomenon with wideranging implications in tissue kinetics. *British Journal of Cancer*, 1972. **26**(4): pp. 239–257.
61. New, S.E., & Aikawa, E., Role of extracellular vesicles in de novo mineralization: an additional novel mechanism of cardiovascular calcification. *Arteriosclerosis, Thrombosis, and Vascular Biology*, 2013. **33**(8): pp. 1753–1758.
62. Abels, E.R., & Breakefield, X.O., Introduction to extracellular vesicles: biogenesis, RNA cargo selection, content, release, and uptake. *Cellular and Molecular Neurobiology*, 2016. **36**(3): pp. 301–312.
63. Bebelman, M.P., et al., Biogenesis and function of extracellular vesicles in cancer. *Pharmacology & Therapeutics*, 2018. **188**: pp. 1–11.
64. Pathirana, R.D., & Kaparakis-Liaskos, M., Bacterial membrane vesicles: biogenesis, immune regulation and pathogenesis. *Cellular Microbiology*, 2016. **18**(11): pp. 1518–1524.
65. Petrovčíková, E., Vičíková, K., & Leksa, V., Extracellular vesicles–biogenesis, composition, function, uptake and therapeutic applications. *Biologia*, 2018. **73**(4): pp. 437–448.

66. Juan, T., & Fürthauer, M., Biogenesis and function of ESCRT-dependent extracellular vesicles. in *Seminars in cell & developmental biology*. 2018. Elsevier.
67. Zhu, X., et al., The multivesicular bodies (MVBs)-localized AAA ATPase LRD6-6 inhibits immunity and cell death likely through regulating MVBs-mediated vesicular trafficking in rice. *PLoS Genetics*, 2016. **12**(9): pp. e1006311.
68. Bobrie, A., et al., Exosome secretion: molecular mechanisms and roles in immune responses. *Traffic*, 2011. **12**(12): pp. 1659–1668.
69. Mathieu, M., et al., Specificities of secretion and uptake of exosomes and other extracellular vesicles for cell-to-cell communication. *Nature Cell Biology*, 2019. **21**(1): pp. 9–17.
70. Cruz-Migoni, A., et al., Structure-based development of new RAS-effector inhibitors from a combination of active and inactive RAS-binding compounds. *Proceedings of the National Academy of Sciences*, 2019. **116**(7): pp. 2545–2550.
71. Shi, L., et al., SNARE proteins: one to fuse and three to keep the nascent fusion pore open. *Science*, 2012. **335**(6074): pp. 1355–1359.
72. Geng, X., et al., Exosomal circular RNA sorting mechanisms and their function in promoting or inhibiting cancer. *Oncology Letters*, 2020. **19**(5): pp. 3369–3380.
73. Kim, K.M., et al., RNA in extracellular vesicles. *Wiley Interdisciplinary Reviews: RNA*, 2017. **8**(4): p. e1413.
74. Di Liegro, C.M., Schiera, G., & Di Liegro, I., Extracellular vesicle-associated RNA as a carrier of epigenetic information. *Genes*, 2017. **8**(10): p. 240.
75. Zaborowski, M.P., et al., Extracellular vesicles: composition, biological relevance, and methods of study. *Bioscience*, 2015. **65**(8): pp. 783–797.
76. Kalra, H., Drummen, S., & Mathivanan, S., Focus on extracellular vesicles: introducing the next small big thing. *International Journal of Molecular Sciences*, 2016. **17**(2): p. 170.
77. Taylor, D.D., & Gercel-Taylor, C., The origin, function, and diagnostic potential of RNA within extracellular vesicles present in human biological fluids. *Frontiers in Genetics*, 2013. **4**: p. 142.
78. Roier, S., et al., A novel mechanism for the biogenesis of outer membrane vesicles in Gram-negative bacteria. *Nature Communications*, 2016. **7**(1): pp. 1–13.
79. Kulkarni, H.M., & Jagannadham, M.V., Biogenesis and multifaceted roles of outer membrane vesicles from Gram-negative bacteria. *Microbiology*, 2014. **160**(10): pp. 2109–2121.
80. Simpson, R.J., Jensen, S.S., & Lim, J.W., Proteomic profiling of exosomes: current perspectives. *Proteomics*, 2008. **8**(19): pp. 4083–4099.
81. Choi, D.S., et al., Proteomics of extracellular vesicles: exosomes and ectosomes. *Mass Spectrometry Reviews*, 2015. **34**(4): pp. 474–490.
82. Morad, G., et al., Tumor-derived extracellular vesicles breach the intact blood–brain barrier via transcytosis. *ACS Nano*, 2019. **13**(12): pp. 13853–13865.
83. Banks, W.A., et al., Transport of extracellular vesicles across the blood-brain barrier: brain pharmacokinetics and effects of inflammation. *International Journal of Molecular Sciences*, 2020. **21**(12): p. 4407.
84. Matsumoto, J., et al., The transport mechanism of extracellular vesicles at the blood-brain barrier. *Current Pharmaceutical Design*, 2017. **23**(40): pp. 6206–6214.
85. Fujita, Y., Yoshioka, Y., & Ochiya, T., Extracellular vesicle transfer of cancer pathogenic components. *Cancer Science*, 2016. **107**(4): pp. 385–390.
86. Witwer, K.W., & Wolfram, J., Extracellular vesicles versus synthetic nanoparticles for drug delivery. *Nature Reviews Materials*, 2021. **6**: pp. 1–4.
87. Maeki, M., et al., Advances in microfluidics for lipid nanoparticles and extracellular vesicles and applications in drug delivery systems. *Advanced Drug Delivery Reviews*, 2018. **128**: pp. 84–100.
88. Mulcahy, L.A., Pink, R.C., & Carter, D.R.F., Routes and mechanisms of extracellular vesicle uptake. *Journal of Extracellular Vesicles*, 2014. **3**(1): p. 24641.
89. Nakase, I., et al., Vectorization of biomacromolecules into cells using extracellular vesicles with enhanced internalization induced by macropinocytosis. *Scientific Reports*, 2016. **6**(1): pp. 1–10.
90. Arteaga-Blanco, L.A., et al., Characterization and internalization of small extracellular vesicles released by human primary macrophages derived from circulating monocytes. *PLoS One*, 2020. **15**(8): p. e0237795.
91. Jiang, L., P. Vader, & Schiffelers, R., Extracellular vesicles for nucleic acid delivery: progress and prospects for safe RNA-based gene therapy. *Gene Therapy*, 2017. **24**(3): pp. 157–166.

92. Parayath, N.N., Padmakumar, S., & Amiji, M.M., Extracellular vesicle-mediated nucleic acid transfer and reprogramming in the tumor microenvironment. *Cancer Letters*, 2020. **482**: pp. 33–43.
93. Ratajczak, M.Z., & Ratajczak, J., Horizontal transfer of RNA and proteins between cells by extracellular microvesicles: 14 years later. *Clinical and Translational Medicine*, 2016. **5**(1): pp. 1–8.
94. Valadi, H., et al., Exosome-mediated transfer of mRNAs and microRNAs is a novel mechanism of genetic exchange between cells. *Nature Cell Biology*, 2007. **9**(6): pp. 654–659.
95. Pegtel, D.M., et al., Functional delivery of viral miRNAs via exosomes. *Proceedings of the National Academy of Sciences*, 2010. **107**(14): pp. 6328–6333.
96. Jungmann, R.A., et al., Glucocorticoid induction of CRE-binding protein isoform mRNAs in rat C6 glioma cells. *Nucleic Acids Research*, 1992. **20**(4): pp. 825–829.
97. Vader, P., et al., New considerations in the preparation of nucleic acid-loaded extracellular vesicles. *Therapeutic Delivery*, 2014. **5**(2): pp. 105–107.
98. O'Brien, K., et al., RNA delivery by extracellular vesicles in mammalian cells and its applications. *Nature Reviews Molecular Cell Biology*, 2020. **21**(10): pp. 585–606.
99. Orefice, N.S., Development of new strategies using extracellular vesicles loaded with exogenous nucleic acid. *Pharmaceutics*, 2020. **12**(8): p. 705.
100. De Jong, O.G., et al., Drug delivery with extracellular vesicles: from imagination to innovation. *Accounts of Chemical Research*, 2019. **52**(7): p. 1761–1770.
101. Maas, S.L., Breakefield, X.O., & Weaver, A.M., Extracellular vesicles: unique intercellular delivery vehicles. *Trends in Cell Biology*, 2017. **27**(3): pp. 172–188.
102. Jeyaram, A., et al., Enhanced loading of functional miRNA cargo via pH gradient modification of extracellular vesicles. *Molecular Therapy*, 2020. **28**(3): pp. 975–985.
103. Villa, F., Quarto, R., & Tasso, R., Extracellular vesicles as natural, safe and efficient drug delivery systems. *Pharmaceutics*, 2019. **11**(11): p. 557.
104. Pomatto, M.A.C., et al., Improved loading of plasma-derived extracellular vesicles to encapsulate antitumor miRNAs. *Molecular Therapy-Methods & Clinical Development*, 2019. **13**: pp. 133–144.
105. Agrawal, A.K., et al., Milk-derived exosomes for oral delivery of paclitaxel. *Nanomedicine: Nanotechnology, Biology and Medicine*, 2017. **13**(5): pp. 1627–1636.
106. Aqil, F., et al., Exosomal formulation enhances therapeutic response of celastrol against lung cancer. *Experimental and Molecular Pathology*, 2016. **101**(1): pp. 12–21.
107. Aqil, F., et al., Exosomes for the enhanced tissue bioavailability and efficacy of curcumin. *The AAPS Journal*, 2017. **19**(6): pp. 1691–1702.
108. Munagala, R., et al., Exosomal formulation of anthocyanidins against multiple cancer types. *Cancer Letters*, 2017. **393**: pp. 94–102.
109. Aqil, F., et al., Exosomal delivery of berry anthocyanidins for the management of ovarian cancer. *Food & Function*, 2017. **8**(11): pp. 4100–4107.
110. Didiot, M.-C., et al., Exosome-mediated delivery of hydrophobically modified siRNA for huntingtin mRNA silencing. *Molecular Therapy*, 2016. **24**(10): pp. 1836–1847.
111. Agarwal, A., et al., Regulatory T cell-derived exosomes: possible therapeutic and diagnostic tools in transplantation. *Frontiers in Immunology*, 2014. **5**: p. 555.
112. Lamichhane, T.N., & Jay, S.M., Production of extracellular vesicles loaded with therapeutic cargo. in *Targeted Drug Delivery*. 2018. Springer. pp. 37–47.
113. Kooijmans, S.A., et al., Electroporation-induced siRNA precipitation obscures the efficiency of siRNA loading into extracellular vesicles. *Journal of Controlled Release*, 2013. **172**(1): pp. 229–238.
114. Lamichhane, T.N., Raiker, R.S., & Jay, S.M., Exogenous DNA loading into extracellular vesicles via electroporation is size-dependent and enables limited gene delivery. *Molecular Pharmaceutics*, 2015. **12**(10): pp. 3650–3657.
115. Fuhrmann, G., et al., Active loading into extracellular vesicles significantly improves the cellular uptake and photodynamic effect of porphyrins. *Journal of Controlled Release*, 2015. **205**: pp. 35–44.
116. Zhang, Y.-F., Shi, J.-B., & Li, C., Small extracellular vesicle loading systems in cancer therapy: current status and the way forward. *Cytotherapy*, 2019. **21**(11): pp. 1122–1136.
117. Pedersen, K.W., Kierulf, B., & Neurauter, A., Specific and generic isolation of extracellular vesicles with magnetic beads. in *Extracellular Vesicles*. 2017. Springer. pp. 65–87.
118. Witwer, K.W., et al., Standardization of sample collection, isolation and analysis methods in extracellular vesicle research. *Journal of Extracellular Vesicles*, 2013. **2**(1): p. 20360.

119. Kenari, A.N., Cheng, L., & Hill, A.F., Methods for loading therapeutics into extracellular vesicles and generating extracellular vesicles mimetic-nanovesicles. *Methods*, 2020. **177**: pp. 103–113.
120. Rayamajhi, S., & Aryal, S., Surface functionalization strategies of extracellular vesicles. *Journal of Materials Chemistry B*, 2020. **8**(21): pp. 4552–4569.
121. Lu, M., et al., Functionalized extracellular vesicles as advanced therapeutic nanodelivery systems. *European Journal of Pharmaceutical Sciences*, 2018. **121**: pp. 34–46.
122. Wan, Y., et al., Enrichment of extracellular vesicles with lipid nanoprobe functionalized nanostructured silica. *Lab on a Chip*, 2019. **19**(14): pp. 2346–2355.
123. Markov, O., Oshchepkova, A., & Mironova, N., Immunotherapy based on dendritic cell-targeted/-derived extracellular vesicles—a novel strategy for enhancement of the anti-tumor immune response. *Frontiers in Pharmacology*, 2019. **10**: p. 1152.
124. Cheng, K., et al., Magnetic targeting enhances engraftment and functional benefit of iron-labeled cardiosphere-derived cells in myocardial infarction. *Circulation Research*, 2010. **106**(10): p. 1570.
125. Mentkowski, K.I., & Lang, J.K., Exosomes engineered to express a cardiomyocyte binding peptide demonstrate improved cardiac retention in vivo. *Scientific Reports*, 2019. **9**(1): pp. 1–13.
126. Martins-Marques, T., et al., Presence of Cx43 in extracellular vesicles reduces the cardiotoxicity of the anti-tumour therapeutic approach with doxorubicin. *Journal of Extracellular Vesicles*, 2016. **5**(1): p. 32538.
127. Piffoux, M., et al., Modification of extracellular vesicles by fusion with liposomes for the design of personalized biogenic drug delivery systems. *ACS Nano*, 2018. **12**(7): pp. 6830–6842.
128. Cabeza, L., et al., Cancer therapy based on extracellular vesicles as drug delivery vehicles. *Journal of Controlled Release*, 2020. 327: pp. 296–315.
129. Kooijmans, S.A., et al., Display of GPI-anchored anti-EGFR nanobodies on extracellular vesicles promotes tumour cell targeting. *Journal of Extracellular Vesicles*, 2016. **5**(1): p. 31053.
130. Susa, F., et al., Engineered extracellular vesicles as a reliable tool in cancer nanomedicine. *Cancers*, 2019. **11**(12): p. 1979.
131. Rufino-Ramos, D., et al., Extracellular vesicles: novel promising delivery systems for therapy of brain diseases. *Journal of Controlled Release*, 2017. **262**: pp. 247–258.
132. Kanada, M., et al., Differential fates of biomolecules delivered to target cells via extracellular vesicles. *Proceedings of the National Academy of Sciences*, 2015. **112**(12): pp. E1433–E1442.
133. Alvarez-Erviti, L., et al., Delivery of siRNA to the mouse brain by systemic injection of targeted exosomes. *Nature Biotechnology*, 2011. **29**(4): pp. 341–345.
134. Li, Z., et al., In vitro and in vivo RNA inhibition by CD9-HuR functionalized exosomes encapsulated with miRNA or CRISPR/dCas9. *Nano Letters*, 2018. **19**(1): pp. 19–28.
135. Guo, S., et al., Intranasal delivery of mesenchymal stem cell derived exosomes loaded with phosphatase and tensin homolog siRNA repairs complete spinal cord injury. *ACS Nano*, 2019. **13**(9): pp. 10015–10028.
136. Ohno, S.-I., et al., Systemically injected exosomes targeted to EGFR deliver antitumor microRNA to breast cancer cells. *Molecular Therapy*, 2013. **21**(1): pp. 185–191.
137. Ren, Y., et al., Tumor-penetrating delivery of siRNA against TNFα to human vestibular schwannomas. *Scientific Reports*, 2017. **7**(1): pp. 1–16.
138. Xin, H., et al., MiR-133b promotes neural plasticity and functional recovery after treatment of stroke with multipotent mesenchymal stromal cells in rats via transfer of exosome-enriched extracellular particles. *Stem Cells*, 2013. **31**(12): pp. 2737–2746.
139. Neviani, P., et al., Natural killer–derived exosomal miR-186 inhibits neuroblastoma growth and immune escape mechanisms. *Cancer Research*, 2019. **79**(6): pp. 1151–1164.
140. Aqil, F., et al., Milk exosomes-Natural nanoparticles for siRNA delivery. *Cancer Letters*, 2019. **449**: pp. 186–195.
141. Fonsato, V., et al., Human liver stem cell-derived microvesicles inhibit hepatoma growth in SCID mice by delivering antitumor microRNAs. *Stem Cells*, 2012. **30**(9): pp. 1985–1998.
142. Zhang, Y., et al., Microvesicle-mediated delivery of transforming growth factor β1 siRNA for the suppression of tumor growth in mice. *Biomaterials*, 2014. **35**(14): pp. 4390–4400.
143. O'Brien, K., et al., Employing mesenchymal stem cells to support tumor-targeted delivery of extracellular vesicle (EV)-encapsulated microRNA-379. *Oncogene*, 2018. **37**(16): pp. 2137–2149.
144. Yu, L., et al., Exosomes derived from microRNA-199a-overexpressing mesenchymal stem cells inhibit glioma progression by down-regulating AGAP2. *Aging (Albany NY)*, 2019. **11**(15): p. 5300.

145. Liu, T., et al., Exosome-transmitted miR-128-3p increase chemosensitivity of oxaliplatin-resistant colorectal cancer. *Molecular Cancer*, 2019. **18**(1): pp. 1–17.
146. Zhang, H., et al., Exosomes serve as nanoparticles to suppress tumor growth and angiogenesis in gastric cancer by delivering hepatocyte growth factor si RNA. *Cancer Science*, 2018. **109**(3): pp. 629–641.
147. Jeong, K., et al., Exosome-mediated microRNA-497 delivery for anti-cancer therapy in a microfluidic 3D lung cancer model. *Lab on a Chip*, 2020. **20**(3): pp. 548–557.
148. Pakravan, K., et al., MicroRNA-100 shuttled by mesenchymal stem cell-derived exosomes suppresses in vitro angiogenesis through modulating the mTOR/HIF-1α/VEGF signaling axis in breast cancer cells. *Cellular Oncology*, 2017. **40**(5): pp. 457–470.
149. Lou, G., et al., Exosomes derived from miR-122-modified adipose tissue-derived MSCs increase chemosensitivity of hepatocellular carcinoma. *Journal of Hematology & Oncology*, 2015. **8**(1): pp. 1–11.

4 Green Synthesis of Nanoparticles in Oligonucleotide Drug Delivery System

Manish P. Patel, Praful D. Bharadia, Kunjan B. Bodiwala and Mustakim M. Mansuri
L. M. College of Pharmacy, Ahmedabad, India

Jayvadan Patel
Sankalchand Patel University, Visnagar, India

CONTENTS

4.1 INTRODUCTION

Approximately 30 million persons in the USA alone have orphan or rare diseases. A low population affected with rare diseases caused a major hurdle toward development of these drugs.

DOI: 10.1201/9781003186083-4

Oligonucleotides are an emerging delivery system toward patient centric treatment of this type of diseases (Kim et al., 2019).

Oligonucleotides are nucleic acid polymers which can be a major impact on treatment or management of wide range of diseases. This therapeutics' particular role is to be the silencing gene. Other roles like splice modulation and gene activation are under investigation.

Oligonucleotides are binding with RNA through DNA base pairing and modulate the function of the targeted RNA. It suggests a wide variety of oligonucleotides that modulate RNA through a different binding mechanisms (Esau et al., 2006). Briefly, these mechanisms can be classified largely as (a) those that bind to RNA and interfere with its function without promoting RNA degradation, such as translation arrest or modulating RNA processing (b) those that promote degradation of the RNA through endogenous enzymes, such as RNase H or argonaute-2 [RNA interference (RNAi)]. Recently, researchers have shown that Anti Sense Oligonucleotides can also be used to increase protein production, either through antagonizing microRNAs, which normally suppress protein production, or through masking upstream open reading frames (Liang et al., 2016). It shows that this drug delivery system has a huge impact in management or treatment of rare diseases.

However, this delivery system has some limitations, so these issues must be overcome for the development of Oligonucleotides. Enzymatic degradations and rapid renal clearance are major issues in this delivery system (Ivanova et al., 2018).

In the attempt to stabilize Oligonucleotides, we can use delivery vehicles for therapeutic agents like (siRNA, DNA) (Le Ouay and Stellacci, 2015). These delivery vehicles are either metal nanoparticles or plant derived components in the synthesis of nanoparticles (Tolaymat et al., 2010). Generally, nanoparticles fall in the range of —one to one hundred nm. Nano medicine is an emerging branch of medicine that uses knowledge of nanomaterials to apply to different nanotechnologies to prevent, diagnose, and treat disease (Satalkar et al., 2016) but synthesis of nanoparticles has concernment toward toxicity in eukaryotic cells and pollution in the atmosphere (Taghavi et al., 2013) so it is important to develop methods that synthesis nanoparticles in such a ways that they cannot cause or reduce toxicity and pollution. Green synthesis is one branch of chemistry where product is produced without causing environmental consequences. Here, environmentally friendly nontoxic reagents are used in the synthesis of nanoparticles on the basis that chemical, and physical methods are prepared as well so these techniques give economic and environmental benefits. In this method, nontoxic, safe reagents that are eco-friendly and biosafe are used. Minimizing waste and providing a sustainable process is the primary aim of this synthesis (Zhu et al., 2019). Various natural resources available in nature, such as plant extracts (Sithara et al., 2017), cyclodextrin (Abou-Okeil et al., 2012), chitosan (Mokhena and Luyt, 2017), and many more, have been studied for the synthesis of metal oxide nanoparticles. As of April 2021, fourteen oligonucleotide drugs have received regulatory approval from the FDA (Stein and Castanotto, 2017). In addition, more than thirty oligonucleotide drugs are in clinical trials or waiting for regulatory approval (Bennett et al., 2017) (Table 4.1).

4.2 TYPES OF OLIGONUCLEOTIDES

4.2.1 Antisense Oligonucleotides (ASO)

They are single stranded oligo deoxy-nucleotides that can modify RNA and can alter the protein expression by mechanism of action. They are further classified into two classes:

4.2.1.1 RNAse H-dependent Oligonucleotide

They hold the process of splicing or translation. These oligonucleotides are target specific. Example: AUG initiation codon. They encourage mRNA degradation. The majority of antisense drugs act through an RNase H-dependent mechanism. They have capacity of down-regulating protein and mRNA expression up to 90 percentage, and target any site of mRNA. Hydrolysis of the RNA strand is caused by this enzyme.

TABLE 4.1
Oligonucleotide Drugs Approved by FDA

	Drug Name	Indication	FDA Approval
1	Fomivirsen	Cytomegalovirus (CMV) Retinitis	1998 (Withdrawn)
2	Pegaptanib	Age-related Macular Degeneration (AMD) of the Retina	2004 (Marketed)
3	Mipomersen sodium	Homozygous Familial Hypercholesterolemia	2013 (Withdrawn)
4	Defibrotide sodium	Hepatic Veno-Occlusive Disease	2016 (Marketed)
5	Eteplirsen	Duchenne Muscular Dystrophy	2016 (Marketed)
6	Nusinersen	Spinal Muscular Atrophy	2016 (Marketed)
7	Hepatitis B vaccine, Adjuvanted	Hepatitis B	2017 (Marketed)
8	Inotersen sodium	Familial Amyloid Neuropathies	2018 (Marketed)
9	Patisiran	Familial Amyloid Neuropathies	2018 (Marketed)
10	Givosiran	Acute Hepatic Porphyria	2019 (Marketed)
11	Golodirsen	Duchenne Muscular Dystrophy	2019 (Marketed)
12	Viltolarsen	Duchenne Muscular Dystrophy	2020 (Marketed)
13	Lumansiran	Primary hyperoxaluria type 1	2020 (Marketed)
14	Casimersen	Duchenne Muscular Dystrophy	2021 (Marketed)

4.2.2 Steric-blocker Oligonucleotides

They hold the process of splicing or translation. They are target specific, i.e., AUG initiation codon (Kole et al., 2012)

4.2.3 Small Interfering RNA or Silencing RNA (siRNA)

They are target specific non coding RNA. siRNA cause gene silencing at the post transcriptional level by providing an interface toward RNAi. Furthermore, mRNA degradation is caused by RNAi, although they have poor stability as well as less delivery (Eberhardt et al., 2007).

4.2.4 Micro RNA (miRNA)

Generally, miRNA are made up of—21 to 25 nucleotides and are single stranded. They interfere with the translation process through accumulation of mRNA in processing bodies (Dua et al., 2017a). They have a major part in different processes, such as breakdown of fats, neuronal patterning, hematopoietic differentiation, cell division, cell death, and immunity (Dua et al., 2017b). They cause degradation of mRNA but have less stability, as well as poor delivery. miRNA should be use in neuro degenerative disorders and cancers also (Lam et al., 2015).

4.2.5 Aptamer

These have a single strain, and their basic structure is made up of DNA or RNA. As well as being very target specific in nature (Röthlisberger and Hollenstein, 2018), they are stable and less immunogenic; scale up of the process is easily placed with reasonable finance. Both DNA and RNA consist of different basic structures, even though they have similar targets (Song et al., 2012).

4.2.6 CpG Oligonucleotides

They are short, single stranded oligo deoxy nucleotides with unmethylated CpG dinucleotides and are classified into four types:

K-type/B-type, D-type/A-type, C-type, P-type (Shirota et al., 2015)

4.3 THERAPEUTIC IMPORTANCE OF OLIGONUCLEOTIDES

Neurodegenerative disorders (Scoles and Pulst, 2018)
Respiratory disorders (Mehta et al., 2019)
Cancer (Mercatante et al., 2001)
Diabetic retinopathy (Hnik et al., 2009)

4.4 DRUG DELIVERY APPROACHES FOR OLIGONUCLEOTIDES

Liposomes (Bheemidi et al., 2011)
Niosomes (Kazi et al., 2010)
Nanoparticles (Audrey and Procter, 2015)

Mucoadhesive targeting of oligonucleotides in respiratory disorders (de Fougerolles and Novobrantseva, 2008)Dendrimers (Conti et al., 2014Micells (Smola et al., 2008)

Among all these approaches, nanoparticles capture significant attraction because of controlled drug release, theranostics, target specificity, and better therapeutic index (Crucho and Barros, 2017). Nanoparticles fall under the size range of —one to one hundred nm. Because of very small size, nanoparticles play a major role as a drug carrier (Florence, 2004). On the basis of this, nanoparticles can be divided into three major parts (Singh and Lillard, 2009).

Polymeric nanoparticles (Soppimath et al., 2001)
Solid-lipid nanoparticles (Manjunath et al., 2005)
Inorganic nanoparticles (Liong et al., 2008)

4.5 GREEN SYNTHESIS OF NANOPARTICLES

Nanoparticles have varied unique properties (optical, magnetic, electrical, etc.) because of their very large specific surface area, high surface energy, and quantum confinement (Hussain et al., 2016). Metallic nanoparticles can be made by different chemical, physical, and biological methods (Iravani et al., 2014). This method falls under two distinguishing approaches:

Top-down approaches: in this method, size reduction of primary materials has taken place.

Bottom-up approaches: in this method, we deal with atoms and molecules to form nanoparticles. (Iqbal et al., 2012)

Physical methods for production of nano particles require lots of time, consume lots of energy, and require thermal stability, as well as raising the environmental temperature around the source material and consuming large spaces in the case of furnaces (Spadavecchia et al., 2005). That is the reason physical methods are not suitable for a greener approach. On the other hand, chemical methods deal with harsh reagents that create toxicity issues along with environmental issues (Gnanasangeetha and SaralaThambavani, 2013). Because of these issues, biological methods are superior over physicochemical methods. In biological methods, bacteria (De Silva et al., 2020), fungi (Guilger-Casagrande and de Lima, 2019b), algie (Khan et al., 2019), and plants (Mohanpuria et al., 2008), but here our concern is a microorganism-based production process which is slow as well as having pathogenicity issues and maintenance on a large scale (Mandal et al., 2006) using plant derivatives or plant parts for reducing metal ions to their elemental forms (Gaur et al., 2019).

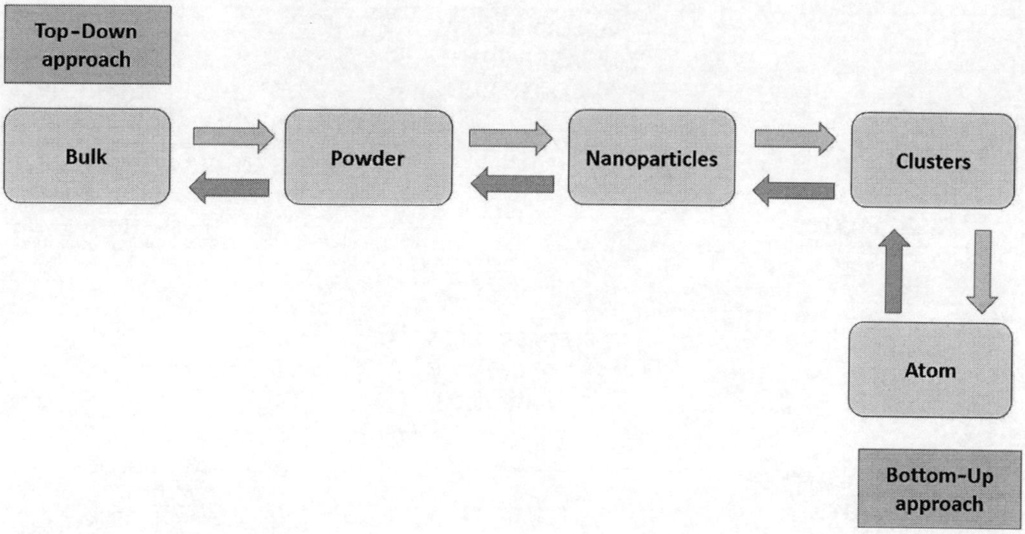

FIGURE 4.1 Synthesis of nanoparticles.

(Abid et al., 2021)

This process can be easily scaled up, simple, efficient, and economical; in addition, there are no worries about biosafety in this process. Silver (Sharma et al., 2009), gold (Ankamwar, 2010a), palladium (Petla et al., 2011), iron (Saif et al., 2016) and zinc oxide (Fakhari et al., 2019b) nanoparticles are generally made up by this synthesis.

4.6 APPROACHES INVOLVED IN THE GREEN SYNTHESIS OF NANOPARTICLES

Generally, two approaches are involved in the green synthesis of nanoparticles:

Top-down approach
Bottom-up approach

In top-down approach, breaking down of bulk material, followed by size reduction, yields nanoparticles (Mignot et al., 2013). In the bottom up approach (Zhigaltsev et al., 2012), atoms or clusters assemble together to form nanoparticles by different biological or chemical methods.

By green synthesis, we can make different types of nanoparticles: metallic, metal oxide-based, alloy based, and magnetic, etc. Silver (Srikar et al., 2016), gold (Ankamwar et al., 2010a, 2010b), platinum (Thirumurugan et al., 2016), palladium (Siddiqi and Husen, 2016), copper (Umer et al., 2014), zinc (Fakhari et al., 2019a), and metallic nanoparticles can be made by synthesis, but all of these silver nanoparticles draw a huge amount of attention because of their low toxicity profile and antibacterial activity, as well as it can be produced by green synthesis. Metal oxides have good thermal stability and are used as an adsorbent. They have metal-oxide-metal or metal-hydroxy-metal structure (Stoimenov et al., 2002). Metallic alloys have also attracted significant attention in the last few years due to their optical and electrical properties (Stoimenov et al., 2002). Magnetic nanoparticles have a magnetic part along with a chemical part (Kodama, 1999). Due to their target specific properties, it's widely used now and is also used for diagnostic purposes.

4.6.1 GREEN SYNTHESIS OF NANOPARTICLES

The above figure shows different approaches toward the green synthesis of nanoparticles. Bacterial strains, fungi spores, actinomycetes, algae, yeast spores, and plants can reduce metal ions and

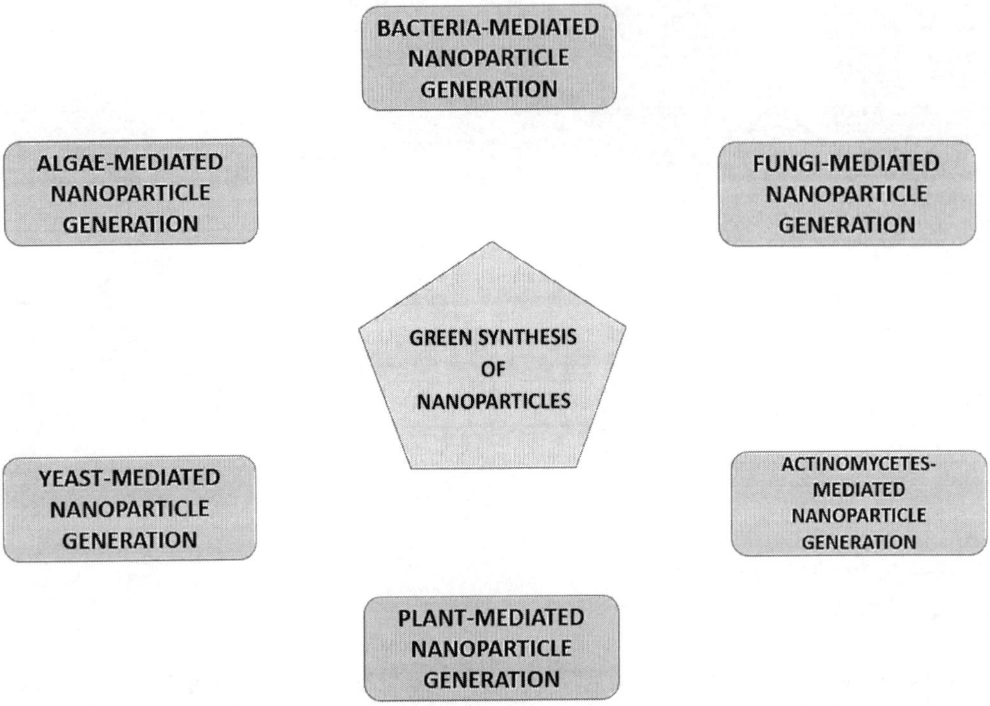

FIGURE 4.2 Green synthesis of nanoparticles.

(Nadaroglu et al. 2017)

biosynthesis metal nanoparticles. Among all of this, bacteria mediated nanoparticles generation has great potential toward biosynthesis of NPs, due to the selective nature of bacterial strains and the amount of nanoparticles, but it has some drawbacks due to problems arising in the scale up procedure and maintenance of the bed of culture mixture. However, the plant mediated approach is clear of all these issues but lacks selectivity potential compared to the bacterial approach.

4.6.2 Bacteria-Mediated Nanoparticle Generation

Synthesis of nanoparticles using a bacterial strain emerged due to the technique's green prospect. These bacteria can mobilize or immobilize metal ions, along with reducing ion concentration and accumulation of these ions in their cell wall. *Pseudomonas stutzeri* AG259 is the first bacterial strain to show silver nanoparticles formation; this strain is isolated from silver mines. After that discovery, significant research has been done in this area, and many bacterial strains are isolated for synthesis of nanoparticles (Prabhu and Poulose, 2012). This approach has immense potential and a bottom up type of synthesis. Magnetotactic bacteria and S layer lattices in prokariyotic cells like bacteria are capable of synthesis of metal ions (xie et al., 2009; Györvary et al., 2004). *Pseudomonas stutzeri* and *Pseudomonas aeruginosa* can get through at even higher concentrations of metal ions (Husseiny et al., 2007). Thiobacillus ferrooxidans, T. thiooxidans, and Sulfolobus acidocaldarius convert Fe^{3+} into Fe^{2+} (Korbekandi et al., 2009). Bacillus cereus, B. subtilis, E. coli, and P. aeruginosa can decrease concentrations of silver ions, cadmium ions, cupric ions, and lanthanum ions (Du and Li, 2016). Bacterial strains or species used in synthesis of nanoparticles are shown in Table 4.2.

TABLE 4.2
Different Metal Nanoparticles Supported by Particular Strain of Bacterial (Iravani et al., 2014)

Type of Bacteria	Metal Nanoparticles
Aeromonas sp. SH10	Silver
Bacillus cereus	Silver
Bacillus megatherium D01	Gold
Bacillus subtilis 168	Gold
Bacillus subtilis	Silver
Clostridium thermoaceticum	Cadmium sulfide
Corynebacterium sp. SH09	Silver
Desulfobacteraceae	Zinc sulfide
Desulfovibrio desulfuricans	Palladium and selenium
Desulfovibrio vulgaris	Gold uranium and chromium
Desulfovibrio magneticus strain RS-1	Magnetite
Enterobacter cloacae	Silver and selenium
Escherichia coli	Cadmium sulfide
Escherichia coli	Silver
Escherichia coli DH5α	Silver
Escherichia coli DH5α	Gold
Escherichia coli MC4100	Gold
Geobacillus sp.	Gold
Geovibrio ferrireducens	Gold
Klebsiella aerogenes	Cadmium sulfide
Klebsiella pneumonia	Silver
Lactobacillus strains	Gold
Lactobacillus strains	Silver
Lactobacillus strains	Silver-gold alloys
Lactobacillus strains	Titanium
Lactobacillus casei subsp. casei	Silver
Magnetospirillum magnetotacticum	Magnetite cluster
Plectonema boryanum UTEX 485	Gold
Pseudomonas aeruginosa	Gold
Pseudomonas aeruginosa Lanthanum	Gold
Pseudomonas fluorescens	Gold
Pseudomonas putida NCIM 2650	Silver
Pseudomonas stutzeri AG259	Silver
Rhodobacter sphaeroides	Zinc sulfide
Rhodopseudomonas capsulate	Gold

4.6.3 Silver Nanoparticles

Cell-free culture supernatants of five psychrophilic bacteria (Phaeocystis antarctica, Pseudomonas proteolytica, Pseudomonas meridiana, Arthrobacter kerguelensis, and Arthrobacter gangotriensis) and two mesophilic bacteria (Bacillus indicus and Bacillus cecembensis) can synthesis silver nanoparticles. This process is dependent upon pH, type of bacteria, and temperature. It is observed that two different type of bacteria cannot produce silver nanoparticles at same temperature (Shivaji et al., 2011). Lactobacillus casei subsp. casei can produce silver nanoparticles at room temperature.

this nanoparticles are spherical and within size limit of 25~50nm. Here nanoparticles are formed at the cytoplasmic cell membrane, in cytoplasm, and extracellular space of the strain, ; this process gives evidence that the enzymes responsible for bioenzymatic reactions are present at the cytosol and cytoplasmic cell membrane (Garmasheva et al. 2016).

4.6.4 Gold Nanoparticle

Au3^{+} gold(III) ion, also known as auric ion, is converted into octahedral gold nanoparticles by Bacillus subtilis 168. These nanoparticles fall in the range of 5~25 nm. With ideal temperature and pressure, AuCl3 converts into nanoparticles with this strain (Beveridge and Murray, 1980; Reddy et al., 2010). Gold nanoparticles can be made in the cell and out of the cell space. Bacillus megaterium D01 can adsorb Au3+, when the biomass of this strain is held along with chloroauric acid nano gold particles, which have a spherical shape formed intracellularly or extracellularly. Here is this dodecanethiol, also added as a capping ligand. Gram-negative bacillus Shewanella alga is a facultative anaerobe bacteria. This bacteria can reduce tetrachloroaurate and form gold nano particle; in this process, hydrogen gas (30 minutes at 25 Celsius within pH range of two to seven) or lactate acts as an electron donor, and the ferric ion acts as electron acceptor. Nanoparticles deposit at the periplasmic space of the cell (Rajeshkumar et al., 2013). Here, pH is a very critical parameter because when we decrease pH below three, NPs deposit extracellularly. At pH below two, smaller nanoparticles deposit in the cell and larger particles deposit extracellularly (Konishi et al., 2007).

For escherichia coli DH5 alpha biosynthesis gold nanoparticles in the size range of 33~17 nm, (Saha et al., 2007) this process leads mostly spherical particles, along with a small amount of trianles and quasi-shaped. When E. coli MC4100 and Desulfovibrio desulfuricans ATCC 29577 are exposed to chloroauric acid, nanoparticles are formed. Below 7 pH, spherical old nano particles in the size range of 10 nm synthesis and at slightly basic pH mixture of bigger and smaller nanoparticles are observed (Deplanche and Macaskie, 2008).

Rhodopseudomonas capsulate also produced nanoparticles of gold under controlled pH. Here, NPs are produced extracellularly when AuCl4- ion is reduced to metal. At a lower concentration of chloroaurate ion, it produces spherical NPs, and at a higher concentration, nanowire is produced. Here, the whole process is enzymatic reduction caused by an NADH dependent enzyme, which is present in R. capsulate (He et al., 2008).

Plectonema boryanum UTEX 485 in aqueous solution of aurochlorate ion and aurothiosulfate ion produce cubic nanoparticles. This bacterial strain is cynobacteria. Interaction of cynobacteria with aurochlorate ion precipitate amorphous AuS at the cell wall with an octahedral shape (Lengke et al., 2006).

4.6.5 Magnetite Nanoparticles

Desulfovibrio magneticus strain RS-1 is capable of reducing sulfates. They accumulate NPs in the cells (Pósfai et al., 2006). Alteromonas putrefaciens oxidize electron donors like pyruvate, lactate, hydrogen and format to reduce the Ferric ion and Manganese (IV) ion. Pelobacter acetylenicus and P. venetianus are also capable of reducing Ferric ion under specific conditions (Lovley et al., 1989) Desulfuromonas acetoxidans is capable of reducing ferric citrate complexed of ferric ion of nitriloacetic acid, pyruvate, and lactate, with the help of butanol as an electron donor (Revati and Pandey, 2011). Pyrobaculum islandicum reduces the ferric ion where hydrogen acts as an electron donor and ferric citrate as an acceptor when the process takes place at 100 celsius. (Knutson, 2011) This strain can also reduce uranium (VI), Technetium (VI), Cobalt (III), Chromium (VI), and Manganese (IV); therefore, this type of strain can also be used in removal of toxic metals from water or metal deposits. Generally, iron nano-sized magnetic particles are used in accurate prognosis, patient centric treatment, cellular interactions of environment, radio therapy, targeted drug delivery, MRI, immunoassays, purification of DNA or RNA (Huber, 2005).

4.6.6 Palladium and Platinum Nanoparticles

Desulfovibrio desulfuricans, which is a sulfate ion reducer, and Shewanella oneidensis, which can reduce the ferric ion, can also reduce Palladium(II) into Palladium(0). Here, reducing agents are lactate, pyruvate, or hydrogen. Shewanella algae can reduce platinum(III) into insoluble platinum.

4.6.7 Selenium and Tellurium Nanoparticles

Stenotrophomonas maltophilia SELTE02 gives significant results toward synthesis of metal selenium from selenite ion. Yield of the above process is found in the cell of the strain or at the extracellular space. Enterobacter cloacae SLD1a-1, Rhodospirillum rubrum, and Desulfovibrio desulfuricans also facilitate this process of yielding selenite nanoparticles from selenite ion. E.coli store selenite in both periplasmic cells and cytoplasm, by reduction of selenite.

This all are the types of nanoparticles that can be use in targeted drug delivery of oligo nucleotides along with this zinc oxide nanoparticle, titanium/titanium dioxide nanoparticles, cadmium sulphide nanoparticles, tellurium nanoparticles, etc. Also, types can be made up or created through biosynthesis by bacterial strains, but they are more prone to use in electronic and mechanical approaches rather than drug delivery vehicles, due to their toxicity and poor compliance with drug molecules (Zoroddu et al., 2014).

4.6.8 Fungi-Mediated Nanoparticle Generation

Fungi have the ability of metal bioaccumulation, high binding capacity, and intake into a cell the same as bacteria. For several years, fungi mediated synthesis has drawn significant attention because of high yields, ease of handling, and less toxicity in the residue.

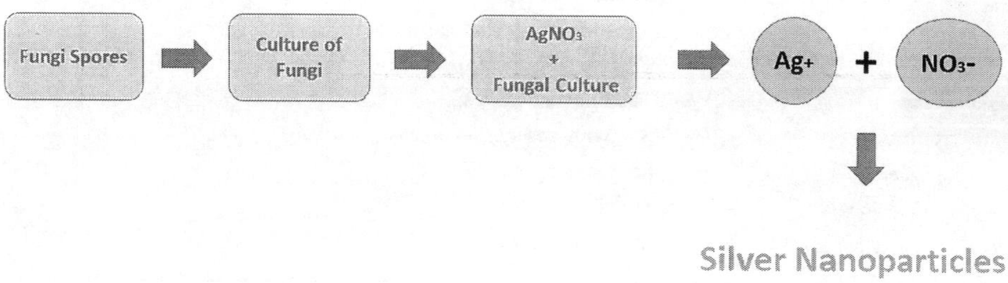

FIGURE 4.3 Fungi meditated nanoparticles.

(Vahabi et al.)

Fusarium sp., Colletotrichum sp., and Phaenerocheate chrysosporium are types of fungus which are used in nanoparticle synthesis. Fungi can produce large amounts of enzymes that are responsible for NPs generation, so that they yield a high amount of product or NPs. Fungus is generally used in biosynthesis of silver nanoparticles. Trichoderma harzianum, Fusarium oxysporum, Colleotrichum sp. ALF2-6, Aspergillus oryzae, Rhizopus stolonifera, Aspergillus fumigatus BTCB10, Fusarium oxysporum, Trichoderma viride, Isaria fumosorosea, Guignardia mangifera, Duddingtonia flagan, Trichoderma longibrachiatum, Penicillium purpurogenum, Epicoccum nigrum, Penicillium oxalicum, Arthroderma fulvum, Sclerotinia sclerotiorum MTCC8785, Guilger-Casagrande, Lima Fungal, Fusarium oxysporum, Fusarium oxysporum, Rhizoctonia solani, and Penicillium oxalicum GRS-1 are some fungi which are widely used in biosynthesis of silver nanoparticles (Guilger-Casagrande and de Lima, 2019a).

TABLE 4.3
Type of Nanoparticles Biosynthesized by Different Actinomycetes (Reddy et al., 2010)

Type of Nanoparticles	Actinomycetes
Silver	Streptomyces sp.
	Streptomyces sp JAR1
	Thermoactinomyces sp.
	Nocardiopsis sp MBRC-1
	Actinomycetes.
	Rhodococcus sp.
	Streptomyces albidoflavus. CNP10
	Streptomyces hygroscopicus. BDUS 49
	Streptomyces spVITPK1.
	Streptomyces rochei.
	Streptomyces sp BDUKAS10.
	Streptomyces sp VITBT7.
	Streptomyces sp I
	Rhodococcus sp Streptomyces aureofaciens. MTCC356
	Actinomycetes sp
	Actinomycetes sp
	Streptomyces glaucus 71MD.
	Streptomyces aureofaciens MTCC356.
	Streptomyces sp JF741876.
	Streptomyces sp ERI-3.
	Streptomyces sp LK3.
Gold	Rhodococcus sp.
	Streptomycetes viridogens HM10
	Thermomonospora spp. 67^{Th}
	Thermomonospora sp.
	Streptomyces hygroscopicus
Zinc	Streptomyces sp.
Zinc/copper	Streptomyces sp.
Zinc/manganese	Streptomyces sp.

4.6.9 Actinomycetes-Mediated Nanoparticle Generation

Thermomonospora sp., Rhodococcus sp., Streptomyces sp., Thermoactinomyces sp., and Rhodococcus sp. are some examples of actinomycetes which produced nanaoparticles. Actinomycetes are mostly used in biosynthesis of silver nanoparticles, though gold, zinc, copper, and manganese are also biosynthesized. Nitrate reductase is present in Streptomyces sp. Reduce agno3 into insoluble silver metal (Table 4.3).

REFERENCES

Abou-Okeil, A., A. Amr, and F. A. Abdel-Mohdy. "Investigation of silver nanoparticles synthesis using aminated β-cyclodextrin." *Carbohydrate Polymers* 89, no. 1 (2012): 1–6.

Abid, Namra, Aqib Muhammad Khan, Sara Shujait, Kainat Chaudhary, Muhammad Ikram, Muhammad Imran, Junaid Haider, Maaz Khan, Qasim Khan, and Muhammad Maqbool. "Synthesis of nanomaterials using various top-down and bottom-up approaches, influencing factors, advantages, and disadvantages: a review." *Advances in Colloid and Interface Science* 300, (2021): 102597.

Ankamwar, Balaprasad. "Biosynthesis of gold nanoparticles (green-gold) using leaf extract of Terminalia catappa." *E-Journal of Chemistry* 7, no. 4 (2010a): 1334–1339.

Ankamwar, Balaprasad. "Biosynthesis of gold nanoparticles (green-gold) using leaf extract of Terminalia catappa." *E-Journal of Chemistry* 7, no. 4 (2010b): 1334–1339.

Audrey, Suzanne, and Sunita Procter. "Employers' views of promoting walking to work: a qualitative study." *International Journal of Behavioral Nutrition and Physical Activity* 12, no. 1 (2015): 1–10.

Bennett, C. Frank, Brenda F. Baker, Nguyen Pham, Eric Swayze, and Richard S. Geary. "Pharmacology of antisense drugs." *Annual Review of Pharmacology and Toxicology* 57 (2017): 81–105.

Beveridge, T. J., and R. G. Murray. "Sites of metal deposition in the cell wall of Bacillus subtilis." *Journal of Bacteriology* 141, no. 2 (1980): 876–887.

Bheemidi, Vijaya Shanti, M. Tiruckovela, and P. Varanasi. "An imperative note on novel drug delivery systems." *Journal of Nanomedicine and Nanotechnology* 2 (2011).

Conti, Denise S., Daniel Brewer, Jordan Grashik, Sumant Avasarala, and Sandro RP da Rocha. "Poly (amidoamine) dendrimer nanocarriers and their aerosol formulations for siRNA delivery to the lung epithelium." *Molecular Pharmaceutics* 11, no. 6 (2014): 1808–1822.

Crucho, Carina IC, and Maria Teresa Barros. "Polymeric nanoparticles: a study on the preparation variables and characterization methods." *Materials Science and Engineering: C* 80 (2017): 771–784.

De Silva, C., A. A. Mohd Noor, M. M. Abd Karim, B. Gunasekaran, S. Abd Gani, M. A. Cabrera, and S. A. Ahmad. "The green synthesis and characterisation of silver nanoparticles from Serratia spp." *Revista Mexicana de Ingeniería Química* 19, no. 3 (2020): 1327–1339.

Deplanche, Kevin, and L. E. Macaskie. "Biorecovery of gold by Escherichia coli and Desulfovibrio desulfuricans." *Biotechnology and Bioengineering* 99, no. 5 (2008): 1055–1064.

Du, Juan, and Tae-Hoo Yi. "Biosynthesis of silver nanoparticles by Variovorax guangxiensis THG-SQL3 and their antimicrobial potential." *Materials Letters* 178 (2016): 75–78.

Dua, Kamal, Nicole G. Hansbro, Paul S. Foster, and Philip M. Hansbro. "MicroRNAs as therapeutics for future drug delivery systems in treatment of lung diseases." *Drug Delivery and Translational Research* 7, no. 1 (2017a): 168–178.

Dua, Kamal, Shakti Dhar Shukla, Terezinha de Jesus Andreoli Pinto, and Philip Michael Hansbro. "Nanotechnology: advancing the translational respiratory research." *Interventional Medicine and Applied Science* 9, no. 1 (2017b): 39–41.

Eberhardt, Wolfgang, Anke Doller, El-Sayed Akool, and Josef Pfeilschifter. "Modulation of mRNA stability as a novel therapeutic approach." *Pharmacology & Therapeutics* 114, no. 1 (2007): 56–73.

Esau, Christine, Scott Davis, Susan F. Murray, Xing Xian Yu, Sanjay K. Pandey, Michael Pear, Lynnetta Watts et al. "miR-122 regulation of lipid metabolism revealed by in vivo antisense targeting." *Cell Metabolism* 3, no. 2 (2006): 87–98.

Fakhari, Shabnam, Mina Jamzad, and Hassan Kabiri Fard. "Green synthesis of zinc oxide nanoparticles: a comparison." *Green Chemistry Letters and Reviews* 12, no. 1 (2019a): 19–24.

Fakhari, Shabnam, Mina Jamzad, and Hassan Kabiri Fard. "Green synthesis of zinc oxide nanoparticles: a comparison." *Green Chemistry Letters and Reviews* 12, no. 1 (2019b): 19–24.

Florence, Alexander T. "Issues in oral nanoparticle drug carrier uptake and targeting." *Journal of Drug Targeting* 12, no. 2 (2004): 65–70.

de Fougerolles, Antonin, and Tatiana Novobrantseva. "siRNA and the lung: research tool or therapeutic drug?." *Current Opinion in Pharmacology* 8, no. 3 (2008): 280–285.

Garmasheva, Inna, Nadezhda Kovalenko, Sergey Voychuk, Andriy Ostapchuk, Olena Livins'ka, and Ljubov Oleschenko "Lactobacillus species mediated synthesis of silver nanoparticles and their antibacterial activity against opportunistic pathogens in vitro." *BioImpacts: BI* 6, no. 4 (2016): 219.

Gaur, Pushpa, Debashish Bhowmik, and Nirmala Soren. "Utilisation of plug-in electric vehicles for frequency regulation of multi-area thermal interconnected power system." *IET Energy Systems Integration* 1, no. 2 (2019): 88–96.

Gnanasangeetha, D., and D. SaralaThambavani. "One pot synthesis of zinc oxide nanoparticles via chemical and green method." *Research Journal of Material Sciences* 2320 (2013): 6055.

Golinska, Patrycja, Magdalena Wypij, Avinash P. Ingle, Indarchand Gupta, Hanna Dahm, and Mahendra Rai. "Biogenic synthesis of metal nanoparticles from actinomycetes: biomedical applications and cytotoxicity." *Applied Microbiology and Biotechnology* 98, no. 19 (2014): 8083–8097.

Gour, Aman, and Narendra Kumar Jain. "Advances in green synthesis of nanoparticles." *Artificial Cells, Nanomedicine, and Biotechnology* 47, no. 1 (2019): 844–851.

Guilger-Casagrande, Mariana, and Renata de Lima. "Synthesis of silver nanoparticles mediated by fungi: a review." *Frontiers in Bioengineering and Biotechnology* 7 (2019a): 287.

Guilger-Casagrande, Mariana, and Renata de Lima. "Synthesis of silver nanoparticles mediated by fungi: a review." *Frontiers in Bioengineering and Biotechnology* 7 (2019b): 287.

Györvary, Erika, Andrea Schroedter, Dimitri V. Talapin, Horst Weller, Dietmar Pum, and Uwe B. Sleytr. "Formation of nanoparticle arrays on S-layer protein lattices." *Journal of Nanoscience and Nanotechnology* 4, no. 1–2 (2004): 115–120.

He, Shiying, Yu Zhang, Zhirui Guo, and Ning Gu. "Biological synthesis of gold nanowires using extract of Rhodopseudomonas capsulata." *Biotechnology Progress* 24, no. 2 (2008): 476–480.

Hnik, Peter, David S. Boyer, Lisa R. Grillone, John G. Clement, Scott P. Henry, and Ellen A. Green. "Antisense oligonucleotide therapy in diabetic retinopathy." *Journal of Diabetes Science and Technology* 3, no. 4 (2009): 924–930.

Huber, Dale L. "Synthesis, properties, and applications of iron nanoparticles." *Small* 1, no. 5 (2005): 482–501.

Hussain, Imtiyaz, N. B. Singh, Ajey Singh, Himani Singh, and S. C. Singh. "Green synthesis of nanoparticles and its potential application." *Biotechnology Letters* 38, no. 4 (2016): 545–560.

Husseiny, M. I., M. Abd El-Aziz, Y. Badr, and M. A. Mahmoud. "Biosynthesis of gold nanoparticles using Pseudomonas aeruginosa." *Spectrochimica Acta Part A: Molecular and Biomolecular Spectroscopy* 67, no. 3–4 (2007): 1003–1006.

Iqbal, Parvez, Jon A. Preece, and Paula M. Mendes. "Nanotechnology: the "top-down" and "bottom-up" approaches." In *Supramolecular chemistry: from molecules to nanomaterials* (2012).edited by: Philip A. Gale

Iravani, Siavach, Hassan Korbekandi, Seyed Vahid Mirmohammadi, and Behzad Zolfaghari. "Synthesis of silver nanoparticles: chemical, physical and biological methods." *Research in Pharmaceutical Sciences* 9, no. 6 (2014): 385.

Iravani, Siavash. "Bacteria in nanoparticle synthesis: current status and future prospects." *International Scholarly Research Notices* 2014 (2014): 1–18.

Ivanova, Nadezhda, Viliana Gugleva, Mirena Dobreva, Ivaylo Pehlivanov, Stefan Stefanov, and Velichka Andonova. "Silver nanoparticles as multi-functional drug delivery systems." In *Nanomedicines*. IntechOpen, 2018: 71–91.

Kazi, Karim Masud, Asim Sattwa Mandal, Nikhil Biswas, Arijit Guha, Sugata Chatterjee, Mamata Behera, and Ketousetuo Kuotsu. "Niosome: a future of targeted drug delivery systems." *Journal of Advanced Pharmaceutical Technology & Research* 1, no. 4 (2010): 374.

Khan, Azhar U., Masudulla Khan, Nazia Malik, Moo Hwan Cho, and Mohammad Mansoob Khan. "Recent progress of algae and blue–green algae-assisted synthesis of gold nanoparticles for various applications." *Bioprocess and Biosystems Engineering* 42, no. 1 (2019): 1–15.

Kim, Jinkuk, Chunguang Hu, Christelle Moufawad El Achkar, Lauren E. Black, Julie Douville, Austin Larson, Mary K. Pendergast et al. "Patient-customized oligonucleotide therapy for a rare genetic disease." *New England Journal of Medicine* 381, no. 17 (2019): 1644–1652.

Knutson, Jacqueline. "Exploring biotic iron transformation by the hyperthermophilic archaeon pyrobaculum islandicum." PhD diss., 2011.

Kodama, R. H. "Magnetic nanoparticles." *Journal of Magnetism and Magnetic Materials* 200, no. 1–3 (1999): 359–372.

Kole, Ryszard, Adrian R. Krainer, and Sidney Altman. "RNA therapeutics: beyond RNA interference and antisense oligonucleotides." *Nature Reviews Drug Discovery* 11, no. 2 (2012): 125–140.

Konishi, Y., T. Tsukiyama, T. Tachimi, N. Saitoh, T. Nomura, and S. Nagamine. "Microbial deposition of gold nanoparticles by the metal-reducing bacterium Shewanella algae." *Electrochimica Acta* 53, no. 1 (2007): 186–192.

Korbekandi, Hassan, Siavash Iravani, and Sajjad Abbasi. "Production of nanoparticles using organisms." *Critical Reviews in Biotechnology* 29, no. 4 (2009): 279–306.

Lam, Jenny KW, Michael YT Chow, Yu Zhang, and Susan WS Leung. "siRNA versus miRNA as therapeutics for gene silencing." *Molecular Therapy–Nucleic Acids* 4 (2015): e252.

Le Ouay, Benjamin, and Francesco Stellacci. "Antibacterial activity of silver nanoparticles: a surface science insight." *Nano Today* 10, no. 3 (2015): 339–354.

Lengke, Maggy F., Bruce Ravel, Michael E. Fleet, Gregory Wanger, Robert A. Gordon, and Gordon Southam. "Mechanisms of gold bioaccumulation by filamentous cyanobacteria from gold (III)– chloride complex." *Environmental Science & Technology* 40, no. 20 (2006): 6304–6309.

Liang, Xue-Hai, Wen Shen, Hong Sun, Michael T. Migawa, Timothy A. Vickers, and Stanley T. Crooke. "Translation efficiency of mRNAs is increased by antisense oligonucleotides targeting upstream open reading frames." *Nature Biotechnology* 34, no. 8 (2016): 875–880.

Liong, Monty, Jie Lu, Michael Kovochich, Tian Xia, Stefan G. Ruehm, Andre E. Nel, Fuyuhiko Tamanoi, and Jeffrey I. Zink. "Multifunctional inorganic nanoparticles for imaging, targeting, and drug delivery." *ACS Nano* 2, no. 5 (2008): 889–896.

Lovley, Derek R., Elizabeth JP Phillips, and Debra J. Lonergan. "Hydrogen and formate oxidation coupled to dissimilatory reduction of iron or manganese by Alteromonas putrefaciens." *Applied and Environmental Microbiology* 55, no. 3 (1989): 700–706.

Mandal, Deendayal, Mark E. Bolander, Debabrata Mukhopadhyay, Gobinda Sarkar, and Priyabrata Mukherjee. "The use of microorganisms for the formation of metal nanoparticles and their application." *Applied Microbiology and Biotechnology* 69, no. 5 (2006): 485–492.

Manjunath, K., J. Suresh Reddy, and V. Venkateswarlu. "Solid lipid nanoparticles as drug delivery systems." *Methods and Findings in Experimental and Clinical Pharmacology* 27, no. 2 (2005): 127–144.

Mehta, Meenu, Devesh Tewari, Gaurav Gupta, Rajendra Awasthi, Harjeet Singh, Parijat Pandey, Dinesh Kumar Chellappan et al. "Oligonucleotide therapy: an emerging focus area for drug delivery in chronic inflammatory respiratory diseases." *Chemico-Biological Interactions* 308 (2019): 206–215.

Mercatante, D. R., P. Sazani, and R. Kole. "Modification of alternative splicing by antisense oligonucleotides as a potential chemotherapy for cancer and other diseases." *Current Cancer Drug Targets* 1, no. 3 (2001): 211–230.

Mignot, Anna, Charles Truillet, François Lux, Lucie Sancey, Cédric Louis, Franck Denat, Frédéric Boschetti et al. "A top-down synthesis route to ultrasmall multifunctional Gd-Based silica nanoparticles for theranostic applications." *Chemistry–A European Journal* 19, no. 19 (2013): 6122–6136.

Mohanpuria, Prashant, Nisha K. Rana, and Sudesh Kumar Yadav. "Biosynthesis of nanoparticles: technological concepts and future applications." *Journal of Nanoparticle Research* 10, no. 3 (2008): 507–517.

Mokhena, Teboho C., and A. S. Luyt. "Electrospun alginate nanofibres impregnated with silver nanoparticles: preparation, morphology and antibacterial properties." *Carbohydrate Polymers* 165 (2017): 304–312.

Nadaroglu, Hayrunnisa, Alayli Güngör Azize, and İ. N. C. E. Selvi. "Synthesis of nanoparticles by green synthesis method." *International Journal of Innovative Research and Reviews* 1, no. 1 (2017): 6–9.

Petla, Ramesh Kumar, Singaravelu Vivekanandhan, Manjusri Misra, Amar Kumar Mohanty, and Nallani Satyanarayana. "Soybean (Glycine max) leaf extract based green synthesis of palladium nanoparticles." *Journal of Biomaterials and Nanobiotechnology* (2011), https://doi.org/10.4236/jbnb.2012.31003.

Pósfai, Mihály, Bruce M. Moskowitz, Balázs Arató, Dirk Schüler, Christine Flies, Dennis A. Bazylinski, and Richard B. Frankel. "Properties of intracellular magnetite crystals produced by Desulfovibrio magneticus strain RS-1." *Earth and Planetary Science Letters* 249, no. 3–4 (2006): 444–455.

Prabhu, Sukumaran, and Eldho K. Poulose. "Silver nanoparticles: mechanism of antimicrobial action, synthesis, medical applications, and toxicity effects." *International Nano Letters* 2, no. 1 (2012): 1–10.

Rajeshkumar, S., C. Malarkodi, M. Vanaja, G. Gnanajobitha, K. Paulkumar, C. Kannan, and G. Annadurai. "Antibacterial activity of algae mediated synthesis of gold nanoparticles from Turbinaria conoides." *Der Pharma Chemica* 5, no. 2 (2013): 224–229.

Reddy, A. Satyanarayana, Chien-Yen Chen, Chien-Cheng Chen, Jiin-Shuh Jean, Hau-Ren Chen, Min-Jen Tseng, Cheng-Wei Fan, and Jung-Chen Wang. "Biological synthesis of gold and silver nanoparticles mediated by the bacteria Bacillus subtilis." *Journal of Nanoscience and Nanotechnology* 10, no. 10 (2010): 6567–6574.

Revati, K., and B. D. Pandey. "Microbial synthesis of iron-based nanomaterials—a review." *Bulletin of Materials Science* 34, no. 2 (2011): 191–198.

Röthlisberger, Pascal, and Marcel Hollenstein. "Aptamer chemistry." *Advanced Drug Delivery Reviews* 134 (2018): 3–21.

Saha, Biswarup, Jaydeep Bhattacharya, Ananda Mukherjee, AnupKumar Ghosh, ChittaRanjan Santra, Anjan K. Dasgupta, and Parimal Karmakar. "In vitro structural and functional evaluation of gold nanoparticles conjugated antibiotics." *Nanoscale Research Letters* 2, no. 12 (2007): 614–622.

Saif, Sadia, Arifa Tahir, and Yongsheng Chen. "Green synthesis of iron nanoparticles and their environmental applications and implications." *Nanomaterials* 6, no. 11 (2016): 209.

Satalkar, Priya, Bernice Simone Elger, and David M. Shaw. "Defining nano, nanotechnology and nanomedicine: why should it matter?." *Science and Engineering Ethics* 22, no. 5 (2016): 1255–1276.

Scoles, Daniel R., and Stefan M. Pulst. "Oligonucleotide therapeutics in neurodegenerative diseases." *RNA Biology* 15, no. 6 (2018): 707–714.

Sharma, Virender K., Ria A. Yngard, and Yekaterina Lin. "Silver nanoparticles: green synthesis and their antimicrobial activities." *Advances in Colloid and Interface Science* 145, no. 1-2 (2009): 83–96.

Shirota, H, D. Tross, D. M. Klinman, CpG oligonucleotides as cancer vaccine adjuvants, Vaccine 3 (2015) 390–407, https://doi.org/10.3390/vaccines3020390.

Shivaji, S., S. Madhu, and Shashi Singh. "Extracellular synthesis of antibacterial silver nanoparticles using psychrophilic bacteria." *Process Biochemistry* 46, no. 9 (2011): 1800–1807.

Siddiqi, Khwaja Salahuddin, and Azamal Husen. "Green synthesis, characterization and uses of palladium/platinum nanoparticles." *Nanoscale Research Letters* 11, no. 1 (2016): 1–13.

Singh, Rajesh, and James W. Lillard Jr. "Nanoparticle-based targeted drug delivery." *Experimental and Molecular Pathology* 86, no. 3 (2009): 215–223.

Sithara, R., P. Selvakumar, C. Arun, S. Anandan, and P. Sivashanmugam. "Economical synthesis of silver nanoparticles using leaf extract of Acalypha hispida and its application in the detection of Mn (II) ions." *Journal of Advanced Research* 8, no. 6 (2017): 561–568.

Smola, Malgorzata, Thierry Vandamme, and Adam Sokolowski. "Nanocarriers as pulmonary drug delivery systems to treat and to diagnose respiratory and non respiratory diseases." *International Journal of Nanomedicine* 3, no. 1 (2008): 1.

Song, Kyung-Mi, Seonghwan Lee, and Changill Ban. "Aptamers and their biological applications." *Sensors* 12, no. 1 (2012): 612–631.

Soppimath, Kumaresh S., Tejraj M. Aminabhavi, Anandrao R. Kulkarni, and Walter E. Rudzinski. "Biodegradable polymeric nanoparticles as drug delivery devices." *Journal of Controlled Release* 70, no. 1–2 (2001): 1–20.

Spadavecchia, J., P. Prete, N. Lovergine, L. Tapfer, and R. Rella. "Au nanoparticles prepared by physical method on Si and sapphire substrates for biosensor applications." *The Journal of Physical Chemistry B* 109, no. 37 (2005): 17347–17349.

Srikar, Sista Kameswara, Deen Dayal Giri, Dan Bahadur Pal, Pradeep Kumar Mishra, and Siddh Nath Upadhyay. "Green synthesis of silver nanoparticles: a review." *Green and Sustainable Chemistry* 6, no. 1 (2016): 34–56.

Stein, Cy A., and Daniela Castanotto. "FDA-approved oligonucleotide therapies in 2017." *Molecular Therapy* 25, no. 5 (2017): 1069–1075.

Stoimenov, Peter K., Rosalyn L. Klinger, George L. Marchin, and Kenneth J. Klabunde. "Metal oxide nanoparticles as bactericidal agents." *Langmuir* 18, no. 17 (2002): 6679–6686.

Taghavi, Sayed Mohammad, Mahdiye Momenpour, Maryam Azarian, Mohammad Ahmadian, Faramarz Souri, Sayed Ali Taghavi, Marzieh Sadeghain, and Mohsen Karchani. "Effects of nanoparticles on the environment and outdoor workplaces." *Electronic Physician* 5, no. 4 (2013): 706.

Thirumurugan, A., P. Aswitha, C. Kiruthika, S. Nagarajan, and A. Nancy Christy. "Green synthesis of platinum nanoparticles using Azadirachta indica–An eco-friendly approach." *Materials Letters* 170 (2016): 175–178.

Tolaymat, Thabet M., Amro M. El Badawy, Ash Genaidy, Kirk G. Scheckel, Todd P. Luxton, and Makram Suidan. "An evidence-based environmental perspective of manufactured silver nanoparticle in syntheses and applications: a systematic review and critical appraisal of peer-reviewed scientific papers." *Science of the Total Environment* 408, no. 5 (2010): 999–1006.

Umer, Asim, Shahid Naveed, Naveed Ramzan, Muhammad Shahid Rafique, and Muhammad Imran. "A green method for the synthesis of copper nanoparticles using L-ascorbic acid." *Matéria (Rio de janeiro)* 19 (2014): 197–203.

Vahabi, Khabat, G. Ali Mansoori, and Sedighe Karimi. "Biosynthesis of silver nanoparticles by fungus Trichoderma reesei (a route for large-scale production of AgNPs)." *Insciences Journal* 1, no. 1 (2011): 65–79.

Xie, Jin, Kai Chen, and Xiaoyuan Chen. "Production, modification and bio-applications of magnetic nanoparticles gestated by magnetotactic bacteria." *Nano Research* 2, no. 4 (2009): 261–278.

Zhigaltsev, Igor V., Nathan Belliveau, Ismail Hafez, Alex K. K. Leung, Jens Huft, Carl Hansen, and Pieter R. Cullis. "Bottom-up design and synthesis of limit size lipid nanoparticle systems with aqueous and triglyceride cores using millisecond microfluidic mixing." *Langmuir* 28, no. 7 (2012): 3633–3640.

Zhu, Xianchun, Kavitha Pathakoti, and Huey-Min Hwang. "Green synthesis of titanium dioxide and zinc oxide nanoparticles and their usage for antimicrobial applications and environmental remediation." In *Green synthesis, characterization and applications of nanoparticles*, pp. 223–263. Elsevier, 2019.edited by: Ashutosh Kumar Shukla, Siavash Iravani.

Zoroddu, Maria Antonietta, Serenella Medici, Alessia Ledda, Valeria Marina Nurchi, Joanna I. Lachowicz, and Massimiliano Peana. "Toxicity of nanoparticles." *Current Medicinal Chemistry* 21, no. 33 (2014): 3837–3853.

5 Development of m-RNA Vaccines in Covid-19 Pandemic Scenario

Disha Jain
University of South Florida, Tampa, USA

Yashwant Pathak
University of South Florida, Tampa, USA
Airlangga University, Surabaya, Indonesia

CONTENTS

Covid-19 has emerged as one of the great pandemics of 21st century after the last one in 1918.

It has eventually resulted in tremendous development for highly effective vaccines to reduce the risk of Covid-19 infection, hospitalizations, and associated infections. Particularly in the United States, the Food and Drug Administration (FDA) authorized emergency administration for two mRNA vaccines BNT162b2 (Pfizer & BioNTech) and mRNA-1273 (Moderna) (Venky Soundararajan et al., 2021). Barajas-Nava (2021) reviewed the scientific literature to identify and evaluate scientific studies on vaccines on the way to development and production.

Few studies involved the mechanism of infections; genomics was excluded. Overall, 141 vaccines were identified, including weakened and inactive viruses, viral vectors, and proteins. Only 13 vaccines were clinically evaluated. Many research teams and organizations developed newer technologies to develop these vaccines. Among those, five vaccines were in Phase I and seven vaccines were in Phase I-II studies. One vaccine by the University of Oxford and AstraZeneca was in clinical evaluation (Phase II-III). The U.S.-based company Moderna revealed excellent results in their studies that led to the rapid design of the mRNA-1273 vaccine. The studies showed positive results to determine the effectiveness of the vaccine. mRNA possesses some unique advantages over other pDNA or viral vector vaccines. One of the greatest attributes includes the power to stimulate the production of neutralizing antibodies, which supported the further progress for the development and approval of the vaccine by the FDA (Leticia B., 2020). Another advantage is mRNA does not generate infectious particles or deliver antigen expression in situ, it can express complex antigens without packaging constraints (Maruggi et al., 2019).

DOI: 10.1201/9781003186083-5

They induce immune responses, such as the production of chemokines, cytokines such as interleukin-12 (IL-12), and tumor necrosis factor (TNF) (Liang et al., 2017).

5.1 HISTORICAL LANDMARKS OUTLINING THE DEVELOPMENT OF mRNA VACCINES

Pitlik (2020) discussed the origins of viral transmission and clinical characteristics of Covid-19 as compared to other pandemic diseases. As discussed in the early 1960s by Almeida, there exist seven different types of human coronavirus, such as HCoV-HKU1, HCV-OC43, HCV-NL63, HCV-229E, SARS-CoV-1, SARS-CoV-2, and MERS-CoV, not just including bats but all other animals (Combs S. 2020).

In the case of COVID-19, pangolins are considered intermediate hosts. These are attributed to multi-curative properties. Various clinical research studies obtained viral sequencing samples from patients affected with SaRS-CoV-2, and it certainly brought the breakthrough power of mRNA vaccines to help decrease infection rates/hospitalizations and provide a nearly accurate representation of viral transmission of SaRS-CoV-2 disease.

Genetic sequencing technology of RNA and DNA has proved to be an invaluable tool in pandemics to trace transmission pathways, detect mutation, and reconstruct the chain of transmission. During the current vaccine development process of Covid-19, many challenges emerged to facilitate the process and approve the efficacious drug to reduce the infection rates.

Pardi et al. (2018) discussed the important pathways and challenges in delivering mRNA vaccine technology against infectious diseases in further detail. mRNA technology has proven to be the most versatile, potent behavior in neutralizing antibody responses, even in low-dose immunizations. mRNA vaccines have been further differentiated as self-amplifying RNA vaccines and non-replicating RNA vaccines. Self-amplifying mRNA vaccines demonstrated strong immunogenicity against lethal infections such as Toxoplasma gondii, influenza, Ebola, and bacterial pathogens, including Streptococcus A and B (.spp). Another form of mRNA vaccine is through dendritic cells.

This type has been utilized mainly for HIV-1 infected individuals, although no clinical benefits have been observed. In addition, mRNA vaccines possess some crucial features, such as increased safety and efficacy power compared to DNA-based vaccines, minimal genetic vector so that anti-vector immunity could be avoided, and repeated administrations of the vaccines and scalable production, owing to in vivo transcription reactions.

These mRNA vaccines are widely used to treat malignant cancers as well and have been extensively reviewed in recent scientific publications also. In addition, they have proven feasible and viable strategies to help combat cancer in numerous clinical studies. These cancer vaccines are mainly therapeutic and induce cell-mediated responses. These cancer vaccines can be administered by various routes, such as intradermal, intramuscular, intranasal, subcutaneous, and other unconventional routes such as intranodal and intrasplenic.

Manufacturing of these mRNA vaccines provides minimal risks and higher safety than other DNA-based vaccines, as they do not require toxic chemicals, or any contaminating microorganisms, live viruses, viral vectors. mRNA vaccines also have some potential safety concerns that need to be evaluated in preclinical and clinical studies, such as local and systemic administrationand type 1 interferon responses. Thus, the future developments of mRNA vaccines will certainly have substantial research for enhancing therapeutic and clinical uses and thereby increasing the immunologic responses.

A retrospective analytic study was conducted among 136, 532 individuals in the Mayo Clinic Health System (Arizona, Florida, Iowa, Minnesota, and Wisconsin). This study compared the clinical outcomes for the vaccinated cohort and unvaccinated control cohort. The significant findings suggest that both the mRNA-based vaccines (BNT162b2 and mRNA-1273) are 88.0% and 93.3% effective in real-world evidence to prevent SaRS-CoV-2 infection (Soundararanjan et al., 2021).

In addition to the above findings, the data showed BNT162b2 and mRNA-1273 were 88.8% and 86.0% effective in preventing COVID-19 associated hospitalizations. Overall, both vaccines mRNA1273 and BNT162b2 demonstrate robust evidence for the assessment of effectiveness in large phase III randomized clinical trials. The meta-analysis with a random-effects model of mRNA vaccines in large observational studies showed a significant protective effect against RT-PCR (reverse transcription-polymerase chain reaction) confirmed SARS-COV-2 infection > 14 days after the first dose, i.e., 55% of vaccine effectiveness with 95% confidence interval and 95% of vaccine effectiveness after the second dose (Kow & Hasan, 2021). These studies are required to investigate vaccine effectiveness in immunocompromised patients and other variants of COVID-19 that can inform better decision-making policies in public health to increase mass vaccination efforts.

Another important aspect of testing these vaccines in clinical trials was to follow-up on vaccinated individuals and evaluate patterns among the people who are at risk of acquiring severe Covid-19 infection.

The Centers for Disease Control & Prevention (CDC) identified "breakthrough" cases that were reported in the United States from early January 2021. As of September 20, 2021, CDC reported 19,136 vaccine breakthrough individuals, resulting in hospitalizations or deaths, and more than 180 million people are fully vaccinated against Covid-19.

5.2 CHALLENGES TOWARDS RATIONALE DESIGN OF mRNA VACCINES

In the early pandemic, the data findings were limited to obtaining and validating the drug manufacturing process or scientific simulations. Various computational approaches have been developed since then to address the information on the molecular level in disease pathology, define a systematic approach towards the rapid vaccine development process, and accelerate the research to find targets for novel anti-SARS-CoV-2 preventive and treatment measures (Ghaebi et al., 2020). Messenger RNA (mRNA) vaccine technology has the innate immunotherapeutic pathway that can treat infectious diseases like COVID-19. Multiple formulations have been under investigation, such as conventional mRNA and self-amplifying mRNA.

There have been several concerns regarding the safety and efficiency of mRNA vaccines over the past few years, but many recent studies suggest that stability concerns have been tackled by modifications of the mRNA backbone structure. Second, optimal carrier development has been ongoing research to enhance the stability and cellular uptake efficiency of these mRNA vaccines. Another important factor to consider that can influence the effect of mRNA vaccines is the innate immunological response of these vaccines. The third but very important factor influencing the effect of mRNA therapeutics is the inherent innate immune-stimulating activity of mRNA, which can either support or hamper the therapeutic outcomes.

5.3 CLINICAL TRIALS EVALUATING THE SAFETY AND IMMUNOGENICITY OF mRNA VACCINES

The National Institute of Allergy and Infectious Diseases (NIAID) have been conducting multiple clinical trials in the research sites to assess the safety, immunogenicity, and reactogenicity of mRNA-1273 (ModernaTX.Inc). The molecule mRNA-1273 is a novel lipid nanoparticle (LNP) encapsulated mRNA-based vaccine. This particular Phase I study is enrolling both males and non-pregnant females ranging from 18 years age and above from three domestic clinical research sites (Georgia, Maryland, and Washington). The study will include up to one hundred and twenty subjects non-randomized in 13 different cohorts based on five dosages (10, 25, 50, 100, 250 mcg). Subjects will receive an intramuscular injection of 0.5 milliliters (mL) on day one and day 29 and will be monitored in follow-up visits through 12 months (one, two, and four weeks post-each vaccination and three to six months post second vaccination (28 days apart in each time point recorded). In addition, the secondary objective is to investigate the immunoglobulin IgG linked immunosorbent assay to SARS-CoV-2 spike protein after two dose vaccination schedules at day 57. This study is active and is estimated to be completed primarily by November 22, 2022.

Jochum et al. (2021) provided an exploratory analysis of the mRNA-1273 phase I trial by assessing the clinical utility of the quantitative Roche Elecsys Anti-SARS-CoV-2 assay (ACOV2S). As for the study, blood samples were collected from 30 healthy participants aged 18-55 years who received two injections of mRNA-1273 at dosages 25 and 100 mcg at baseline (day one, first vaccination), 15, (day 29, second vaccination), 43, and 57. These samples were analyzed, and serum testing was performed as well. The samples were compared with antibody responses from individuals infected with polymerase-chain reaction (PCR) lab confirmed SARS-CoV-2 in two time points, i.e., zero to 15 days and 16 to 35 post-PCR diagnosis, then further analyzed for the presence of SARS-CoV-2 antibodies, using ACOV2S.

The results depicted strong correlations of ACOV2S measurements with those from the receptor binding domain (RBD) ELISA. There was notable increase in antibody levels in the second vaccination from the first vaccination by approximately 10-100 times in ACOV2S induced vaccination, as compared with natural SARS-CoV-2 antibodies. This data suggested that ACOV2S assay should be considered as a significant, feasible, and accessible tool to help assess and quantify the antibodies from receptor binding protein and provide us enhanced understanding of building robust immunological pathways against COVID-19.

5.4 mRNA VACCINES IN ADOLESCENTS AND OLDER ADULTS

Jackson et al. (2020) published a preliminary report regarding phase I, open-label trial mRNA-1273 trial including healthy adults 18-55 years of age. These individuals will receive mRNA-1273 doses of 25, 100, and 250 mcg as part of the vaccination research.

The results displayed after the second vaccination; few systemic adverse events were reported in individuals receiving the highest dose (graded as mild to moderate). Overall, the mRNA-1273 vaccine induced anti-SARS-CoV-2 immune responses in all the participants and no safety concerns were reported (Jackson et al., 2020). The safety profile has been similar to reports in trials of avian influenza (influenza A/H1N08) and influenza A/H7N9.

5.5 mRNA TECHNOLOGY: A PROMISING ALTERNATIVE FOR FUTURE IMPLICATIONS AND SUPPORTING DATA REVIEW

The mRNA technology has greatly transformed the impact of vaccines globally in terms of the rapid development, low-cost manufacturing process, and completely synthetic nature to generate highly potent, safe immunogenicity. It has the potential to provide the adjuvant activity for dendritic cell maturation and stimulate T cell and B cell immune responses (Pardi et al., 2018).

Nucleic acid vaccines, including mRNA, pDNA (plasmid DNA), and other viral vectors, have easy adaptability and flexibility. The manufacturing process is independent of encoded antigens, which allows different vaccines to utilize the same production and purification methods.

This also eliminates the possibility of any infectious pathogen by specific pattern recognition receptors (PRRs) to promote increase in adjuvant activity and any innate humoral and T cell responses (Maruggi et al., 2019). Nucleoside base modification (modified mRNA) is an emerging approach to increase the potency of mRNA (Maruggi et al., 2019).

Another critical factor is the efficient mRNA delivery to achieve therapeutic relevance in vaccination. *Ex vivo DC* loading has proven to allow high targeted delivery of injections, though it seems a costlier and more labor-intensive approach than vaccine delivery. Various forms of physical, genomic deliveries have been proposed to essentially increase the uptake of vaccine delivery and improve substantial progress in humans.

5.6 SAFETY AND EFFICACY OF mRNA-1273 AND BNT162B2 VACCINES (PHASE III CLINICAL TRIAL)

In the Phase III randomized, placebo-controlled trial of SARS-CoV-2 vaccine mRNA-1273, the participants were enrolled in 1:1 ratio to determine the safety and efficacy of the vaccine (Baden et al., 2021). The results suggested that more than 95% of the participants were administered with both the injections, and 2.2 % had serologic SARS-CoV-2 infection at the baseline.

The analysis at the primary endpoint reveals that the efficacy of mRNA1273 is 30% or less. The findings from secondary analyses displayed a higher level of efficacy of 95.2% (95% CI). The efficacy endpoints were consistent across all the subgroups and indicated that the magnitude of vaccine efficacy is higher than other vaccines for respiratory diseases, such as influenza vaccine, for symptomatic, confirmed disease in adults. Similarly, the studies suggested that efficacy rate of mRNA-1273 is like BNT162b2 (95% effective). Systematic reactogenicity was observed as mild or moderate in older adults as compared to younger adults (Polack et al., 2020). With these changes in some severity in few vaccine trials, it also presents us with some limitations.

5.7 CONCLUSION

Many multi-phase clinical studies involving mRNA vaccines were executed to assess safety and effectiveness in real-world evidence data. Since the early pandemic, there was a certain lack of data, such as the infection/hospitalizations rate in people, including some immunocompromised, co-morbid conditions. As the situation progressed, there were some robust developments toward understanding Covid-19 transmission pathways through a mixed, random-effects model, laboratory results, and case-control studies in different patient cohorts. Currently, studies are on-going regarding the SaRS-CoV-2 variants, and data has revealed in many studies that it has helped tremendously to prevent SaRS-CoV-2 incidence coverage across all age groups (5-12 years/12-18-year-olds/ >18-65 years old). This is important for us as public health professionals to educate and spread awareness in the community about vaccines, and help promote to increase vaccine coverage.

REFERENCES

Baden, L. R., El Sahly, H. M., Essink, B., Kotloff, K., Frey, S., Novak, R., Diemert, D. et al. COVE Study Group (2021). Efficacy and Safety of the mRNA-1273 SARS-CoV-2 Vaccine. *The New England Journal of Medicine*, 384(5), 403–416. https://doi.org/10.1056/NEJMoa2035389

Barajas-Nava, L. A. (2021). Development of SARS-CoV-2 Vaccines. *Boletín Médico del Hospital Infantil de México*, 78(1), 66–74. https://doi.org/10.24875/BMHIM.20000217

Brito, L. A., Kommareddy, S., Maione, D., Uematsu, Y., Giovani, C., Berlanda Scorza, F., Otten, G. R. et al. (2015). Self-Amplifying mRNA Vaccines. *Advances in Genetics*, 89, 179–233. https://doi.org/10.1016/bs.adgen.2014.10.005

Ghaebi, M., Osali, A., Valizadeh, H., Roshangar, L., & Ahmadi, M. (2020). Vaccine Development and Therapeutic Design for 2019-nCoV/SARS-CoV-2: Challenges and Chances. *Journal of Cellular Physiology*, 235(12), 9098–9109. https://doi.org/10.1002/jcp.29771

Huang, Q., Zeng, J., & Yan, J. (2021). COVID-19 mRNA Vaccines. *Journal of Genetics and Genomics*, 48(2), 107–114. https://doi.org/10.1016/j.jgg.2021.02.006

Hekele, A., Bertholet, S., Archer, J., Gibson, D. G., Palladino, G., Brito, L. A., Otten, G. R. et al. (2013). Rapidly Produced SAM(®) Vaccine Against H7N9 Influenza is Immunogenic in Mice. *Emerging Microbes & Infections*, 2(8), e52. https://doi.org/10.1038/emi.2013.54

Jackson, L. A., Anderson, E. J., Rouphael, N. G., Roberts, P. C., Makhene, M., Coler, R. N., McCullough, M. P. et al. mRNA-1273 Study Group (2020). An mRNA Vaccine Against SARS-CoV-2: Preliminary Report. *The New England Journal of Medicine*, 383(20), 1920–1931. https://doi.org/10.1056/NEJMoa2022483

Jochum, S., Kirste, I., Hortsch, S., Grunert, V. P., Legault, H., Eichenlaub, U., Kashlan, B. et al. (2021). Clinical Utility of Elecsys Anti-SARS-CoV-2 S Assay in COVID-19 Vaccination: An exploratory Analysis of the mRNA-1273 Phase I Trial. *medRxiv: The Preprint Server for Health Sciences*, 2021. https://doi.org/10.1101/2021.10.04.21264521

Kow, C. S., & Hasan, S. S. (2021). Real-World Effectiveness of BNT162b2 mRNA Vaccine: A Meta-Analysis of Large Observational Studies. *InflammoPharmacology*, 29(4), 1075–1090. https://doi.org/10.1007/s10787-021-00839-2

Liang, F., Lindgren, G., Lin, A., Thompson, E. A., Ols, S., Röhss, J., John, S. et al. (2017). Efficient Targeting and Activation of Antigen-Presenting Cells In Vivo after Modified mRNA Vaccine Administration in Rhesus Macaques. *Molecular Therapy: The Journal of the American Society of Gene Therapy*, 25(12), 2635–2647. https://doi.org/10.1016/j.ymthe.2017.08.006

Maruggi, G., Zhang, C., Li, J., Ulmer, J. B., & Yu, D. (2019). mRNA as a Transformative Technology for Vaccine Development to Control Infectious Diseases. *Molecular Therapy: The Journal of the American Society of Gene Therapy*, 27(4), 757–772. https://doi.org/10.1016/j.ymthe.2019.01.020

Polack, F. P., Thomas, S. J., Kitchin, N., Absalon, J., Gurtman, A., Lockhart, S., Perez, J. L. et al. C4591001 Clinical Trial Group (2020). Safety and Efficacy of the BNT162b2 mRNA Covid-19 Vaccine. *The New England Journal of Medicine*, 383(27), 2603–2615. https://doi.org/10.1056/NEJMoa2034577

Pardi, N., Hogan, M. J., Porter, F. W., & Weissman, D. (2018). mRNA Vaccines: A New Era in Vaccinology. *Nature Reviews. Drug Discovery*, 17(4), 261–279. https://doi.org/10.1038/nrd.2017.243

Pitlik S. D. (2020). COVID-19 Compared to Other Pandemic Diseases. *Rambam Maimonides Medical Journal*, 11(3), e0027. https://doi.org/10.5041/RMMJ.10418.

6 Gene Therapy for Cardiovascular Diseases

Clinical Evidences

Dhwani Thakkar
Nirma University, Ahmedabad, India

Vandit Shah
L. M. College of Pharmacy, Ahmedabad, India

Jigna Shah
Nirma University, Ahmedabad, India

CONTENTS

6.1 INTRODUCTION

Cardiovascular diseases (CVD) are the group of disorders mainly related to the blood vessels and heart.[1] CVD is the major cause of death worldwide, killing more people each year than cancer and respiratory diseases combined. From 271 million in the 1990s to 523 million in 2019, the number of CVD cases has risen dramatically. The number of deaths has continuously climbed from 12.1 million in the 1990s to 18.6 million in 2019.[2] The Western world is the most impacted, with one out of every three Americans suffering from cardiovascular disease. The ageing and population growth factors are predicted to be important drivers for the increase in the number of CVD patients.[3] Efforts to promote practical and accessible methods for CVD prevention and control, as well as to monitor the results, are desirable.

The limitations associated with available therapies have prompted extensive research on new therapies. Researchers now have the capabilities to target specific genes that play a critical role in cardiovascular illnesses by virtue of gene transfer vectors and delivery systems. Gene therapy has emerged as a promising treatment option for heart disease. Several clinical investigations have shown positive results in terms of efficacy and safety. Myocardial gene transfer can improve angiogenesis

DOI: 10.1201/9781003186083-6

with VEGF or FGF, boost myocardial contractility, and minimize arrhythmia vulnerability with sarcoplasmic reticulum Ca^{2+} adenosine triphosphatase, and trigger heart repair with stromal-derived factor-1, according to preclinical study (SDF-1).[4] Connexins 40 and 43, the skeletal muscle sodium channel SCN4a, or a dominant-negative mutation of the KCNH2-G628S, a rapid component of the delayed rectifier potassium channel, minimize the likelihood of atrial fibrillation and ventricular tachycardia in atrial fibrillation, according to preclinical studies.[5]

The field of cardiac genetic manipulation is vast due to the extremely sophisticated and multiple disease pathways. Several gene products can be targeted to improve clinical quality, and several delivery vectors can be used in the clinic. This chapter looks at the most common cardiac illnesses that are now being studied to see if gene therapy can help. Also, a description of the genes that are being altered, as well as the current situation of preclinical and clinical trials, are discussed.

6.2 TARGETS FOR CARDIOVASCULAR GENE THERAPY

6.2.1 Ischemic Heart Diseases

Lack of oxygen leads to a condition known as an ischemia. In mid 1990s, gene therapy was explored to induce therapeutic angiogenesis in coronary and peripheral arteries in patients suffering with ischemic heart disease.[6] This concept is based on a perception that new blood vessels are formed by a cytokine-driven process and expression of some angiogenic cytokines, which are capable of solving acute conditions of ischemia in animal models. However, there is not a single successful application to achieve therapeutic angiogenesis in humans after over hundreds of clinical trials.

The formation of new blood vessels from pre-existing vessels is aided by a variety of angiogenic growth factors, as shown in Table 6.1. VEGF, or vascular endothelial growth factor, is one of the most significant and commonly used factors related with angiogenesis. Members of the VEGF family include VEGF A, B, C, D, and E [7, 8] and placental growth factors. Alternative splicing produces a variety of VEGF isoforms, with varying numbers of amino acids, including as VEGF-121, 145, 165, 189, and 206.[9] They exert their biological effects through several membrane tyrosine kinase receptors. The recombinant proteins of VEGF-A or VEGF-C, along with the expression of plasmids or

TABLE 6.1
Tool Box Targeted Genes for Their Therapeutic Targets

Therapeutic Targets	Targeted Genes
Thrombosis	Hirudin, thrombomodulin, COX, TFPI, tPA
Vulnerable plaques	COX, TIMPs, soluble VCAM
Therapeutic angiogenesis	VEGF-A, B, C, D, E, Angiopoetin 1, 2, HGF, PDGF, MCP-1, FGF-1, 2, 4, 5, eNOS, iNOS
Systemic hypertension	Adrenomedullin, tissue kallikrein, eNOS, ANP, adrenoceptor, angiotensin II type-I receptor, ACE, Angiotensin gene activating element, carboxypeptidase Y, CYP4A1 antisense oligonucleotides, c-fos, TRH receptor, TRH
Pulmonary hypertension	ANP, VEGF-A, eNOS, prostacyclin synthase, prepro-calcitonin gene related peptide
Atherosclerosis, hyperlipidemia	Soluble VCAM or ICAM, SOD, Il-10, PAF-AH, Lipoprotein lipase, soluble scavenger-receptor decoy, hepatic lipase, apoB, lipid transfer protein, LDL receptor, apo A1,VLDL receptor, Lp(a) inhibition, LCAT, Lipid transfer protein
Graft failure in veins and restenosis	FAS ligand, PCNA antisense oligonucleotide ribozyme, gax, c-myc, cecropine A, cdk-2, ras, blocking PDGF or TGFβ expression or their receptors, iNOS, eNOS, gax, CyA, VEGF-A, TK, p16, p27, p21, p53, COX, TIMPS, VEGF-C, TK, E2F decoys, cdc-2, Gβ Rb, sdi-NFkB, C-myb

adenovirus, have shown favourable effects on myocardium, injured arteries, and ischemic tissue in various animal models. Other biological factors, such as angiopoietin-1[10] and angiopoietin-2, influence the stability and development of new vasculature, which helps to reshape the response to VEGF gene therapy. In addition, angiogenesis can also be induced with the help of fibroblast growth factors (FGFs).[11–13] *In-vivo*, the most prevalent FGFs, such as FGF-1, FGF-2, and FGF-5, affect a variety of cells. These receptors, unlike the VEGF family, are mostly found on monocytes, macrophages, and endothelial cells. Other growth factors and cytokines, which indirectly induce angiogenesis by reviving the production of VEGF from cells *in-vivo*, are platelet derived growth factors, monocyte chemotactic protein-1, and hepatocyte growth factor HGF[14], on which clinical trials have begun, along with its intra-myocardial plasmid delivery. As numerous transitory expressions of growth factors are required for the sprouting of new blood vessels, gene therapy may be the most effective technique to create therapeutic angiogenesis. Angiogenesis stimulation has yet to be proven to result in the creation of functional collateral arteries, and other factors, such as pericyte activation and continuous flow, may be necessary. It's unknown whether diabetic or elderly patients have lower angiogenic responses. We also don't know if ischemia is necessary for the production of angiogenic factors. These considerations are essential, since ischemia does not occur on a continuous basis, even in severe peripheral artery disease, but only during workouts.[15]

The risks associated with therapeutic angiogenesis include hemangioma, stimulation of angiogenesis in the tumor, development of non-functional leaky blood vessels, and excessive new blood vessel formation in atherosclerotic lesions that can lead to plaque rupture and intraplaque hemorrhage. These obstacles can be overcome by increasing the tissue specificity of gene constructs and promoting them, because it can specifically control gene expression. In addition to gene therapy, local genes can also be delivered to the heart through various techniques as described above to reduce the risk to other organs. A possible way to stimulate angiogenesis is physical exercise, through electrical stimulation or through pharmacological agents.[15, 16]

6.2.2 Atherogenesis and Thrombosis

Atherosclerotic lesions gear up their formation in the second decade of life and proceed for the next 20 to 40 years as clinically important lesions. Pathogenesis of atherosclerosis involves several genes and environmental factors. Thus, atherosclerosis is not treated with the help of single gene or local gene transfer. In order to make the gene therapy a successful venture in atherosclerosis, many hereditary diseases and single gene defects that are susceptible to atherosclerosis need to be addressed. Here we discuss a few single gene transfers that facilitate to treat atherosclerosis. For example, lecithin cholesterol acetyl transferase (LCAT) or lipid transfer protein gene transfer can be used to treat certain dyslipoproteinemias. Gene transfer of very low density lipoprotein (VLDL) and low density lipoprotein (LDL) to the liver is an efficient treatment for low density lipoprotein receptor dysfunction.[17, 18] Gene transfer of the catalytic subunit of the ApoB editing enzyme-apobec-1 can suppress elevated levels of atherogenic apolipoprotein (Apo) B100.[19] Similarly, apoE gene transfer helps to decrease the lipoprotein levels that is used to treat type III hyperlipoproteinemia. In ApoA1 deficient patients, ApoA1 gene transfer is used to promote the reverse cholesterol transport. Patients with a lack of enzymes like hepatic lipase and lipoprotein lipase benefit from gene transfer since these enzymes are necessary for lipoprotein metabolism. Class A soluble scavengers and class B soluble scavenger receptors gene transfer also benefits by decreasing the lipid accumulation in macrophages and by altering high-density lipoprotein levels (HDL) respectively.

In early stages of atherosclerosis, endothelial dysfunction occurs due to decrease in the NO bioavailability, which can be treated using VEGF and eNOS genes.[20, 21] However, in advanced atherosclerosis cases, increased production of NO is not beneficial. Transferring genes for platelet activating factors acetyl hydrolase (PAF-AH), Interlukin-10 (IL-10), and dominant negative RhoK can also help to reduce atherosclerosis.[22, 23] Rho family GTPases are involved in the regulation of the actin cytoskeleton and cell adhesion.[24]

A thrombosis can be caused by a ruptured plaque in an atherosclerotic artery, resulting in an acute ischemic episode, such as unstable angina and myocardial infarction. Transfer of the tissue inhibitor of metalloproteinase (TIMP) gene can aid in the stabilization of unstable plaques.[25] Many studies have shown that transferring genes for biological indicators, including hirudin, thrombodulin, tissue plasminogen activator, cyclooxygenase, and tissue factor pathway inhibitor, lowers thrombotic events.[15]

6.2.3 Restenosis, In-Stent Restenosis, Graft Failure

Post-angioplasty and in-stent restenosis are important targets for gene therapy since both illnesses cause balloon-dilated arteries to become occluded within six months of the treatment in 20% of patients. However, the process of restenosis is slow in grafts which are occluded in five years within 50% of patients. There are various factors, such as cell proliferation, remodeling thrombosis, matrix deposition, smooth muscle, as well as platelet and leukocyte adhesion, which play a crucial role in the development of arterial restenosis after stenting, in-vein graft diseases, and angioplasty. Gene therapy is more focused toward inhibition of proliferation and migration of smooth muscle cells, undesirable growth factor effects, and formation of connective tissues. Construction of decoy oligonucleotide ribozymes and antisense against NF-κB, E2F, *c-myb, c-myc, cdc-2, cdk-2, ras, bcl-x*, and antigen of nuclear proliferating cells decreased the intimal thickening in a restenosis experiment.[26–29] *p21, p27, p53, gax* and non-phosphorylated retinoblastoma gene is reported to inhibit cell cycles in an animal model. When HSV-TK, cytosine deaminase, fas ligand, and preprocecropine A genes were transferred into several animal models, there was a reduction in smooth muscle cell migration and cellular proliferation in blood vessels.[30–32] The VEGF gene was transferred to the vessel wall, which reduced the creation of neointima and encouraged endothelial regeneration in wounded blood vessels. It is thought that substances such as C-type natriuretic peptide, prostacyclin I2, and nitric oxide are secreted during the rapid regeneration of endothelial cells, and that gene transfer of nitric oxide synthase and tissue inhibitor of metalloproteinase (TIMP-1) inhibited the formation of neointima in animal models. In numerous animal models, gene transfer of recombinant hirudin (a thrombin inhibitor) and dominant negative Rho kinase suppressed the formation of neointima.[33] In human trial, successful ex-vivo gene transfer of E2F decoy in vein graft reduced the graft failure rate.[15, 34]

6.2.4 Systemic Hypertension

The prevalence of hypertensive patients with genetic abnormalities is minor. Nevertheless, systemic hypertension is the major risk factor involved in cardiac arrhythmia, ischemic heart disease, progressive renal damage, and peripheral vascular diseases. It is easily controlled by pharmacological agents, and these drugs are very successful in controlling systemic hypertension. However, long term management of systemic hypertension using gene transfer can improve the patient compliance. The viability of gene therapy is questioned because of its raveled and multifactorial etiology. Still there are certain mediators of gene therapy that effectively lower the systemic blood pressure. One of the methods of gene transfer is to increase the vasodilator protein by transferring the genes encoding atrial natriuretic peptide (ANP)[35], adrenomedullin[36], nitric-oxide synthase (e-NOS), and tissue kallikrein,[37] which successfully reduced blood pressure in animal models. Another method uses anti-sense oligonucleotides against angiotensin converting enzyme (ACE), angiotensin gene activating element, angiotensin type 1 receptor, adrenoreceptors, carboxypeptide Y, CYP4A1, thyrotropin releasing hormone (TRH), TRH receptor, and c-fos focus on decreasing vasoconstrictor proteins. Both of the methods have shown promising results in the pre-clinical stage.[34] However, no clinical trials are reported to investigate the effect of gene transfer on systemic hypertension until this date.

6.2.5 Pulmonary Hypertension

Progressive increase in pulmonary artery pressure is known as pulmonary hypertension. Etiology of primary pulmonary hypertension (PPH) is unknown; however, mutation in Bone Morphogenic Protein Receptor II (BMPR-II) is reported in scattered cases of PPH.[38] Secondary pulmonary hypertension is caused due to chronic thrombotic or embolic diseases, chronic hypoxia, collagen vascular disease, congenital heart diseases, chronic obstructive pulmonary diseases, and exposure to certain drugs. Also, right ventricle hypertrophy, plexogenic pulmonary arteriopathy, and vascular remodeling in small- and medium-sized pulmonary vessels is observed in many cases. Prostacyclin, inhaled nitric oxide, endothelin receptor antagonists, and calcium channel blockers have all been used to treat pulmonary hypertension, but they have all failed to treat PPH, since it is progressive and malignant. Various genes, such as e-NOS, prostacyclin synthase, MCP-1, prepro-calcitonin gene related peptide, VEGF, and atrial natriuretic peptide, are utilized to reduce pulmonary artery vasospasm and cellular proliferation, although these strategies are only successful in animal models.[35, 39, 40]

6.2.6 Heart Failure

Some of the most essential targets in the system can help failing cardiomyocytes regain their function. Its primary purpose is to restore functions once heart failure has been induced in animal models. It should be associated with increase in the expression of the gene of interest to improve its function, and it is not associated with established gene-dose effects or arrhythmogenesis. Development of heart failure takes place because of deregulation at multiple levels in excitation-contraction coupling. Thus, various transporters, critical proteins, and various channels can be targeted and the contractile functions can be restored by gene editing.[41] The two important targets for heart failure are discussed below:

6.2.6.1 Ca^{2+} Protein Cycling as a Target

HF is caused by a number of abnormalities in the Ca^{2+} processing protein involved in excitation-contraction coupling. These effects can be reversed with the help of gene therapy, using some of the novel techniques described below:

a. Overexpression of SERCA2a:
Abnormality in calcium cycling due to the partial decrease in SERCA2a was reported in human heart failure twenty years ago. Gene transfer of SARCA2 showed improvement in cardiac contractility in many experimental models of HF. AAV mediated SERCA2a showed long term overexpression, preserved systolic function, and improved ventricular remodeling.[42] It also restored the energetic state of the heart in terms of utilization and supply, enhancing coronary flow by activating e-NOS in endothelial cells and decreased ventricular arrhythmia.[43]

b. Inhibition of phospholamban (PLN):
Inhibition of PLN is another technique for improving Ca^{2+} handling. Reduced PLN levels in human cardiac myocytes resulted in improved relaxation velocities and contractions, similar to SERCA2a gene transfer.[44] An animal model (sheep) showed positive results in activity of SERCA, as evident from improved function of systolic and diastolic LV function by silencing PLN expression. In addition to these therapies, RNAi therapy is also used in various HF animal models to suppress Phospholamban (PLN) expression.[45]

c. Active I-1 and inhibition of protein phosphatase 1 (PP1):
In mouse hearts, increased PPI activity induces dephosphorylation of PLN, upregulation of PPI, and excision of I-1, which is linked to a reduction in adrenergic receptor-related contractile response, premature mortality, and poor cardiac function, all of which lead to heart failure. Recent research on transgenic mice with constitutively active I-1 expression found

that inhibiting PPI, which increased the phosphorylation of PLN, enhanced cardiac contractility. It also confirmed a relationship between SERCA2a activity and phosphorylation of PLN. I-1 expression improved the contractile recovery and reduced the infract size by refining the reperfusion-induced injury or ischemia, in addition to decreasing the levels of biomarkers of apoptosis and endoplasmic reticulum stress response.[46–48]

d. SUMO1:
The cytoplasmic protein small ubiquitin-like modifier type 1 (SUMO1)[49] influences the activity and levels of SERCA2a in cardiomyocytes. They are members of the peptide SUMO family, which changes the capabilities of other proteins in cells after translation modification, a process known as sumoylation. This process enhances the stability and increases the activity of SERCA2a. Levels of SUMO1 were increased with the help of AAV9 gene transfer in failing human ventricles, as well as heart failure models of pig and murine, wherein the hemodynamic performance was improved, mortality was reduced in animals with HF, and SERCA2a levels were restored.[41]

e. S100A1:
S100A1, a superabundant isoform of the S100 protein found in the heart, is a calcium modulated protein implicated in intracellular regulatory actions. It enhances the activity of both SERCA2a and RYRs, which stimulates cardiac relaxation and contractile function.[50] Recently, AAV-6 mediated S100A1 showed encouraging results for LV remodeling and dysfunction in a rat model for heart failure. In a pre-clinical model of induced cardiomyopathy, S100A1 demonstrated a significant increase in contractility, indicating that human clinical trials of S100A1 gene therapy are necessary in the upcoming years.[51, 52]

6.2.6.2 Targeting of Beta-Adrenergic System

Signalling of adrenergic system is adversely affected leading to down regulation and de-sensitization of β-adrenergic receptors. Signalling abnormalities of β-AR are seen because of the up-regulation of GRK 2 protein. Several gene-based experiments have hypothesized that cardiac function can be enhanced by genetic modification of β-AR system.

a. Overexpression of β-AR:
Transgenic mice that overexpressed β1-AR had severe cardiomyopathy,[53] overexpression of 2-AR increased myocardial adenylate cyclase activity and improved left ventricular function in mice.[54] The intracoronary and direct myocardial delivery of AAV harboring the 2-AR transgene improved cardiac function in mammalian and rodent models.[55, 56]

b. Cardiac Adenylyl Cyclase (AC) expression activation:
Significant components of the -adrenergic system were employed to increase cAMP expression, but the results were not satisfactory. However, an activation of Adenylyl cyclase (AC) type IV had a beneficial profile. The up-regulation of AC IV in transgenic mice improved adrenergic-stimulated heart function and increased cAMP production in isolated cardiomyocytes. AC IV has no influence on normal heart function and has no interaction with structural abnormalities of the heart.[57] In various animal models, AC IV intracoronary delivery improved the LV function and remodeling in response to increased cAMP generating capacity.[58] Successful results of pre-clinical studies using gene AC IV will promote the start of human trials in patients with HF.[59]

c. G-protein coupled receptor kinase (GRKs) inhibition:
Kinases regulate the interactions between β-ARs and G-protein that modulate the receptors by phosphorylation. The family of GRKs mediates the agonist dependent desensitization which results in functional uncoupling. The most visible GRK in the heart is GRK2, which binds to the Gbg subunit of activated G protein phosphorylated β-ARs and then to β-arrestin, an

inhibitory protein with detrimental action in failing hearts due to faulty cardiac β-AR signaling.[60] Several studies involving myocardial infarction induced HF showed that selective excision of GRK2 post-infraction halted the ventricular remodeling, enhanced the cardiac contractile, and increased the survival in mice.[61] Transgene intracoronary adenovirus mediated distribution of peptide ARKct after myocardial infraction established a marked reversal of ventricular dysfunction and is another approach that inhibits GRK2 mediated -AR desensitization. Thus, further studies are focusing on large animal models.[62] Other targets such as targeting cell death and homing of stem cells are also used to treat heart failure.[41]

6.3 CLINICAL STUDIES ON GENE THERAPY FOR CVDS

As we have studied in previous sections about various genes selected as a target for treating different cardiovascular diseases using various vectors and modes of delivery. This section will focus on the recent, ongoing, or planned clinical trials for delivery of genes for various cardiovascular diseases. Despite the failures of clinical trials for various diseases, in the past few years the vision is restored because of the new vectors which are developed for accurate and efficacious delivery of genes in diseases such as infectious diseases, cancer, monogenetic diseases, and CVDs,.

Insufficient blood supply to the myocardium is caused by ischemic heart disease and atherosclerotic stenosis. The fundamental concept of angiogenesis and formation of blood vessels, which has already been addressed, can be used to cure this. Several growth factors, including fibroblast growth factor (FGF), vascular endothelial growth factor (VEGF), and hepatocyte growth factor (HGF), are employed in gene therapy procedures combined with cell treatment. Several clinical trials, such as APIRE[63], AWARE, KAT 301,[64] and many more as mentioned in Table 6.2, have been put into effect in the last 10 years using plasmid or adenovirus as a vector to achieve success. However, phase II/III trials have yielded no clinically relevant evidence yet.[65] There are several new targets, such as thymosin B4, GalNac-antisense oligonucleotide, promoter activating siRNA, exosomes, and hypoxia inducible factor 1a, which help to target ischemic myocardium. However, currently no clinical trials are focusing on the new target for treatment.[66]

The very first clinical trial for heart failure, named CUPID, was launched in 2007 in the United States.[67] The SERCA2a gene transfer employing intracoronary administration was evaluated in a multicentric trial to determine the biological effects and safety profile.[68] There are currently various trials ongoing or in the implementing stages to rescue the failing myocardium by targeting various pathways. The recent clinical trials include the target adenylyl cyclase type 6, SDF-1, and SERCA2a (Refer Table 6.2). The most recent trial for congestive heart failure is NAN-CS101, which gives the intracoronary administration of an active I-1 transgene (AA 1-65 with T35D), and the second is the vector BNP116.,

In hyperlipidemia, gene therapy for atherosclerosis is potentially used for the treatment of lipoprotein metabolism. GLYBERA was the very first gene-based medication granted in the Western world for lipoprotein lipase deficiency and severe pancreatitis attacks. It was reported for lipoprotein lipase deficiency and severe pancreatitis attacks.[69] Recently, various clinical trials using gene therapy are ongoing, with focus on lowering the plasma concentration level and homozygous familial hypercholesterolemia by targeting LDL receptors, lipoprotein (a) as mentioned in Table 6.2.[66, 70–72]

6.4 CONCLUSION AND FUTURE PERSPECTIVE

In summary, different types of vectors and gene delivery methods are currently in use. Different types of genetic targets are available for the treatment of cardiovascular disease. A quick look at the latest clinical trials in this area will help researchers develop sensible therapies. Gene therapy has not been successfully applied to the cardiovascular field since its inception for 30 years. Although the success of gene therapy in the past 30 years has opened up new and innovative therapies for

TABLE 6.2
Clinical Trials for Gene Therapy of CVD

Trial Name	Therapeutic Agent	Vector	Delivery	Diseases	Phase	Main Results	Primary Endpoint	Reference
VEGF peripheral trial	VEGF-A	Ad, PI	Percutaneous injection at angioplasty site	PAD	II, RCT	Positive	Increase in vascularity	—[73]
KAT 301	VEGF-D$\Delta^N\Delta^C$	Ad	PET/NOGA guided percutaneous intramuscular injections	CAD	I, RCT	Positive	Perfusion reserve in ischemic area	—[64]
ReGenHeart	VEGF-D$\Delta^N\Delta^C$	Ad	PET/NOGA guided percutaneous intramuscular injection	CAD	II,RCT	NA	Perfusion reserve in ischemic area,6 min walking test	NCT03039751
VEGR-A116A	VEGF-A116A	Ad	Thoracotomy, intramuscular injection	CAD	I/II, Open label	NA	Depression on ETT, time to 1mm ST	NCT01757223
HGF-X7	HGF	Ad	Percutaneous, intra muscular injections	CAD	I	NA	Safety	—[74]
HGF-X7 (VM202)	HGF	PI	Intra muscular injection	PAD	II	NA	Visual analogue scale	NCT01064440
HGF-X7 (NL003)	HGF	PI	Intra-muscular injection	PAD	II	NA	Visual analogue scale	NCT01548378
Neovasculgen	VEGF-A165	PI	Intra-muscular injection	PAD	II/III	Positive, pain free walking increased by 167% after a year	Pain free walking distance	—[75]
KAT-PAD101	VEGF- D$\Delta^N\Delta^C$	Ad	Intra-muscular injection	PAD	I,RCT	NA	Perfusion in treated area and safety	Eudra CT 001019-22
AWARE	FGF-4	Ad	Percutaneous, intra-cutaneous injection	CAD	III/RCT	NA	ECG changes on ETT	NCT00438867
ASPIRE	FGF-4	Ad	Percutaneous, intra-coronary injection	CAD	III/ open-label	NA	Reversible perfusion defect	—[63]
JVS-100	SDF-1	PI	Intramuscular injection	PAD	II	NA	Amputation rate, safety	Nct01410331
MULTIGENE ANGIO	VEGF-A165	RV	Percutaneous intra-arterial injection	PAD	I, open label	Positive Amputation-rate at 1 year was 72%	Amputation -free survival, safety	—[76]
AGENT-HF	SERCA-2a	AAV	Percutaneous, intra-coronary injection	Heart-failure	II, RCT	NA	End-systolic volume changes in left ventricle	NCT01966887
STOP HF	SDF-1	PI	Percutaneous intra-muscular helical infusion catheter-mediated injection	Heart-failure	II	Negative	6-min walking distance	—[77]
CUPID 2	SERCA2a	AAV 1	Percutaneous intra-coronary injection	Heart-failure	II, RCT	Negative	Time to recurrent cardiovascular event	—[67]
SERCA-LVAD	SERCA2a	AAV 1	Percutaneous intra-coronary injection	Chronic heart failure	II,RCT	NA	Feasibility and safety	NCT00534703

AC6	Adenylyl cyclase type 6	Ad	Percutaneous intra-coronary injection	Heart failure	I/II, RCT	NA	Combined cardiac function and ETT before and during dobutamine stress	NCT00787059
RETRO-HF	SDF-1	PI	Percutaneous retrograde injection via coronary vein	Heart failure	I/II, open label part and RCT part	NA	6-min walking distance	NCT019617266
Familial hypercholesterolemia trial	LDL receptor	RV	Ex vivo gene transfer to patient hepatocytes that were then injected back into the patient	Homozygous familial hypercholesterolemia because of LDL receptor mutation	I, open label	3 patients showed moderate decrease in cholesterol level	Decreased in plasma LDL cholesterol level, safety	—[70]
apoC-III	As-oligonucleotide against apolipoprotein C-III	As-oligonucleotide	Sub-cutaneous injection (weekly)	Hypertriglyceridemia	II, RCT	Positive, plasma levels of apoC decreased by 40%-80%	Change in plasma apoC-III level	—[72]
PCSK9	siRNA against PCSK9 mRNA	siRNA	Sub-cutaneous injections (monthly/ weekly)	High LDL cholesterol level	I, RCT	Positive, reduction by 51% to 60% in plasma LDL cholesterol	Change plasma LDL concentration	—[78]
Lp(a)	As-oligonucleotide against Lp(a)	As-oligonucleotide	Sub-cutaneous injections (weekly)	High lipoprotein level	II, RCT	Positive, reduction in plasma Lp(a) concentration by 67%	Change in plasma lipoprotein concentration	—[72]
AAV FH	LDL receptor	AAV	Liver directed delivery	Homozygous familial hypercholesterolemia because of LDL receptor mutation	I, open label	NA	Decrease in plasma LDL cholesterol, safety	NCT02651675
Lp(a)- I	Liver targeted asoligonucleotide against Lp(a)	GalNac conjugated as-oligonucleotide	Sub-cutaneous injection (weekly)	High lipoprotein level	I/II, RCT	Positive, plasma Lp(a) reduced by 66%–92%	Changes in plasma Lp(a) concentration	—[72]
NANCS-101	Active I-1 (AA-165 AND T53d)	BNP 116	Intra-coronary administration	Congestive Heart failure	I, open label	NA	Adverse event, all,-cause mortality and HF hospitalization	NCT04179643

Note: CAD, coronary heart disease; PAD, peripheral vascular disease; RCT, randomized controlled trial; ECG, electrocardiogram; RV, retrovirus; Pl, plasmid; ETT, exercise tolerance test; S-T segment; NA, not available, HF- heart failure, BNP- Brain natriuretic peptide.

CVD patients, improvements are still needed to overcome obstacles such as DNA vector efficacy, existing gene delivery methods, and development of genetic targets, gene size selection, vector and vector tolerance, duration of gene expression, and patient death on gene therapy. As understanding of cardiovascular pathophysiology increases, new targets for gene therapy will be discovered. Pharmacogenomics will make it possible to advance in the search for the best gene therapy methods and tools to adjust dysfunction pathways and reverse the course of quality of life of patients. For the purpose of delivery, we need to develop viral and non-viral vectors, which can overcome the obstacles and routes of systemic administration. As we enter the world of personalized medicine, with the development of delivery and packaging of genetic material, gene therapy is becoming increasingly suitable for clinical medicine. To improve cardiovascular therapy through gene delivery, genetic engineers, delivery scientists, and biologists must work together. By diagnosing new genetic targets and developing new gene delivery technologies, cardiac gene therapy is expected to become an emerging field that will ultimately enhance clinical success.

REFERENCES

1. Ashworth MT. The cardiovascular system. *Keeling's Fetal Neonatal Pathol.* 2015:481–529. doi:10.1007/978-3-319-19207-9_19
2. Roth GA, Mensah GA, Johnson CO, et al. Global burden of cardiovascular diseases and risk factors, 1990–2019: Update from the GBD 2019 study. *J Am Coll Cardiol* 2020;76(25):2982–3021. doi:10.1016/j.jacc.2020.11.010
3. Thomas H, Diamond J, Vieco A, et al. Global atlas of cardiovascular disease 2000–2016: The path to prevention and control. *Glob Heart* 2018;13(3):143–163. doi:10.1016/j.gheart.2018.09.511
4. Wolfram JA, & Donahue JK. Gene therapy to treat cardiovascular disease. *J Am Heart Assoc* 2013;2(4):1–11. doi:10.1161/JAHA.113.000119
5. Donahue JK. Current state of the art for cardiac arrhytmia gene therapy. *Physiol Behav* 2017;176(3):139–148. doi:10.1016/j.pharmthera.2017.06.005.
6. Cannatà A, Ali H, Sinagra G, & Giacca M. Gene therapy for the heart lessons learned and future perspectives. *Circ Res* 2020;126(10):1394–1414. doi:10.1161/CIRCRESAHA.120.315855
7. Ferrara N, & Bunting S. Vascular endothelial gowth factor, a specific regulator of angiogenesis. *Curr Opin Nephrol Hypertens* 1996;5(1):35–44.
8. Verheul HMW, Pinedo HM, & Neufeld G. Vascular endothelial growth factor and its inhibitors. *Drugs of Today* 2003;39:81–93.
9. Ferrara N. VEGF-A: A critical regulator of blood vessel growth. *Eur Cytokine Netw* 2009;20(4):158–163. doi:10.1684/ecn.2009.0170
10. Shyu KG, Manor O, Magner M, Yancopoulos GD, & Isner JM. Direct intramuscular injection of plasmid DNA encoding angiopoietin-1 but not angiopoietin-2 augments revascularization in the rabbit ischemic hindlimb. *Circulation* 1998;98(19):2081–2087. doi:10.1161/01.CIR.98.19.2081
11. Unger EF, Banai S, Shou M, et al. Basic fibroblast growth factor enhances myocardial collateral flow in a canine model. *Am J Physiol Heart Circ Physiol* 1994;266(4):H1588–H1595. doi:10.1152/ajpheart.1994.266.4.h1588
12. Lopez JJ, Edelman ER, Stamler A, et al. Angiogenic potential of perivascularly delivered aFGF in a porcine model of chronic myocardial ischemia. *Am J Physiol Heart Circ Physiol* 1998;274(3):H930–H936. doi:10.1152/ajpheart.1998.274.3.h930
13. Safi J, Gloe TR, Riccioni T, Kovesdi I, & Capogrossi MC. Gene therapy with angiogenic factors: A new potential approach to the treatment of ischemic diseases. *J Mol Cell Cardiol* 1997;29(9):2311–2325. doi:10.1006/jmcc.1997.0459
14. Van Belle E, Witzenbichler B, Chen D, et al. Potentiated angiogenic effect of scatter factor/hepatocyte growth factor via induction of vascular endothelial growth factor: The case for paracrine amplification of angiogenesis. *Circulation* 1998;97(4):381–390. doi:10.1161/01.CIR.97.4.381
15. Ylä-Herttuala S, & Martin JF. Cardiovascular gene therapy. *Pharm Gene Deliv Syst* 2003;355:345–362. doi:10.1201/9780824758608.

16. Lau M, Gundwekar S, Lefer David J. Gene therapy for ischemic heart disease. *Bone* 2012;23(1):1–7. doi:10.1016/j.yjmcc.2010.06.007.Gene
17. Kozarsky KF, McKinley DR, Austin LL, Raper SE, Stratford-Perricaudet LD, & Wilson JM. In vivo correction of low density lipoprotein receptor deficiency in the Watanabe heritable hyperlipidemic rabbit with recombinant adenoviruses. *J Biol Chem* 1994;269(18):13695–13702. doi:10.1016/s0021-9258(17)36886-2
18. Pakkanen TM, Laitinen M, Hippeläinen M, et al. Enhanced plasma cholesterol lowering effect of retrovirus-mediated LDL receptor gene transfer to WHHL rabbit liver after improved surgical technique and stimulation of hepatocyte proliferation by combined partial liver resection and thymidine kinase-ganci. *Gene Ther* 1999;6(1):34–41. doi:10.1038/sj.gt.3300796
19. Greeve J, Jona VK, Roy Chowdhury N, Horwitz MS, & Roy Chowdhury J. Hepatic gene transfer of the catalytic subunit of the apolipoprotein B mRNA editing enzyme results in a reduction of plasma LDL levels in normal and Watanabe heritable hyperlipidemic rabbits. *J Lipid Res* 1996;37(9):2001–2017. doi:10.1016/s0022-2275(20)37565-9
20. Wever RMF, Lüscher TF, Cosentino F, & Rabelink TJ. Atherosclerosis and the two faces of endothelial nitric oxide synthase. *Circulation* 1998;97(1):108–112. doi:10.1161/01.CIR.97.1.108
21. Sato J, Mohácsi T, Noel A, et al. In vivo gene transfer of endothelial nitric oxide synthase to carotid arteries from hypercholesterolemic rabbits enhances endothelium-dependent relaxations. *Stroke* 2000;31(4):968–975. doi:10.1161/01.STR.31.4.968
22. Von Der Thüsen JH, Kuiper J, Fekkes ML, De Vos P, Van Berkel TJ, & Biessen EA. Attenuation of atherogenesis by systemic and local adenovirus-mediated gene transfer of interleukin-10 in LDLr-/- mice. *FASEB J* 2001;15(14):2730–2732. doi:10.1096/fj.01-0483fje
23. Quarck R, De Geest B, Stengel D, et al. Adenovirus-mediated gene transfer of human platelet-activating factor-acetylhydrolase prevents injury-induced neointima formation and reduces spontaneous atherosclerosis in apolipoprotein E-deficient mice. *Circulation* 2001;103(20):2495–2500. doi:10.1161/01.CIR.103.20.2495
24. Morishige K, Shimokawa H, Eto Y, Hoshijima M, Kaibuchi K, & Takeshita A. In Vivo gene transfer of dominant-negative Rho-kinase induces regression of coronary arteriosclerosis in pigs. *Ann N Y Acad Sci* 2001;947:407–411. doi:10.1111/j.1749-6632.2001.tb03974.x
25. George SJ. Expert opinion on investigational drugs therapeutic potential of matrix metalloproteinase inhibitors in atherosclerosis. *Expert Opin Investig Drugs* 2000;9(5):993–1007.
26. Morishita R, Gibbons GH, Ellison KE, et al. Single intraluminal delivery of antisense cdc2 kinase and proliferating- cell nuclear antigen oligonucleotides results in chronic inhibition of neointimal hyperplasia. *Proc Natl Acad Sci U S A* 1993;90(18):8474–8478. doi:10.1073/pnas.90.18.8474
27. Indolfi C, Awedimento E V., Rapacciuolo A, et al. Inhibition of cellular ras prevents smooth muscle cell proliferation after vascular injury in vivo. *Nat Med* 1995;1(6):541–545. doi:10.1038/nm0695-541
28. Burgess TL, Fisher EF, Ross SL, et al. The antiproliferative activity of c-myb and c-myc antisense oligonucleotides in smooth muscle cells is caused by a nonantisense mechanism. *Proc Natl Acad Sci U S A* 1995;92(9):4051–4055. doi:10.1073/pnas.92.9.4051
29. Soto C, Sigurdsson EM, Morelli L, Kumar RA, Castano EM, & Frangione B. β-sheet breaker peptides inhibit fibrillogenesis in a rat brain model of amyloidosis: Implications for Alzheimer's therapy. *Nat Med* 1998;4(7):822–826. doi:10.1038/nm0798-822
30. Guzman RJ, Hirschowitztt EA, Brody SL, & Ronald G. In vivo suppression of injury-induced vascular smooth muscle cell accumulation using adenovirus-mediated transfer of the herpes simplex virus thymidine kinase gene. *Nature* 1994;91:10732–10736.
31. Sata M, Perlman H, Muruve DA, et al. Fas ligand gene transfer to the vessel wall inhibits neointima formation and overrides the adenovirus-mediated T cell response. *Proc Natl Acad Sci U S A* 1998;95(3):1213–1217. doi:10.1073/pnas.95.3.1213
32. Nikol S, Huehns TY, Krausz E, et al. Needle injection catheter delivery of the gene for an antibacterial agent inhibits neointimal formation. *Gene Ther* 1999;6(5):737–748. doi:10.1038/sj.gt.3300888
33. Eto Y, Shimokawa H, Hiroki J, et al. Gene transfer of dominant negative Rho kinase suppresses neointimal formation after balloon injury in pigs. *Am J Physiol-Heart Circ Physiol* 2021;278(6):1744–1750.
34. Gruchala M, Roy H, Bhardwaj S, & Yla-Herttuala S. Gene therapy for cardiovascular diseases. *Curr Pharm Des* 2004;10:407–423. doi:10.1016/B978-0-12-801888-0.00029-1

35. Lin K, Chao J, & Chao Lee. Atrial natriuretic peptide gene delivery attenuates hypertension, cardiac hypertrophy, and renal injury in salt-sensitive rats. *Hum Gene Ther* 1998;9(10):1429–1438.
36. Chao J, Jin L, Lin KF, & Chao L. Adrenomedullin gene delivery reduces blood pressure in spontaneously hypertensive rats. *Hypertens Res* 1997;20(4):269–277. doi:10.1291/hypres.20.269
37. Chao J, & Chao L. Kallikrein gene therapy: A new strategy for hypertensive diseases. *Immunopharmacology* 1997;36(2–3):229–236. doi:10.1016/S0162-3109(97)00026-X
38. Atkinson C, Stewart S, Upton PD, et al. Primary pulmonary hypertension is associated with reduced pulmonary vascular expression of type II bone morphogenetic protein receptor. *Circulation* 2002;105(14):1672–1678. doi:10.1161/01.CIR.0000012754.72951.3D
39. Suhara H, Sawa Y, Fukushima N, et al. Gene transfer of human prostacyclin synthase into the liver is effective for the treatment of pulmonary hypertension in rats. *J Thorac Cardiovasc Surg* 2002;123(5):855–861. doi:10.1067/mtc.2002.118687
40. Champion HC, Bivalacqua TJ, Greenberg SS, Giles TD, Hyman AL, & Kadowitz PJ. Adenoviral gene transfer of endothelial nitric-oxide synthase (eNOS) partially restores normal pulmonary arterial pressure in eNOS-deficient mice. *Proc Natl Acad Sci U S A* 2002;99(20):13248–13253. doi:10.1073/pnas.182225899
41. Tilemann L, Ishika K, Weber T, & Hajjar RJ. Gene therapy for HEART failure. *Bone* 2008;23(1):1–7. doi:10.1161/CIRCRESAHA.111.252981.Gene
42. Kawase Y, Ly HQ, Prunier F, et al. Reversal of cardiac dysfunction after long-term expression of SERCA2a by gene transfer in a pre-clinical model of heart failure. *J Am Coll Cardiol* 2008;51(11):1112–1119. doi:10.1016/j.jacc.2007.12.014
43. Hadri L, Bobe R, Kawase Y, et al. SERCA2a gene transfer enhances eNOS expression and activity in endothelial cells. *Mol Ther* 2010;18(7):1284–1292. doi:10.1038/mt.2010.77
44. del Monte F, Harding SE, Dec GW, Gwathmey JK, & Hajjar RJ. Targeting phospholamban by gene transfer in human heart failure. *Circulation* 2002;105(8):904–907.
45. Suckau L, Fechner H, Chemaly E, et al. Long-term cardiac-targeted RNA interference for the treatment of heart failure restores cardiac function and reduces pathological hypertrophy. *Circulation* 2009;119(9):1241–1252. doi:10.1161/CIRCULATIONAHA.108.783852
46. Nicolaou P, Rodriguez P, Ren X, et al. Inducible expression of active protein phosphatase-1 inhibitor-1 enhances basal cardiac function and protects against ischemia/reperfusion injury *Circ Res* 2010;104(8):1012–1020. doi:10.1161/CIRCRESAHA.108.189811.Inducible
47. Pathak A, Baldwin B, & Kranias EG. Key protein alterations associated with hyperdynamic cardiac function: Insights based on proteomic analysis of the protein phosphatase 1 inhibitor-1 overexpressing hearts. *Hell J Cardiol* 2007;48(1):30–36.
48. Pathak A, Del Monte F, Zhao W, et al. Enhancement of cardiac function and suppression of heart failure progression by inhibition of protein phosphatase. *Circ Res* 2005;96(7):756–766. doi:10.1161/01.RES.0000161256.85833.fa
49. Kho C, Lee A, Jeong D, et al. SUMO1-dependent modulation of SERCA2a in heart failure. *Nature* 2012;477(7366):601–605. doi:10.1038/nature10407.SUMO1-dependent
50. Most P, Pleger ST, Völkers M, et al. Cardiac adenoviral S100A1 gene delivery rescues failing myocardium. *J Clin Invest* 2004;114(11):1550–1563. doi:10.1172/JCI21454
51. Pleger ST, Most P, Boucher M, et al. Stable myocardial-specific AAV6-S100A1 gene therapy results in chronic functional heart failure rescue. *Circulation* 2007;115(19):2506–2515. doi:10.1161/CIRCULATIONAHA.106.671701
52. Most P, & Koch WJ. S100A1: A calcium-modulating inotropic prototype for future clinical heart failure therapy. *Futur Cardiol* 2007;3(1):5–11. doi:10.2217/14796678.3.1.5
53. Engelhardt S, Hein L, Dyachenkow V, Kranias EG, Isenberg G, & Lohse MJ. Altered calcium handling is critically involved in the cardiotoxic effects of chronic β-adrenergic stimulation. *Circulation* 2004;109(9):1154–1160. doi:10.1161/01.CIR.0000117254.68497.39
54. Milano CA, Allen LF, Rockman HA, et al. Enhanced myocardial function in transgenic mice overexpressing the β2-adrenergic receptor. *Science* 1994;264(5158):582–586. doi:10.1126/science.8160017
55. Maurice JP, Hata JA, Shah AS, et al. Enhancement of cardiac function after adenoviral-mediated in vivo intracoronary β2-adrenergic receptor gene delivery. *J Clin Invest* 1999;104(1):21–29. doi:10.1172/JCI6026

56. Shah AS, Lilly RE, Kypson AP, et al. Intracoronary adenovirus-mediated delivery and overexpression of the beta(2)- adrenergic receptor in the heart: Prospects for molecular ventricular assistance. *Circulation* 2000;101:408–414.
57. Gao MH, Lai NC, Roth DM, et al. Adenylylcyclase increases responsiveness to catecholamine stimulation in transgenic mice. *Circulation* 1999;99(12):1618–1622. doi:10.1161/01.CIR.99.12.1618
58. Lai NC, Roth DM, Gao MH, et al. Intracoronary adenovirus encoding adenylyl cyclase VI increases left ventricular function in heart failure. *Circulation* 2004;110(3):330–336. doi:10.1161/01.CIR.0000136033.21777.4D
59. Lai NC, Roth DM, Gao MH, et al. Intracoronary delivery of adenovirus encoding adenylyl cyclase VI increases left ventricular function and cAMP-generating capacity. *Circulation* 2000;102:2396–2401. doi:10.1161/01.CIR.102.19.2396
60. Hata JA, Williams ML, & Koch WJ. Genetic manipulation of myocardial β-adrenergic receptor activation and desensitization. *J Mol Cell Cardiol* 2004;37(1):11–21. doi:10.1016/j.yjmcc.2004.03.014
61. Raake PW, Vinge LE, Gao E, et al. G protein-coupled receptor kinase 2 ablation in cardiac myocytes before or after myocardial infarction prevents *Heart Failure* 2009;103(4):413–422. doi:10.1161/CIRCRESAHA.107.168336.G
62. Katz MG, Fargnoli AS, Swain JD, et al. AAV6-βARKct gene delivery mediated by molecular cardiac surgery with recirculating delivery (MCARD) in sheep results in robust gene expression and increased adrenergic reserve. *J Thorac Cardiovasc Surg* 2012;143(3):720–726.e3. doi:10.1016/j.jtcvs.2011.08.048
63. Kaski JC, & Consuegra-Sanchez L. Evaluation of ASPIRE trial: A phase III pivotal registration trial, using intracoronary administration of Generx (Ad5FGF4) to treat patients with recurrent angina pectoris. *Expert Opin Biol Ther* 2013;13(12):1749–1753. doi:10.1517/14712598.2013.827656
64. Hartikainen J, Hassinen I, Hedman A, et al. Adenoviral intramyocardial VEGF-DDNDC gene transfer increasesmyocardial perfusion reserve in refractory angina patients: A phase I/IIa study with 1-year follow-up. *Eur Heart J* 2017;38(33):2547–2555. doi:10.1093/eurheartj/ehx352
65. Ylä-Herttuala S, Bridges C, Katz MG, & Korpisalo P. Angiogenic gene therapy in cardiovascular diseases: Dream or vision? *Eur Heart J* 2017;38(18):1365–1371. doi:10.1093/eurheartj/ehw547
66. Ylä-Herttuala S, & Baker AH. Cardiovascular gene therapy: Past, present, and future. *Mol Ther* 2017;25(5):1095–1106. doi:10.1016/j.ymthe.2017.03.027
67. Jaski BE, Jessup ML, Mancini DM, et al. Calcium upregulation by percutaneous administration of gene therapy in cardiac disease (CUPID trial), a first-in-human phase 1/2 clinical trial. *Physiol Behav* 2016;176(1):100–106. doi:10.1161/CIRCULATIONAHA.111.022889.Calcium
68. Hajjar RJ, Zsebo K, Deckelbaum L, et al. Design of a phase 1/2 trial of intracoronary administration of AAV1/SERCA2a in patients with heart failure. *J Card Fail* 2008;14(5):355–367. doi:10.1016/j.cardfail.2008.02.005
69. Ylä-Herttuala S. Endgame: Glybera finally recommended for approval as the first gene therapy drug in the European union. *Mol Ther* 2012;20(10):1831–1832. doi:10.1038/mt.2012.194
70. Grossman M, Rader DJ, Muller DWM, et al. A pilot study of ex vivo gene therapy for homozygous familial hypercholesterolaemia. *Nat Med* 1995;1(11):1148–1154. doi:10.1038/nm1195-1148
71. Carpentier AC, Frisch F, Labbe SM, et al. Effect of alipogene tiparvovec (aav1-lpls447x) on post-prandial chylomicron metabolism in Lipoprotein lipase-deficient patients. *J Clin Endocrinol Metab* 2012;97(5):1635–1644. doi:10.1210/jc.2011-3002
72. Gaudet D, Alexander VJ, Baker BF, et al. Antisense inhibition of apolipoprotein C-III in patients with hypertriglyceridemia. *N Engl J Med* 2015;373(5):438–447. doi:10.1056/nejmoa1400283
73. Mäkinen K, Mannine H, Hedman M, et al. Increased vascularity detected by digital subtraction angiography after VEGF gene transfer to human lower limb artery: A randomized, placebo-controlled, double-blinded phase II study. *Mol Ther* 2002;6(1):127–133. doi:10.1006/mthe.2002.0638
74. Yang ZJ, Zhang YR, Chen B, et al. Phase I clinical trial on intracoronary administration of Ad-hHGF treating severe coronary artery disease. *Mol Biol Rep* 2009;36(6):1323–1329. doi:10.1007/s11033-008-9315-3
75. Deev RV, Bozo IY, Mzhavanadze ND, et al. PCMV- vegf165 intramuscular gene transfer is an effective method of treatment for patients with chronic lower limb ischemia. *J Cardiovasc Pharmacol Ther* 2015;20(5):473–482. doi:10.1177/1074248415574336
76. Flugelman MY, Halak M, Yoffe B, et al. Phase Ib safety, two-dose study of multigeneangio in patients with chronic critical limb ischemia. *Mol Ther* 2017;25(3):816–825. doi:10.1016/j.ymthe.2016.12.019

77. Chung ES, Miller L, Patel AN, et al. Changes in ventricular remodelling and clinical status during the year following a single administration of stromal cell-derived factor-1 non-viral gene therapy in chronic ischaemic heart failure patients: The STOP-HF randomized Phase II trial. *Eur Heart J* 2015;36(33):2228–2238. doi:10.1093/eurheartj/ehv254
78. Fitzgerald K, White S, Borodovsky A, Bettencourt BR, Strahs A, Clausen V, Wijngaard P et al. A highly durable RNAi therapeutic inhibitor of PCSK9. *N Engl J Med* 2018;376(1):41–51. doi:10.1056/NEJMoa1609243.A

7 Current Application of CRISPR/Cas9 Gene-Editing Technique to Eradication of HIV/AIDS

Prachi Pandey
Babaria Institute of Pharmacy, Vadodara, India

Jayvadan Patel
Sankalchand Patel University, Visnagar, India.

Samarth Kumar
Sun Pharmaceutical Industries Ltd., Vadodara, India

CONTENTS

7.1 INTRODUCTION

Human immunodeficiency virus/acquired immunodeficiency syndrome (HIV/AIDS) is among the major health problems worldwide. As per the statistical data from World Health Organization (WHO), about **37.6** million people were living with HIV by the end of 2020, Among these, 35.9 million people were adults and 1.7 million were children (<15 years old). Surprisingly, only 59% of infected patients receive highly active antiretroviral therapy (HAART), which still remains the main therapeutic option for HIV-1 patients. This therapy can reduce the morbidity and mortality of HIV-related disease. However, HAART cannot totally eliminate latent viral reservoirs, thereby making HIV-1/AIDS a chronic and incurable disease. With recent studies in HIV/AIDS prevention, diagnosis, and treatment, the morbidity and mortality of HIV/AIDS has been decreased significantly. It is caused by HIV infection and induces immune destruction [1]. There are two different types of HIV: HIV1 and HIV-2. They both have many similarities and both can lead to AIDS [2]. Compared to HIV-1, HIV-2 has lower transmissibility and is less pathogenic. HIV-1 is recognized as the major cause of AIDS and becomes the main target to prevent and cure AIDS. HIV-1 is a retrovirus with an RNA genome of about 9.8 kb. The complete genome is flanked by two long terminal repeat (LTR) sequences, and it encodes 10 viral proteins, including gag, pol, vif, vpr, vpu, env, tat, rev, nef, and

DOI: 10.1201/9781003186083-7

the antisense protein (ASP) [3], which have different functions in virus invasion and replication. Three main routes that spread HIV-1 are sex, intravenous injection, and vertical transmission [4, 5].

All HIV-1 infected patients have three stages, acute HIV infection, chronic HIV infection (clinical latency), and clinical disease (AIDS) [6]. Mechanistically, HIV1 invades host cells by binding its gp120 envelope protein to the CD4 receptor on the membrane of the target cell, and then interacts with the CCR5 or CXCR4 coreceptor, which depends on the tropism of viral strain. The host cells mainly correspond to T cells, monocytes, and dendritic cells (and even microglial cells, astrocytes, and perivascular macrophages) in the central nervous system. The life cycle of HIV-1 is complex, which contributes to ineffective virus elimination. HIV-1 entry into cells will establish two types of infection, including latent infection and active infection. Latent infection occurs in the early stage of infection within a few cells, while active infection appears in most cells [7]. For active infection, the provirus is active and produces viral particles, which make infected cells bud new progeny virions. The establishment of latent infection may be mediated by complex mechanisms, including RNA interference [8], chromatin environment [9, 10], transcription factors [11], and HIV-1 provirus integration sites [9].

Latent infection results in the generation of latent reservoirs, which contain infected resting CD4+ T cells [7, 12], astrocytes [13], macrophages [14], and microglial cells [15]. The latent reservoirs are often located in gastrointestinal tracts [7, 14], brains [14, 15], and lymphoid tissues [7], which are difficult to reach by antiviral drugs. Once the latently infected cells are reactivated by stimuli, newly generated virus will be produced and infect neighboring cells, then a new latent reservoir will re-establish. Therefore, HIV-1 latent reservoirs are the major challenge for an effective cure for HIV-1/AIDS, due to which HIV/AIDS still remains an incurable, chronic disease because of the presence of multiple HIV latent reservoirs in the body of patients.

The ultimate therapy for HIV/AIDS is based on the removal or disruption of integrated HIV provirus from latently infected cells or the elimination of these latent cells completely. However, the gene therapy for HIV/AIDS has progressed very slowly in past years. The recent developments in gene-editing technology have increased the hopes for eradication of HIV, as the provirus can be removed from the host cell genome through the newly developed technique of clustered regularly interspaced short palindromic repeat (CRISPR)/CRISPR associated protein 9 (Cas9) gene-editing [16–19]. Also, CRISPR/Cas9 system can be useful to induce apoptosis in reservoir cells [20, 21].

The CRISPR/Cas9 system has emerged as a popular gene-editing system with proven efficacy in mammalian cells. Even in year 2013, soon after the introduction of CRISPR/Cas9 technique, its potential applications for treating HIV/AIDS were significantly considered [28]. The use of the CRISPR/Cas9 system is now considered as a powerful gene-editing tool to delete or disrupt HIV provirus from latently infected cell genomes or to knock out CXCR4 and CCR5 receptors for HIV infection treatment and prevention. The elimination or disruption of HIV provirus in reservoir cells or the elimination of reservoir cells might result in a cure for HIV/AIDS.

7.2 OVERVIEW OF CRISPR/CAS9 TECHNOLOGY

The ultimate goal of gene-editing is gene repair. Several gene editing technologies recently demonstrated strong potential for therapeutic uses.

In the recent studies, the three main nuclease-mediated gene editing tools include transcription activator-like nucleases (TALENs), zinc finger nucleases (ZFNs), and clustered regularly interspaced short palindromic repeat (CRISPR)/CRISPR-associated nuclease 9 (Cas9), which are been widely used in HIV-1/AIDS treatment studies.

In principle, ZFNs and TALENs fuse DNA nuclease and DNA-binding domains together to break DNA at specific loci, while Cas9 nucleases work with guide RNA to break DNA at specific loci. Both ZFNs and TALENs techniques have significant applications and studies in HIV/AIDS prevention and therapy. [22–25]. Compared to the initial CRISPR/Cas9 system in bacteria, the current CRISPR/Cas9 system has been simplified, modified, and adapted for mammalian genome editing and bioengineered for better nucleus localization and mammalian cell expression The current CRISPR/Cas9 system includes a single guide RNA (sgRNA) consisting of crRNA and tracRNA and Cas9, which excises the target DNA. [20, 26, 27].

Cas9 is a nuclease of 1,368 amino acids, with two nuclease activity domains named HNH and RuvC. [59, 64–66] Each domain can cleave a DNA strand directed by a sgRNA complementary to the target DNA sequence (generally 20 nucleotides long). The prerequisite to be a target sequence is the presence of an NGG sequence (N: any nucleotides; G: guanine nucleotide, also called protospacer adjacent motifs (PAMs)) downstream (3′ end) of the target site. [59,64–66] In the nucleus, Cas9, sgRNA and target DNA form a complex. Then, HNH and RuvC domains each cleave a DNA strand [28–30]. The double strand DNA breaks are subsequently repaired by two different strategies. One is non-homologous end-joining (NHEJ) when there is no template, and the other is homology-directed repair (HDR) when there is a homologous nucleotide template present, such as ssDNA or dsDNA. (Figure 7.1).

Such DNA template must carry homology sequences to the regions that flank the Cas9 excision site. When the two homologous regions flank an accurate sequence, this may be used as a template to 'repair' the genomic sequences. NHEJ usually cause an insertion or deletion in DNA sequence, in total called indel, while HDR leads to correct repair as directed by the template. [31] CRISPR/Cas9 system can even be employed in biological functions aside from gene-editing. as an example, catalytically inactive Cas9 (i.e., dCas9 has no active nuclease domains, so is often applied to be a site-specific DNA-binding factor) [32]. The fusion protein of dCas9 and transcription activator or repressor domain are often utilized to regulate gene expressions [32, 33]. In summary, CRISPR/Cas9 system could be a fully developed, multifunction gene-editing and regulating system that has unlimited potential. RISPR/Cas9 technology cannot only mediate gene editing effectively, but can also exhibit biological functions. Mutation of Cas9 in two nuclease domains induces inactive Cas9 (dCas9), which is employed as locus-specific DNA-binding protein [16].

This dCas9 fused with a transcription activator or repressor domain can regulate gene expression. [17]. Moreover, other gene editing technologies, like CRISPR/Cpf1, single RNA-guided endonuclease without tracrRNA, and CRISPRCas13a/C2c2, an RNA-guided RNA-targeting effector that induces siRNA cleavage, are developed supporting the muse of CRISPR/Cas9 technology [18].

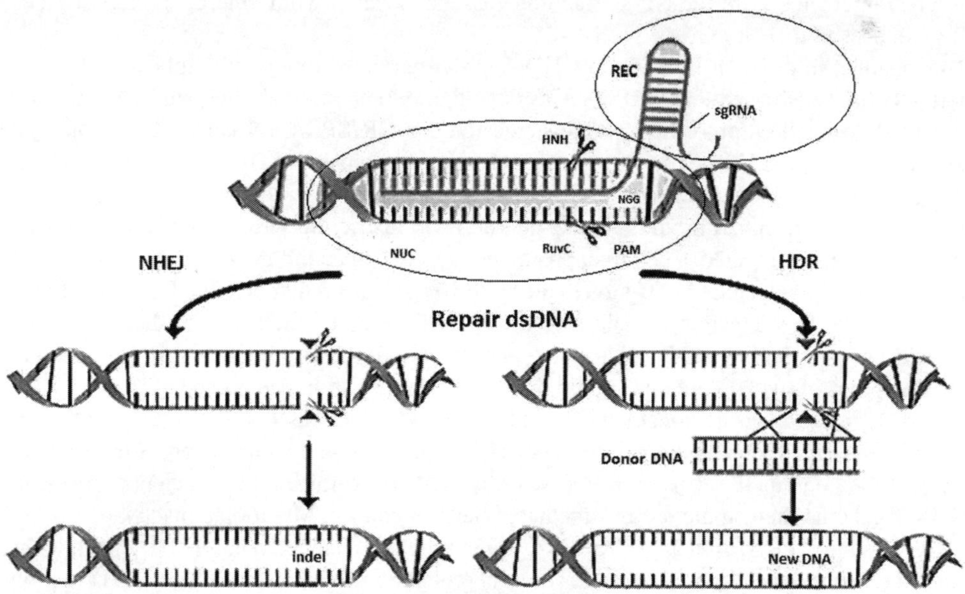

FIGURE 7.1 CRISPR/Cas9 technology.

A study on trial with ZFN mediated C-C chemokine receptor type five (CCR5) editing in autologous CD4 T cells had been successfully tried in HIV-1 infected patients, which showed that ZFNs-CCR5 modification is also effective and safe in human AIDS therapy [19, 23]. CRISPR/Cas9 application in HIV-1/AIDS therapy, thanks to the costly and time-consuming construction of TALENs and ZFNs, the latest gene-editing technique, CRISPR/Cas9 has been developed rapidly with the benefits of precise insertion, deletion, and replacement of target DNA sequences [34]. The primary CRISPR/Cas9 application within the prevention of HIV-1 infection was performed in 2013 by disruption of latent HIV-1 provirus [21]. From then on, numerous studies on HIV-1/AIDS gene therapy by CRISPR/Cas9 technology are reported. Therefore, CRISPR/Cas9 could be a powerful gene editing tool, with unlimited potential in biomedical research field.

7.3 APPLICATION OF CRISPR/CAS9 SYSTEM TO HIV/AIDS PREVENTION AND TREATMENT

CRISPR/cas9 technology has been widely applied in HIV1/AIDS research using experimental laboratory adapted HIV1 strains within the past few years due to its characteristics of straightforward, high efficiency, and limited off-target effect [35]. The targets include HIV-1 genome (Table 7.1) and host factors (Table 7.2) as explained.

7.3.1 Inactivation and Elimination of HIV-1 Provirus by CRISPR/Cas9 Technology

The main obstacle for HIV/AIDS treatment is efficient targeting of latent viral reservoirs. HAART can control HIV-1 replication in patients, but it cannot completely eliminate the provirus in resting CD4+ T cells, leading to the steadiness of latent viral reservoirs for several years [12].

In early studies, researchers used Tre-recombinases to specifically target HIV-1 LTR, which resulted within the excision of HIV-1 provirus in HeLa cells [51]. In a study, the researchers generated ZFNs to target TAR region in HIV-1 LTR, which led to efficient cleavage of integrated HIV-1 proviral DNA in Jurkat T cells and also the latent cell line C11 [52]. The CRISPR/Cas9-based approach was first tested in HIV1/AIDS treatment in 2013. Researchers have successfully used CRISPR/Cas9 to suppress the expression of HIV-1 genes in Jurkat cell lines by targeting HIV-1 LTR. The target sites were the NF-κB binding cassettes located within the U3 region of LTR and TAR sequences in R region, respectively.

This resulted in efficient inhibition of HIV-1 provirus transcription and replication [21]. More importantly, it also showed that CRISPR/Cas9 could eliminate internal integrated viral genes from the infected host cell chromosome, which suggested that CRISPR/Cas9 is also a possible tool for HIV-1/AIDS treatment [21]. Soon afterwards, research on excision of HIV-1 genome via CRISPR/Cas9 was conducted [36]. They used Cas9/gRNA to focus on conserved sites in the HIV1 LTR U3 region, leading to inactivating viral gene expression and restricting virus replication in an exceedingly HIV-1 latently infected T cell line, pro-monocytic cell line, and microglial cell line with little genotoxicity, and no detectable off-target editing. It has also been demonstrated that targeting multiple sites of the HIV-1 genome could increase the efficiency of excision and disruption of non-integrated proviral genome [37]. additionally, combination of two effective sgRNAs to focus on different regions of the HIV genome could prevent viral replication and escape [40].

Recently, it had been demonstrated that Staphylococcus aureus Cas9 (SaCas9)/gRNAs in an all-in-one lentiviral vector could excise latent HIV-1 provirus and suppress provirus reactivation. Moreover, the combined SaCas9/gRNAs was observed to exhibit higher efficiency in disrupting the HIV-1 genome than single sgRNA mediated SaCas9 editing. Mutational inactivation of HIV-1 provirus by single sgRNA induced CRISPR/Cas9 editing has also been reported [39]. While targeting the LTR sequence and essential genes for viral replication by single sgRNA, the HIV-1 provirus was found to be inactivated by mutation of the target site. Thus, the virus can escape single gRNA

TABLE 7.1
CRISPR/cas9 Systems Target HIV-1 Provirus for Excision and Elimination

CRISPR Cas	Delivery	Target Region	Cell Type/Organism	Targeting Locus	Target Sequence	Efficiency (%)	References
Sp Cas9	Transfection	LTR (U3 region)	293T, Hela, Jurkat	465–484	GTTAGACCAGATCTGAGCCT	30–90	[21]
Sp Cas9	Transfection	LTR (U3 region)	CHME5, TZM-Bl, U937	101–127 312–341	GCCAGGGATCAGATATCCACTGACCTT GAGTACTTCAAGAACTGCTGACATCGAGCT	30–90	[36]
Sp Cas9	Lentivirus	LTR (R region)	293T-CD4-CCR5, 293 Primary T cells, hPSC	464–486 485–507	GGTTAGACCAGATCTGAGCCTGG GGGAGCTCTCTGGCTAACTAGGG	20–90	[37]
Sp Cas9	Transfection	Rev (the second exon)	JLat10.6	8513–8532	GGTGGTAGCTGAAGAGGCAC	30	[38]
Sp Cas9	Lentivirus	Gag/Pol/Rev/Env	SupT1	2249–2277 8497–8525	TCAGATCACTCTTTGGCAGCGAC GTGCCTCTTCAGCTACCACCGCT	30–90	[39]
Sp Cas9	Lentivirus	LTR (U3 and R region)	J.Lat FL,SupT1	300–408 463–482	GCCACTCCCCAGTCCCGCCC GCTCAGATCTGGTCTAACCA	35–98	[40]
Sa Cas9	Lentivirus AAV	LTR and gag (U3 region)	Tg26 transgenic mouse	83–103 380–399 1061–1081	GCAGAACTACACACCAGGGCC GTGTGGCCTGGGCGGGACTG GGATAGATGTAAAAGACACCA	20–80	[41]

TABLE 7.2
CRISPR/Cas9 Systems Target Co-Receptor CCR5 or CXCR4 and Restriction Factors

CRISPRCas9	Delivery	Target Gene/ Restriction Factors	Cells	Strategy	Target Sequence	Efficiency (%)	References
SpCas9	Transfection	CCR5	K562	Disruption	TGACATCAATTATTATACAT	13	[42]
SpCas9	Transfection	CCR5	iPSC	CCR5132	GATACAGTCAGTATCAATTC	33–100	[43]
SpCas9	Lentivirus	CCR5	TZM-Bl, CEMss,	Disruption	GCTTGTGACACGGACTCAAG GGTCCTGCCGCTGCTTGTCA GTAAACTGAGCTTGCTCGCT	10.8–67.7	[40]
SpCas9	Lentivirus Adenovirus	CCR5	TZM-Bl, CHO, C8166, primary CD4+ T	Disruption	TCACTATGCTGCCGCCCAGT CAATGTGTCAACTCTTGACA	32–75	[44]
SpCas9	Transfection	CCR5	K562, CD34+ HSPC	Disruption	ACTGGGCGGCAGCATAGTGA CCCAGAAGGGGACAGTAAGA	19–46	[45]
SpCas9	Lentivirus	CXCR4	Ghost, Jurkat, primary CD4+ T	Disruption	GCTTCTACCCCAATGACTTG GTTCCAGTTTCAGCACATCA	10–45	[46]
SaCas9	Lentivirus AAV	CXCR4	TZM-Bl, Ghost, Jurkat, primary CD4+ T	Disruption	CCTGGTATTGTCATCCTGTCC TCCTGCTATTGCATTATCATC	8.5–80	[39]
SpCas9	Lentivirus	Restriction factors	293T, CEMss	APOBEC3G(A3G) and APOBFC3B(A3B)	-	50–90	[47]
-	-	Restriction factors	-	SERINC, HUSH, NONO	-	-	[32, 48–50]

mediated cleavage. This viral breakthrough may be alleviated by a combinatorial CRISPR/Cas9 gene-editing approach [40].

In a study, the researchers used CRISPR/Cas9 target LTR of provirus in HIV-1 latently infected T lymphocyte lines. Sequencing analysis demonstrated mutations occur at each target site, leading to reduction of viral gene expression and viral production while on tumor necrosis factor alpha (TNFα) treatment [38]. Several studies also performed targeting on incoming virus by CRISPR/Cas9 system. They confirmed that dual sgRNAs have higher cleavage efficiency on a non-integrated HIV-1 reporter plasmid than single sgRNA [39, 40].

Recent work also suggested that CRISPR/Cas9 could cleave the non-integrated HIV-1, leading to a 3–4-fold reduction of integrated HIV-1 provirus. Surprisingly, the NHEJ mediated DNA repair mechanism also responds for the non-integrated HIV-1 provirus. Thus, the CRISPR/Cas9 functions in both HIV-1 and proviral DNA in latent cells, of which makes it more potential and promising in HIV-1/AIDS treatment. Excision of HIV-1 proviral DNA in animal models was reported in 2017 [41].

Researchers demonstrated the feasibility and efficiency of disrupting HIV-1 provirus using an all-in-one adeno-associated virus (AAV) combined with multiplex sgRNAs and SaCas9 in three different animal models. Quadruplex sgRNA/SaCas9 AAV-DJ/8 intravenously injected into Tg26 mice could cleave HIV-1 proviral DNA and significantly reduce virus replication. Additionally, after intravenous injection of quadruplex sgRNA/SaCas9 AAV-DJ/8 in humanized bone marrow/liver/thymus (BLT) mice infected with HIV-1, the cleavage of provirus was detected in brain, colon, spleen, heart, and lung. this method could also inhibit HIV-1 replication in Eco-HIV acutely infected mice [41]. This successful application of excision and elimination of HIV-1 proviral DNA by SaCas9/gRNA in vivo delivered via AAV lays the basis for the planning of clinical trials in humans.

7.3.2 Disruption of Co-receptors CCR5 and CXCR4 by CRISPR/Cas9 Technology

For targeting HIV-1 genome, CRISPR/Cas9 technology also can be utilized to block HIV-1 entry by editing of co-receptors. HIV-1 enters host cells through binding to the CD4 receptor and CCR5 or CXCR4 co-receptors [53]. As CD4 is critical for a functional system, disruption of CD4 is an unadvisable strategy in preventing HIV-1 infection. It has been reported that individuals with a homozygous 32-bp deletion of the CCR5 gene (CCR5132) could live healthily and resist R5-tropic HIV-1 infection naturally [54].

Transplantation of CCR5132 HSPCs to a patient with acute chronic myelocytic leukemia (AML) and HIV-1 infection accidently prevented HIV-1 replication and rebound to this point [55]. This is often an encouraging case, but because of the few homozygous CCR5132 donors, broad application of this strategy to HIV-1/AIDS patients is significantly limited. Therefore, co-receptors CCR5 and CXCR4 become the potential targets for HIV-1/AIDS gene therapy. Gene editing of CCR5 by ZFNs has been successfully employed in CD4+ T cells and CD34+ HSPCs, which could resist HIV-1 infection [56]. The studies have been conducted for disruption of CCR5 through ZFN delivered by adenoviral vector in CD4+ T cells from HIV patients and reinfused autologously after ex- vivo expansion, resulting in significant decrease of HIV DNA and RNA to undetectable levels in most patients [23]. This important trial (NCT00842634) was followed by additional studies showing that CCR5-modified cells may well be detected in HIV-1 patients for several months [23]. For the other co-receptor, CXCR4 could even be modified by ZFN to resist to X4-tropic HIV-1 infection [57]. While, simultaneous editing of CXCR4 and CCR5 by ZFN in Sup T1-R5 and primary CD4+ T cells was also reported. Interestingly, the in vivo study showed a significant survival within the presence of HIV-1 infection in an NSG mouse transferred with modified primary CD4+ T cells compared to the unmodified cohort [58]. Compared with ZFN technology, the CRISPR/Cas9 approach provides appropriate target sites with simple design and plasmid construction. CRISPR/Cas9 has been widely employed in disrupting CCR5 and CXCR4 expression.

The observations during a study showed that silencing of CCR5 by CRISPR/Cas9 is often successfully achieved in human embryonic kidney (HEK) 293T cells by transfecting Cas9 and sgRNAs

[42]. Not long ago, researchers conducted a combined CRISPR/Cas9 or TALENs with the piggy-Bac technology to perform the homozygous CCR5132 mutation in induced pluripotent stem cells (iPSCs) seamlessly. These CCR5-modified iPSCs could normally differentiate into monocytes/macrophages, which were resistant against HIV-1 infection [43]. The studies also used adenovirus-delivered CRISPR/Cas9 combined with sgRNAs targeting the fourth exon of CCR5 to disrupt CCR5 expression. Two specific sgRNAs can induce 60% cutting efficiency in TZM-BL cells. The results were also confirmed in a Chinese hamster ovary (CHO) and human T cell lines. Finally, they used a chimeric Ad5/F35 adenovirus vector to deliver the CRISPR/Cas9 system to human CD4+ T cells to silence CCR5 expression, which protected cells from HIV-1 infection with high efficiency and small off-target effects [44]. Another study used the CRISPR/Cas9 system to focus on the CCR5 gene in human CD34+ HSPCs and achieved long-term CCR5 disruption in vivo, which resulted in inhibition of HIV-1 infection. Silencing of CCR5 expression was stable within the secondary repopulating hematopoietic stem cells (HSCs), which provides a basis for the long run of developing an HIV-1/AIDS cure for clinical application by transplanting CCR5-modified HSCs [45].

Meanwhile, several studies have reported CXCR4 gene disruption by CRISPR/Cas9. Some researchers efficiently disrupted CXCR4 expression in human CD4+ T cells and rhesus macaque CD4+ T cells using CRISPR/Cas9 delivered by lentivirus and two sgRNAs that specifically target CXCR4 conserved sequences. Reduced p24 production was observed within the human CXCR4 knockout (KO) CD4+ T cells after infection with X4-tropic HIV-1, which interprets that CXCR4-modified cells resist HIV1 infection. Moreover, CXCR4-edited CD4+ T cells by CRISPR/Cas9 behaved normally in propagation and there was no detectable off-target effect and toxicity in CD4+ T cells [46].

Other researchers also disrupted CXCR4 expression by Cas9/gRNA ribonucleoprotein (Cas9 RNPs) in human CD4+ T cells, which reduced about 40% of CXCR4 expression on the cell surface. When given a repair template, the cellular repair machineries can lead to the required modification in human T cells. Although the CRISPR-Cas9 induced NHEJ process was a stochastic event, the deep sequencing revealed the knocking efficiency up to twenty percent [59].

Further, a Staphylococcus aureus (SaCas9), which is 1 kb shorter compared to SpCas9, was utilized to disrupt CXCR4 in human primary CD4+ T cells delivered by AAV and guarded cells from HIV-1 infection with low toxicity and small off-target effects [60]. As CXCR4 and its ligand CXC12 (SDF-1) play pivotal roles in hematopoietic cells development and thymic differentiation [61]. It is important to consider the safety and side effects of clinical application by targeting CXCR4 in HSPCs. However, CXCR4-deficient human T cells remain functional in a mouse model. The fact that we also used CRISPR-Cas9 and piggyBac recombinant technologies to make a CXCR4 P191A mutant with HIV-1 infection inhibition function and without deficiency of CXCR4 function [62, 63] means it is possible for CRISPRCas9 editing of CXCR4 in human mature post-thymic CD4+ T cells for the purpose of HIV-1/AIDS therapy.

7.3.3 Reactivation of Latent HIV-1 Virus by CRISPR/Cas9 Technology

To eradicate latent HIV-1 reservoirs, it is necessary to reactivate dormant virus within the host cells and induce cell killing by HAART and activation of antiviral immune responses. This strategy is thought of as "shock and kill" [8, 11]. Several drugs are shown to reactivate viral gene expression, like the histone deacetylase (HDAC) inhibitor [64], which introduces the acetylation and remodeling of chromatin, thus leading to the enhancement of HIV-1 RNA expression within the latent reservoirs. The drugs only induce the transcription of latent HIV1 in cells, but don't kill the virus or cause necrobiosis [33, 47].

The combination use of latent reversing agents (LRAs) might overcome several side effects in patients under HAART and cause a much better efficiency of purging the HIV-1 latent reservoir [65]. However, this approach might not be ready to target all viral reservoirs, and so might not be very efficient. Considering the important side effects observed in patients treated with HAART and/or HDAC inhibitors, additional strategies to reactivate HIV-1 reservoirs have to be developed. The CRISPR/Cas9 technology could also be a possible tool for the activation of latent HIV-1 viral reservoirs. [66]

Many researchers have used deficient Cas9 (dCas9) fusion protein combined with sgRNAs specific to effector domains of target DNA sequences to activate or repress gene transcription [16, 17]. The catalytically inactive dCas9 fused with transcription activator domains can activate viral gene expression in HIV-1 latent reservoirs, which can improve the "shock and kill" strategy. Some researchers designed 20 sgRNAs to focus on the LTR-U3 region of HIV-1 promoter and screened two target sites located near or at NF-κB binding sites with high specificity and efficiency. Those specific target sgRNAs could induce reactivation of HIV-1 provirus in HIV-1 latent cell lines like TZM-Bl epithelial cells, CHME5 microglial cells, and Jurkat T lymphocytic cells. They also found that this reactivation could induce suicide death in CHME5 microglial cells and Jurkat T lymphocytic cells, but not in TZM-Bl because of the buildup of toxic viral proteins, with no effect on the non-reactivated cells. [47, 67]

Researchers also designed 23 sgRNAs to focus on the LTR U3 region of HIV1 provirus and located the robust activation sites also near NFκB binding sequences. This activation system worked more efficiently than latency reversing compounds like SAHA and prostratin in numerous latent T cells models [68]. The 7 sgRNAs has been designed to focus on the key functional elements of HIV-1 LTR, including U3 region, NF-κB and Sp1 binding sites, R domain, and U5 region. All 7 sgRNAs could induce gene activation from HIV-1 LTR promoter and only two sgRNAs, overlapping with the NF-κB binding sites and transactivation response elements, significantly stimulated latent HIV-1 gene expression. Moreover, combining latency breaking reagents, including prostratin and SAHA, with CRISPR activators could increase latent HIV-1 re-activation. Therefore, CRISPR/Cas9-mediated activation of viral transcription may provide an alternate approach to focus on and activate viral gene expression in latent HIV-1 reservoirs. [69]

7.3.4 Reactivation of Host Restriction Factors During HIV-1 Infection

Several proteins in mammalian cells function as restriction factors during infection by HIV-1 and other viruses. However, these host factors are often weakly expressed in infected cells. Simultaneous activation of the expression of restriction factors may represent an alternate strategy to stop HIV-1 replication. The Cas9-based approach has been utilized to induce the expression of the restriction factors APOBEC3G (A3G) and APOBFC3B (A3B) in human cells. They also found that using two sgRNAs had higher efficiency than single sgRNA, and both activated proteins could block Vif-deficient HIV-1 infection by inducing dC residues to dU residues (dC-to-Du) editing of HIV-1 genome [70]. However, studies on the applying of CRISPR/Cas9 technology in activating cellular host factors to inhibit HIV-1 infection are very limited.

Recently discovered restriction factors, like serine incorporator five (SERINC5) [48], human silencing hub (HUSH) [32], and NONO [49], could also be new targets to be considered for this application. SERINC5 inhibits viral infection by preventing virus and cell fusion. The HIV-1 accessory protein Nef counteracts the function of SERINC5 by redirecting it to a Rab7-positive endosomal compartment, thus preventing its incorporation in newly generated virions. [48]. The HUSH complex, composed by TASOR, MPP8, and periphilin, may be degraded by the viral protein Vpxvia, a DCAF1-dependent proteasomal pathway in primary T cells and HIV-2 infected cells [32].

It is often also observed that the HUSH complex can be degraded by Vpx and Vpr from HIV-1, HIV-2, and SIV to counteract HUSH-induced repression of provirus transcription [50]. The studies suggested that the HUSH complex could be a critical host factor in HIV infection. NONO has been identified as a capsid-binding factor for Cyclic GMP-AMP synthase (cGAS)-mediated immune activation in macrophages and dendritic cells after HIV-1/2 infection employing a two-hybrid yeast screening. NONO directly interacted with HIV-1/2 capsid proteins to extend DNA sensing mediated by cGAS, but had little effect on HIV infection, which suggest the important role of NONO in cGAS-mediated immune activation after infection with HIV [49]. Since these representative restriction factors can inhibit HIV infection via different mechanisms, the CRISPR/Cas9 technology may be utilized to simultaneously activate their expression in infected cells so as to focus on different phases of the viral life cycle.

This approach may provide new strategies for the treatment of HIV-1 infection, and more efforts are needed to further develop this toolset. Activation of host restriction by CRISPR/Cas9 is typically delivered by lentivirus with high transduction efficiency and simple virus production. Additionally, specific sgRNAs are essential for the activation of the target gene. Two sgRNAs will enhance the expression of the target gene by increasing the target specificity to assure the entire activation of gene promoter [70].

As the restriction factors act on HIV-1 at different stages within the life cycle, and therefore the expression of some restriction factors is merely induced by infection, better understanding of the molecular mechanisms of HIV-1 counteraction with restriction factors will provide more options and rationale for the look of CRISPR/Cas9. Additionally, several restriction factors have dual functions in immunomodulation and different expression levels in various cells, which makes its more complex within the activation of gene expression by CRISPR/Cas9. However, the enforced long-term activation of restriction factors expression has side effects in vivo and desires further investigation. For instance, the virus restriction factor APOBEC3G, which is induced by IFNs, has an antiviral effect on HIV-1 and HBV. These are different expression levels in various cells. But their activated expression in several cells over an extended period of your time isn't bound to exhibit deleterious effects in vivo just because it has not been reported at the present. [71]

7.3.5 CRISPR/CAS9 System Delivery Approach

The major challenge of curing HIV/AIDS lies within the persistence of multiple reservoirs in several drug-inaccessible parts of the physical body. Moreover, while current HIV/AIDS patients can take medications to suppress symptoms, the side effects of those drugs may contribute to high morbidity in these patients. The whole cure is the elimination or disruption of HIV provirus from the patient's reservoir cells' genomes or the elimination of patient reservoir cells. As stated above, gene-editing technology makes it possible to manipulate DNA at the cell genomic level. Since its discovery, CRISPPR/Cas9 system has been a premier tool in gene editing.

As mentioned before, many experimental efforts and clinical trials are underway for the appliance of the CRISPR/Cas9 system for eradication of HIV/AIDS. The foremost substantive attain date that has been reported are ex vivo studies. Future efforts are going to be directed towards efficient delivery of CRISPR/Cas9 system components to latently infected cells. Currently, the delivery methods of the CRISPR/Cas9 system are divided into viral delivery and non-viral delivery. The CRISPR/Cas9 system is extremely flexible and may be delivered within the style of DNA, RNA, or protein/RNA complex. The foremost popular approach is that the DNA form, collectively or several plasmids, within the viral delivery methods, lentivirus, baculovirus81 and recombinant adenoassociated virus (rAAV), are tried. [72, 73]

Among these virus vectors, rAAV has the benefits of high safety, flexible administration routes, and low or no reaction, but it has the disadvantage of limited cargo size. The main administration approaches of rAAV include stereotactic injection, intranasal and intratracheal administration, IV Injection, intraperitoneal injection, and injection. Within the non-viral delivery approach, vectors like cationic polymer polyethyleneimine (PEI), liposomes, lipid nanoparticles, virus-like particles derived from bacteriophage, and self-assembled DNA nanoparticles are applied for the delivery of the CRISPR/Cas9 system. Since viral and non-viral approaches have relative advantages and drawbacks, some investigators started trying the combined approach. For instance, lipid nanoparticle-mediated delivery of Cas9 mRNA has been combined with adeno-associated viruses encoding a sgRNA and a repair template. [74–78]

This delivery strategy successfully repaired a disease gene in an adult mouse model. Regardless of which delivery strategy is getting used, the emphases are biocompatible and targetable carriers, controllable reagents release, and low-invasive administration. As for HIV/AIDS eradication, so as to ascertain more efficient evaluation methods, HIV/AIDS primate models are the middle of the following phase of study. Drug brain delivery is the key for the eradication of the HIV reservoir in patients' CNS systems. The obstacle for drug brain delivery of the CRISPR/Cas9 system is the existence of the blood– brain barrier (BBB). The tight junction between capillary endothelial cells blocks transportation of huge molecules, and only limited molecules with mass smaller than 400 Da and appropriate lipophilicity can cross.

All large molecules and most of the little molecules cannot cross the BBB. Great efforts are made to enhance BBB permeability for therapeutic agents for the treatment of varied brain diseases and HAND. However, currently available procedures are mainly invasive, including intracerebroventricular (ICV) infusion and intra-cerebral injection. Although these approaches have shown successful disruption of the BBB, the invasiveness of the technique and also the risk of harm to the brain remain significant concerns. To beat these problems, the utilization of nanomaterials as drug delivery carriers to the brain has been gaining appreciation. [79]

The most popular approach for brain targeting using nanomaterials is receptor-mediated transcytosis. The endothelial cells that form the BBB are known to specific receptors, including the transferrin receptor and insulin receptor. Ligand functionalized nanoparticles can promote efficient brain delivery through the binding of ligands to the receptors on endothelial cells. Various nanoparticles are developed for brain targeting using this receptor-mediated transcytosis, like transferrin, lactoferrin, monoclonal antibody-conjugated liposomes, polymer nanoparticles, gold nanoparticles, and iron oxide nanoparticles. Magnetic targeting is another approach to boost BBB transmigration of nanoparticles.

There is important progress in brain delivery using magnetic nanoparticles and magneto-electric nanoparticles. Magnetoelectric nanoparticles are used as carrier molecules, which could cross the BBB and mediate controlled release of the loaded drugs. These nanoparticles were ready to cross the BBB using magnetic attraction induced on the nanoparticles by the magnetic field gradient. So far, these nanoparticles haven't been applied to the delivery of the CRISPR/Cas9 system, nor have they been fully tested in animal models. However, the potential for these nanoparticles to deliver the CRISPR/Cas9 system and target the HIV reservoir in the brain seems high because of the nanoparticles' capacities of specific targeting, crossing the BBB, constant or controlled release, and non-invasive application. [80–87]

7.4 CONCLUSION

Currently, HAART remains the key strategy for treatment of HIV-1/AIDS patients within the clinic. It can reduce HIV-1 to an undetectable level and make AIDS a chronic disease. Recently, broadly neutralized antibodies showed promising results but still have an extended path to transfer from bench to bedside. With the development of gene editing technologies, like ZFN, TALEN, and CRISPR/Cas9, more and newer work focuses on using these new strategies to eliminate the virus in patients. ZFN, with the dimensions of $\sim$1 kb, is simpler to deliver. Nevertheless, the limitation of the target site and high off-target effects make it difficult to be applied within the HIV-1/AIDS gene therapy field. TALEN is more flexible in DNA target design and has lower off target effects compared with ZFN. However, the time-consuming and expensive construction of the recognition site of TALEN for a DNA target hampers the event of this gene editing tool. For CRISPR/Cas9, with more convenient and efficient design of target sites, less laborious vector construction, and limited off target effects, it may be applied quickly in every research field, not only in HIV-1/AIDS therapy.

For different target DNA, it needs only a change of sgRNA to search out the foremost effective site. although the big size of SpCas9 decreases the efficiency of delivery, the optional SaCas9 will overcome this limitation to some extent. The CRISPR/Cas9 system is indeed a promising gene editing tool applied in gene therapy fields, however, the high target efficiency and effective delivery are essential for successful application in trial. Moreover, the low off-target effect and safety must be the prerequisites of consideration. The research concerning successful application of gene editing tools in vitro and mouse models to inhibit HIV-1 infection allows clinical trials to be true. ZFN had been utilized in clinical HIV-1/AIDS therapy and completed with promising results.

Disruption of CCR5 by ZFN in autologous CD4+ T cells provided long-term HIV-1 resistant when reinfusing these cells into patients. Another clinical test about ZFN gene-editing of CCR5 in HSPCs followed autologous engraftment were conducted in 2015. This phase I clinical trial study will run to 2019 to estimate the potential of CCR5-disrupted HSPCs in HIV-1 resistance in AIDS patients. The CRISPR/Cas9 mediated run was first conducted in 2016. The modified human T cells were reinfused into a person with metastatic non-small cell carcinoma, which is meant to own a promising result.

Therefore, gene editing of autologous HSPCs provides an option for resistance of HIV-1 infection within the future. For the successful editing of HSPCs by CRISPR/Cas9 in HIV-1/AIDS treatment, it's important to style and screen effective sgRNAs to scale back the off-target effects. The security and delivery efficiency of CRISPR/Cas9 to HSPCs also need consideration, since long run expression of Cas9/sgRNA may induce non-specific injury to the host genome and immune reaction. For the gene editing therapy in clinical treatment, the ethics of animal and human experimentation and therefore the rationale are always the first consideration before its application.

Recently, the birth of gene-edited babies in 2018 has aroused widespread criticism around the scientific fields. This is the time that the CCR5 gene edited by CRISPR-Cas9 human embryos were implanted into women to have HIV-1 resistant babies. This work would make a permanent change to the germ line, which might be passed on to the longer term generations. Obviously, this experiment didn't meet the moral guideline concerning germline and embryo editing. Additionally, it lacks the rationale to switch CCR5 in human embryos, since HAART can inhibit HIV-1 replication. Also, the experimental couple can have a healthy baby. For HIV-positive mothers, cesarean delivery can protect babies from HIV infection. Even an HIV-positive father would not have any risk to transmit to the babies. Moreover, a CCR5 edited baby cannot resist all HIV strains, since the virus can evolved to utilize CXCR4 as an alternative co-receptor. Finally, CRISPR/Cas9 technology has limitations in application, like off-target effects. The safety of heritable germline editing should be monitored and evaluated. The side effects of permanent editing of the CCR5 gene haven't been reported, yet some researches have showed that CCR5 deficiency increases the chance of symptomatic West Nile viral infection.

The major obstacle to cure HIV-1/AIDS is the existence of latent reservoirs. The "shock and kill" strategy is meant to clear the HIV-1 reservoirs. Latency reversing agents (LRAs) are utilized to reverse HIV-1 latency. However, the dearth of specificity, and therefore the heterogeneous and dynamic nature of those drugs, make this pharmacologic strategy less safe and inefficient. CRISPR-based shock strategy with the characteristic of sequence specificity has advantages over pharmacological strategies, but its potential drawbacks must be evaluated, like off-target effect and fewer efficient delivery methods.

The low metabolic activity in HIV-1 latently infected cells also inhibits the function of CRISPR/Cas9 reagents; however, the stimulation of specific cytokines and HIDAC inhibitors can enhance the CRISPR/Cas9 reagents mediated restricting of HIV-1 infection in latent cells. CRISPR/Cas9 technology could be a powerful gene editing tool and has been widely applied in experimental HIV-1/AIDS gene therapy researches.

Moreover, this technology also has great potential to be applied in various areas, like medical screening and gene ontology analysis. Its emergence brings hope for 36.9 million individuals with HIV-1 infection, but it's worth noting that the negative effects like off target and viral escape must be considered. Therefore, successful cure of HIV-1/AIDS still has a great distance to travel.

Clinical trials of CRISPR/Cas9 in HIV-1 treatment remains a challenge, and ethics should be put in first place. The restrictions and difficulties of this technology suggest that several aspects must be improved for future applications: (1) Exploration of recent vehicles to deliver CRISPR/Cas9 compound safely and effectively; (2) For activating the latent viral reservoirs by CRISPR/Cas9, several specific agents will be combined to boost immune responses to eliminate the virus; (3) Design specific sgRNAs and explore new strategies to decrease off-target effects; (4) Understanding the precise mechanism of the CRISPR/Cas9-induced effects is critical for the adoption of simpler strategies (5) and to focus on optimization of animal models of HIV1/AIDS. The mix of the CRISPR/Cas9 technology with other strategies may help overcoming these limitations, thus resulting in exciting and promising progress within the HIV1/AIDS field.

REFERENCES

1. Bowers, N. L., Helton, E. S., Huijbregts, R. P., Goepfert, P. A., Heath, S. L., and Hel, Z. Immune suppression by neutrophils in HIV-1 infection: role of PD-L1/PD-1 pathway. *PLoS Pathog* 2014; 10: e1003993.
2. Nyamweya, S., Hegedus, A., Jaye, A., Rowland-Jones, S., Flanagan, K. L., and Macallan, D. C. Comparing HIV-1 and HIV-2 infection: lessons for viral immune pathogenesis. *Rev Med Virol* 2013; 23: 221–240.
3. Cassan, E., Arigon-Chifolleau, A. M., Mesnard, J. M., Gross, A., and Gascuel, O. Concomitant emergence of the antisense protein gene of HIV-1 and of the pandemic. *Proc Natl Acad Sci U S A* 2016; 113: 11537–11542.
4. Hladik, F., and McElrath, M. J. Setting the stage: host invasion by HIV. *Nat Rev Immunol* 2008; 8: 447–457.
5. Cohen, M. S., Shaw, G. M., McMichael, A. J., and Haynes, B. F. Acute HIV-1 infection. *N Engl J Med* 2011; 364: 1943–1954.
6. Sharp, P. M., and Hahn, B. H. Origins of HIV and the AIDS pandemic. *Cold Spring Harb Perspect Med* 2011; 1: a006841.
7. Chun, T. W., Carruth, L., Finzi, D., Shen, X., DiGiuseppe, J. A., Taylor, H., et al. Quantification of latent tissue reservoirs and total body viral load in HIV-1 infection. *Nature* 1997; 387: 183–188.
8. Huang, J., Wang, F., Argyris, E., Chen, K., Liang, Z., Tian, H., et al. Cellular micro RNAs contribute to HIV-1 latency in resting primary CD4+ T lymphocytes. *Nat Med* 2007; 13: 1241–1247.
9. du Chene, I., Basyuk, E., Lin, Y. L., Triboulet, R., Knezevich, A., Chable-Bessia, C., et al. Suv39H1 and HP1gamma are responsible for chromatinmediated HIV-1 transcriptional silencing and post-integration latency. *EMBO J* 2007; 26: 424–435.
10. Gallastegui, E., Millan-Zambrano, G., Terme, J. M., Chavez, S., and Jordan, A. Chromatin reassembly factors are involved in transcriptional interference promoting HIV latency. *J Virol* 2011; 85: 3187–3202.
11. Barboric, M., Nissen, R. M., Kanazawa, S., Jabrane-Ferrat, N., and Peterlin, B. M. NF-kappaB binds P-TEFb to stimulate transcriptional elongation by RNA polymerase II. *Mol Cell* 2001; 8: 327–337.
12. Siliciano, J. D., Kajdas, J., Finzi, D., Quinn, T. C., Chadwick, K., Margolick, J. B. et al. Longterm follow-up studies confirm the stability of the latent reservoir for HIV-1 in resting CD4+ T cells. *Nat Med* 2003; 9: 727–728.
13. Churchill, M. J., Wesselingh, S. L., Cowley, D., Pardo, C. A., McArthur, J. C., Brew, B. J., et al. Extensive astrocyte infection is prominent in human immunodeficiency virusassociated dementia. *Ann Neurol* 2009; 66: 253–258.
14. Smith, P. D., Meng, G., Salazar-Gonzalez, J. F., and Shaw, G. M. Macrophage HIV-1 infection and the gastrointestinal tract reservoir. *J LeukoBiol* 2003; 74: 642–649.
15. Bagasra, O., Lavi, E., Bobroski, L., Khalili, K., Pestaner, J. P., Tawadros, R., et al. Cellular reservoirs of HIV-1 in the central nervous system of infected individuals: identification by the combination of in situ polymerase chain reaction and immunohistochemistry. *AIDS* 1996; 10: 573–585.
16. Gilbert, L. A., Horlbeck, M. A., Adamson, B., Villalta, J. E., Chen, Y., Whitehead, E. H., et al. Genome-scale CRISPR-mediated control of gene repression and activation. *Cell* 2014; 159: 647–661.
17. Konermann, S., Brigham, M. D., Trevino, A. E., Joung, J., Abudayyeh, O. O., Barcena, C., et al. Genome-scale transcriptional activation by an engineered CRISPR-Cas9 complex. *Nature* 2015; 517: 583–588.

18. Gootenberg, J. S., Abudayyeh, O. O., Lee, J. W., Essletzbichler, P., Dy, A. J., Joung, J., et al. Nucleic acid detection with CRISPR-Cas13a/C2c2. *Science* 2017; 356: 438–442.
19. Xiao, Q., Guo, D., and Chen, S. Application of CRISPR/Cas9-based gene editing in HIV-1/AIDS therapy. *Front Cell Infect Microbiol* 2019; 9: 69.
20. Hsu, P. D., Lander, E. S., and Zhang, F. Development and applications of CRISPR-Cas9 for genome engineering. *Cell* 2014; 157: 1262–1278.
21. Ebina, H., Misawa, N., Kanemura, Y., and Koyanagi, Y. Harnessing the CRISPR/Cas9 system to disrupt latent HIV-1 provirus. *Sci Rep* 2013; 3: 2510.
22. Shi, B., Li, J., Shi, X., Jia, W., Wen, Y., Hu, X., et al. TALEN-mediated knockout of CCR5 confers protection against infection of human immunodeficiency virus. *J Acquir Immune Defic Syndr* 2016; 74(2): 229–241.
23. Tebas, P., Stein, D., Tang, W. W., Frank, I., Wang, S. Q., Lee, G. et al. Gene editing of CCR5 in autologous CD4 T cells of persons infected with HIV. *N Engl J Med* 2014; 370: 901–910.
24. Maier, D. A., Brennan, A. L., Jiang, S., Binder-Scholl, G. K., Lee, G., Plesa, G., et al. Efficient clinical scale gene modification via zinc finger nuclease-targeted disruption of the HIV co-receptor CCR5. *Hum Gene Ther* 2013; 24: 245–258.
25. Strong, C. L., Guerra, H. P., Mathew, K. R., Roy, N., Simpson, L. R., and Schiller, M. R. Damaging the integrated HIV proviral DNA with TALENs. *PLoS One* 2015; 10: e0125652.
26. Cong, L., Ran, F. A., Cox, D., Lin, S., Barretto, R., Habib, N., et al. Multiplex genome engineering using CRISPR/Cas systems. *Science* 2013; 339: 819–823.
27. Mali, P., Yang, L., Esvelt, K. M., Aach, J., Guell, M., DiCarlo, J. E., et al. RNA-guided human genome engineering via Cas9. *Science* 2013; 339: 823–826.
28. Anders, C., Niewoehner, O., Duerst, A., and Jinek, M. Structural basis of PAM-dependenttarget DNA recognition by the Cas9 endonuclease. *Nature* 2014; 513: 569–573.
29. Jinek, M., Jiang, F., Taylor, D. W., Sternberg, S. H., Kaya, E., Ma, E., et al. Structures of Cas9 endonucleases reveal RNA-mediated conformational activation. *Science* 2014; 343: 1247997.
30. Nishimasu, H., Ran, F. A., Hsu, P. D., Konermann, S., Shehata, S. I., Dohmae, N., et al. Crystal structure of Cas9 in complex with guide RNA and target DNA. *Cell* 2014; 156: 935–949.
31. Ran, F. A., Hsu, P. D., Wright, J., Agarwala, V., Scott, D. A., and Zhang, F. Genome engineering using the CRISPR-Cas9 system. *Nat Protoc* 2013; 8: 2281–2308.
32. Chougui, G., Munir-Matloob, S., Matkovic, R., Martin, M. M., Morel, M., Lahouassa, H., et al. HIV-2/SIV viral protein X counteracts HUSH repressor complex. *Nat Microbiol* 2018; 3: 891–897.
33. Archin, N. M., Liberty, A. L., Kashuba, A. D., Choudhary, S. K., Kuruc, J. D., Crooks, A. M., et al. Administration of vorinostat disrupts HIV-1 latency in patients on antiretroviral therapy. *Nature* 2012; 487: 482–485.
34. Hsu, P. D., Lander, E. S., and Zhang, F. Development and applications of CRISPR-Cas9 for genome engineering. *Cell* 2014; 157: 1262–1278.
35. Duan, J., Lu, G., Xie, Z., Lou, M., Luo, J., Guo, L., et al. Genome-wide identification of CRISPR/Cas9 off-targets in human genome. *Cell Res* 2014; 24: 1009–1012.
36. Hu, W., Kaminski, R., Yang, F., Zhang, Y., Cosentino, L., Li, F., et al. RNA-directed gene editing specifically eradicates latent and prevents new HIV-1 infection. *Proc Natl Acad Sci U S A* 2014; 111: 11461–11466.
37. Liao, H. K., Gu, Y., Diaz, A., Marlett, J., Takahashi, Y., Li, M., et al. Use of the CRISPR/Cas9 system as an intracellular defense against HIV-1 infection in human cells. *Nat Commun* 2015; 6: 6413.
38. Zhu, W., Lei, R., Le Duff, Y., Li, J., Guo, F., Wainberg, M. A., et al. The CRISPR/Cas9 system inactivates latent HIV-1 proviral DNA. *Retrovirology* 2015; 12: 22.
39. Wang, G., Zhao, N., Berkhout, B., and Das, A. T. CRISPRCas based antiviral strategies against HIV-1. *Virus Res* 2018; 244: 321–332.
40. Lebbink, R. J., de Jong, D. C., Wolters, F., Kruse, E. M., van Ham, P. M., Wiertz, E. J., et al. A combinational CRISPR/Cas9 gene-editing approach can halt HIV replication and prevent viral escape. *Sci Rep* 2017; 7: 41968.
41. Yin, H., Song, C. Q., Dorkin, J. R., Zhu, L. J., Li, Y., Wu, Q., et al. Therapeutic genome editing by combined viral and non-viral delivery of CRISPR system components in vivo. *Nat Biotechnol* 2016; 34: 328–333.

42. Cho, S. W., Kim, S., Kim, J. M., and Kim, J. S. Targeted genome engineeringin human cells with the Cas9 RNA-guided endonuclease. *Nat Biotechnol* 2013; 31: 230–232.
43. Ye, L., Wang, J., Beyer, A. I., Teque, F., Cradick, T. J., Qi, Z., et al. Seamless modification of wild-type induced pluripotent stemcells to the natural CCR5Delta32 mutation confers resistance to HIV infection. *Proc Natl Acad Sci U S A* 2014; 111: 9591–9596.
44. Li, C., Guan, X., Du, T., Jin, W., Wu, B., Liu, Y., et al. Inhibition of HIV-1 infection of primary CD4+ T-cells by gene editing of CCR5 using adenovirus-delivered CRISPR/Cas9. *J Gen Virol* 2015; 96: 2381–2393.
45. Xu, L., Yang, H., Gao, Y., Chen, Z., Xie, L., Liu, Y., et al. CRISPR/Cas9- Mediated CCR5 ablation in human hematopoietic stem/progenitor cells confers HIV-1 resistance in vivo. *Mol Ther* 2017; 25: 1782–1789.
46. Hou, P., Chen, S., Wang, S., Yu, X., Chen, Y., Jiang, M., et al. Genomeediting of CXCR4 by CRISPR/cas9 confers cells resistant to HIV-1 infection. *Sci Rep* 2015; 5: 15577.
47. Kim, S., Koo, T., Jee, H. G., Cho, H. Y., Lee, G., Lim, D. G., et al. CRISPR RNAs trigger innate immune responses in human cells. *Genome Res* 2018; 28: 367–373.
48. Rosa, A., Chande, A., Ziglio, S., De Sanctis, V., Bertorelli, R., Goh, S. L., et al. HIV-1 Nef promotes infection by excluding SERINC5 from virion incorporation. *Nature* 2015; 526: 212–217.
49. Lahaye, X., Gentili, M., Silvin, A., Conrad, C., Picard, L., Jouve, M., et al. NONO detects the nuclear HIV capsid to promote cGAS-mediated innate immune activation. *Cell* 2018; 175: 488–501.
50. Yurkovetskiy, L., Guney, M. H., Kim, K., Goh, S. L., McCauley, S., Dauphin, A., et al. Primate immunodeficiency virus proteins Vpx and Vpr counteract transcriptional repression of proviruses by the HUSH complex. *Nat Microbiol* 2018; 3: 1354–1361.
51. Sarkar, I., Hauber, I., Hauber, J., and Buchholz, F. HIV-1 proviral DNA excision using an evolved recombinase. *Science* 2007; 316: 1912–1915.
52. Qu, X., Wang, P., Ding, D., Li, L., Wang, H., Ma, L., et al. Zinc-fingernucleases mediate specific and efficient excision of HIV-1 proviral DNA from iinfected and latently infected human T cells. *Nucleic Acids Res* 2013; 41: 7771–7782.
53. Cocchi, F., DeVico, A. L., Garzino-Demo, A., Arya, S. K., Gallo, R. C., and Lusso, P. Identification of RANTES, MIP-1 alpha, and MIP-1 beta as the major HIV-suppressive factors produced by CD8+ T cells. *Science* 1995; 270: 1811–1815.
54. Samson, M., Libert, F., Doranz, B. J., Rucker, J., Liesnard, C., Farber, C. M., et al. Resistance to HIV-1 infection in caucasian individuals bearing mutant alleles of the CCR-5 chemokine receptor gene. *Nature* 1996; 382: 722–725.
55. Hutter, G., Nowak, D., Mossner, M., Ganepola, S., Mussig, A., Allers, K., et al. Long-term control of HIV by CCR5 Delta32/Delta32 stem-cell transplantation. *N Engl J Med* 2009; 360: 692–698.
56. Perez, E. E., Wang, J., Miller, J. C., Jouvenot, Y., Kim, K. A., Liu, O., et al. Establishment of HIV-1 resistance in CD4+ T cells by genome editing using zinc-finger nucleases. *Nat Biotechnol* 2008; 26: 808–816.
57. Wilen, C. B., Wang, J., Tilton, J. C., Miller, J. C., Kim, K. A., Rebar, E. J., et al. Engineering HIV-resistant human CD4+ T cells with CXCR4-specific zincinger nucleases. *PLoSPathog* 2011; 7: e100202.
58. Didigu, C. A., Wilen, C. B., Wang, J., Duong, J., Secreto, A. J., Danet-Desnoyers, G. A., et al. Simultaneous zinc-finger nuclease editing of the HIV coreceptors ccr5 and cxcr4 protects CD4+ T cells from HIV-1 infection. *Blood* 2014; 123: 61–69.
59. Schumann, K., Lin, S., Boyer, E., Simeonov, D. R., Subramaniam, M., Gate, R. E., et al. Generation of knock-in primary human T cells using Cas9 ribonucleoproteins. *Proc Natl Acad Sci U S A* 2015; 112: 10437–10442.
60. Wang, Q., Chen, S., Xiao, Q., Liu, Z., Liu, S., Hou, P., et al. Genome modification of CXCR4 by Staphylococcus aureus Cas9 renders cells resistance to HIV-1 infection. *Retrovirology* 2017; 14: 51.
61. Nagasawa, T., Nakajima, T., Tachibana, K., Iizasa, H., Bleul, C. C., Yoshie, O., et al. Molecular cloning and characterization of a murine pre-B-cell growthstimulating factor/stromal cell-derived factor 1 receptor, a murine homolog of the human immunodeficiency virus 1 entry coreceptor fusin. *Proc Natl Acad Sci U S A* 1996; 93: 14726–14729.
62. Chung, S. H., Seki, K., Choi, B. I., Kimura, K. B., Ito, A., Fujikado, N., et al. CXC chemokine receptor 4 expressed in T cells plays an important roin the development of collagen-induced arthritis. *Arthritis Res Ther* 2010; 12: R188.

63. Liu, Q., Lai, Y. T., Zhang, P., Louder, M. K., Pegu, A., Rawi, R., et al. Improvement of antibody functionality by structure-guided paratope engraftment. *Nat Commun* 2019; 10: 721.
64. Manson McManamy, M. E., Hakre, S., Verdin, E. M., and Margolis, D. M. Therapy for latent HIV-1 infection: the role of histone deacetylase inhibitors. *Antivir Chem Chemother* 2014; 23: 145–149.
65. Bullen, C. K., Laird, G. M., Durand, C. M., Siliciano, J. D., and Siliciano, R. F. New ex vivo approaches distinguish effective and ineffective single agents for reversing HIV-1 latency in vivo. *Nat Med* 2014; 20: 425–429.
66. Rasmussen, T. A., Tolstrup, M., Brinkmann, C. R., Olesen, R., Erikstrup, C., Solomon, A., et al. Panobinostat, a histone deacetylase inhibitor, for latent-virus reactivation in HIV-infected patients on suppressive antiretroviral therapy: a phase 1/2, single group, clinical trial. *Lancet HIV* 2014; 1: e13–21.
67. Zhang, Y., Yin, C., Zhang, T., Li, F., Yang, W., Kaminski, R., et al. CRISPR/gRNA-directed synergistic activation mediator (SAM) induces specific, persistent and robust reactivation of the HIV-1 latent reservoirs. *Sci Rep* 2015; 5: 16277.
68. Saayman, S. M., Lazar, D. C., Scott, T. A., Hart, J. R., Takahashi, M., Burnett, J. C., et al. Potent and targeted activation of latent HIV-1 using the CRISPR/dCas9 activator complex. *Mol Ther* 2016; 24: 488–498.
69. Limsirichai, P., Gaj, T., and Schaffer, D. V. CRISPR-mediated activation of latent HIV-1 expression. *Mol Ther* 2016; 24: 499–507.
70. Bogerd, H. P., Kornepati, A. V., Marshall, J. B., Kennedy, E. M., and Cullen, B. R. Specific induction of endogenous viral restriction factors using CRISPR/Cas-derived transcriptional activators. *Proc Natl Acad Sci U S A* 2015; 112: E7249–56.
71. Wang, F. X., Huang, J., Zhang, H., Ma, X., and Zhang, H. APOBEC3G upregulation by alpha interferon restricts human immunodeficiency virus type 1 infection in human peripheral plasmacytoid dendritic cells. *J Gen Virol* 2008; 89(Pt 3): 722–730.
72. Kaminski, R., Bella, R., Yin, C., Otte, J., Ferrante, P., Gendelman, H. E., et al. Excision of HIV-1 DNA by gene editing: a proof-of-concept in vivo study. *Gene Ther* 2016; 23: 696.
73. Wang, G., Zhao, N., Berkhout, B., and Das, A. T. CRISPR-Cas9 can inhibit HIV-1 replication but NHEJ repair facilitates virus escape. *Mol Ther* 2016; 24: 522–526.
74. Zuckermann, M., Hovestadt, V., Knobbe-Thomsen, C. B., Zapatka, M., Northcott, P. A., Schramm, K., et al. Somatic CRISPR/Cas9-mediated tumour suppressor disruption enables versatile brain tumour modelling. *Nat Commun* 2015; 6: 7391.
75. Wang, L., Hao, Y., Li, H., Zhao, Y., Meng, D., Li, D., et al. Co-delivery of doxorubicin and siRNA for glioma therapy by a brain targeting system: angiopep-2-modified poly (lactic-co-glycolic acid) nanoparticles. *J Drug Target* 2015; 23: 832–846.
76. Wang, M., Zuris, J. A., Meng, F., Rees, H., Sun, S., Deng, P., et al. Efficient delivery of genome-editing proteins using bioreducible lipid nanoparticles. *Proc Natl Acad Sci U S A* 2016; 113: 2868–2873.
77. Qazi, S., Miettinen, H. M., Wilkinson, R. A., McCoy, K., Douglas, T., and Wiedenheft, B. Programmed self-assembly of an active P22-Cas9 nanocarrier system. *Mol Pharm* 2016; 13: 1191–1196.
78. Sun, W., Ji, W., Hall, J. M., Hu, Q., Wang, C., Beisel, C. L., et al. Self-assembled DNA nanoclews for the efficient delivery of CRISPR-Cas9 for genome editing. *Angew Chem Int Ed Engl* 2015; 54: 12029–12033.
79. Yin, H., Song, C. Q., Dorkin, J. R., Zhu, L. J., Li, Y., Wu, Q., et al. Therapeutic genome editing by combined viral and non-viral delivery of CRISPR system components in vivo. *Nat Biotechnol* 2016; 34: 328–333.
80. Clark, A. J., and Davis, M. E. Increased brain uptake of targeted nanoparticles by adding an acid-cleavable linkage between transferrin and the nanoparticle core. *Proc Natl Acad Sci U S A* 2015; 112: 12486–12491.
81. Ding, H., Sagar, V., Agudelo, M., Pilakka-Kanthikeel, S., Atluri, V. S., Raymond, A., et al. Enhanced blood-brain barrier transmigration using a novel transferrin embedded fluorescent magneto-liposome nanoformulation. *Nanotechnology* 2014; 25: 055101.
82. De Luca, M. A., Lai, F., Corrias, F., Caboni, P., Bimpisidis, Z., Maccioni, E., et al. Lactoferrin and antitransferrin-modified liposomes for brain targeting of the NK3 receptor agonist senktide: preparation and in vivo evaluation. *Int J Pharm* 2015; 479: 129–137.
83. Tomitaka, A., Arami, H., Gandhi, S., and Krishnan, K. M. Lactoferrin conjugated iron oxide nanoparticles for targeting brain glioma cells in magnetic particle imaging. *Nanoscale* 2015; 7: 16890–16898.

84. Fornaguera, C., Dols-Perez, A., Caldero, G., Garcia-Celma, M. J., Camarasa, J., and Solans, C. PLGA nanoparticles prepared by nano-emulsion templating using low-energy methods as efficient nanocarriers for drug delivery across the blood-brain barrier. *J Control Release* 2015; 211: 134–143.
85. Huang, Y., Zhang, B., Xie, S., Yang, B., Xu, Q., and Tan, J. et al. Modified with Tween 80 pass through the intact blood-brain barrier in rats under magnetic field. *ACS Appl Mater Interfaces* 2016; 8: 11336–11341.
86. Kaushik, A., Jayant, R. D., Nikkhah-Moshaie, R., Bhardwaj, V., Roy, U., Huang, Z., et al. Magnetically guided central nervous system delivery and toxicity evaluation of magneto-electric nanocarriers. *Sci Rep* 2016; 6: 25309.
87. Nair, M., Guduru, R., Liang, P., Hong, J., Sagar, V., and Khizroev, S. Externally controlled on-demand release of anti-HIV drug using magneto-electric nanoparticles as carriers. *Nat Commun* 2013; 4: 1707.

8 siRNA Delivery for Therapeutic Applications Using Nanoparticles

Sonia Barua
Chung-Ang University, Seoul, South Korea

Yashwant Pathak
University of South Florida, Tampa, USA
Airlangga University, Surabaya, Indonesia

CONTENTS

8.1 INTRODUCTION

RNA interference is a process of RNA degradation in which RNA molecules induce degradation of corresponding messenger RNAs (mRNA), thus blocking the expression of protein synthesis. This process is also called RNAi [1]. RNAi is initiated by exposing cells to long dsRNA via transfection or endogenous expression. dsRNAs proceed into smaller fragments of (usually 21–23 nucleotide) of small interfering RNAs, which further form a complex with the RNA induced silencing complexes [2]. Introduction of siRNA into mammalian cells causes downregulation of target genes without triggering interferon responses [3]. Moreover, siRNA based therapeutic application has shown a great potential to combat diseases such as various cancers, viral infections, and genetic disorders [3].

DOI: 10.1201/9781003186083-8

Over the past decades, a tremendous effort has been made toward efficient delivery of siRNA in the biological systems [1, 2]. The major obstacle of siRNA application is its polar nature. They don't readily pass through the cell membrane, due to their size and negative charges. Several approaches have been implemented to enhance the efficiency of siRNA to reach the desired site. Among the several strategies cationic liposomes-based carrier systems are considered as a promising tool in cellular delivery of siRNA [4]. However, many problems associated with lipid based siRNA delivery, such as rapid clearance by the liver, lack of target tissue specificity, instability, aggregation, and precipitate out of siRNA. Nonviral gene delivery systems are highly attractive in gene therapy because they are relatively safe and easier to produce than the viral vectors. Various side effects are arising in viral carriers, such as insertional mutagenesis and immunogenicity. Currently, nanotechnology has made significant advances in siRNA based formulation with efficient therapeutic activity. Different kinds of nanoparticle based siRNA delivery are extensively studied, such as lipid complexes, conjugated polymers, cationic polymers, and inorganic nanoparticles [4].

8.2 MECHANISM OF GENE SILENCING

The gene silencing mechanism is initiated by the RNAi process in which the enzyme dicer cleaves the double stranded RNAs into short double-stranded siRNAs of 21 to 25 nt [5]. The siRNA passenger strand is then unwound, and the guide strand of siRNA is loaded into the RNA-induced silencing (RISC) complex paired with the mRNA complementary sequence, causing cleavage of target mRNAs by Argonaute 2 (Ago2) (Figure 8.1). This important mechanism has allowed to open novel therapeutic approaches by designing oligonucleotide molecules through using mRNA transcripts sequences found in the existing human genomic data. Therefore, a careful sequence selection and synthesis of tailored siRNAs may have enormous repercussions in therapy, as almost all genes might be down-regulated, while splice variants, separate transcripts, or mutations might also be specifically targeted. As a consequence, this powerful approach might help circumvent the limitations exhibited by small molecule drugs in conventional cancer therapy treatments, leading to drug development processes based on gene functionality. Therefore, the development of this therapeutic strategy may have a high impact on modern medicine [5–7].

Recently, the potential application of RNA interference (RNAi) has drawn central attention in siRNA delivery. RNAi is the term given to the ability of a double- RNA-induced silencing (RISC) complex stranded RNA (dsRNA) containing a homologous sequence to a specific gene, leading to sequence-specific gene silencing [8]. RNAi is an endogenous post-transcriptional regulation

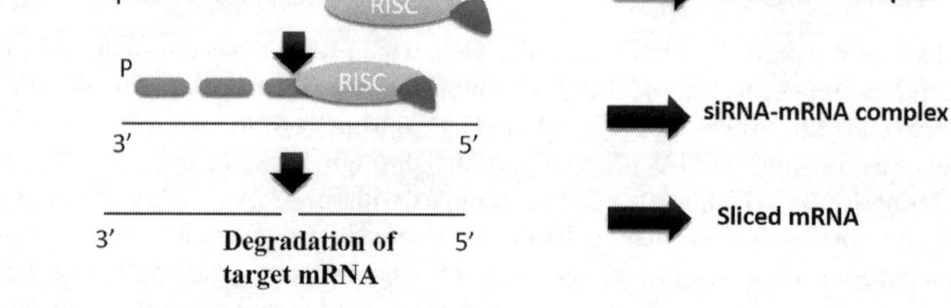

FIGURE 8.1 Mechanism of siRNA gene silencing.

process that consists of small regulatory RNAs, including microRNAs (miRNAs) or small interfering RNAs (siRNAs) that are able to silence target messenger RNAs (mRNAs) in a sequence-specific procedure. After the discovery of RNAi in *Caenorhabditis elegans* and subsequent demonstration of siRNA activity in mammalian cells, RNAi has received considerable attention as an effective therapy for multiple diseases like cancer and viral infections, particularly for those diseases with "undruggable" molecular targets [9, 10].

8.3 NANOPARTICLES IN siRNA DELIVERY

However various challenges have been reported, such as cellular uptake, long-term stabilities, and off-target effects. These issues have been observed in vitro and in vivo after testing in animal models. To overcome these limitations, nanoparticles have received much attention to stabilize the siRNA based therapeutic approaches. The current trends of siRNA based nanoparticles under clinical investigations are enlisted in Table 8.1.

8.4 siRNA CONJUGATION WITH PEPTIDES OR POLYMERS (LESS THAN 10 NM IN SIZE)

Conjugation of small molecules, such as peptides or polymers, with the sense strands of siRNA produces the smallest siRNA nanoparticles at 10 nm size [11]. Modification of sense strands of siRNA maintain the degradation effect on mRNA because the recognition of mRNA involves the antisense strand. Studies show that siRNA conjugation with CPPs and PEG has increased gene transfer in

TABLE 8.1
List of siRNA Based Nanoparticle Formulations in Clinical Trial

Drugs	Targeted Delivery	Vehicles	Diseases	Clinical Trial	Company
siRNA-EphA2-DOPC	EphA2	LNP	Late stage cancer	I	M.D. Anderson Cancer Center
Lumasiranz	Primary hyperoxaluria type (PH1)	GAlNAc-siRNA conjugate	HAO1	III	Alnylam pharmaceutical
TKM-080301	PLK1	LNP	Multiple Cancers	I	National Cancer Institute (NCI)
CALAA-01	RRM2	Cyclodextrin NP	Cancer Solid Tumor	I	Calando Pharmaceuticals
RXi-109	CTGF	Self-delivering RNAi compound	Cicatrix scar prevention	I	RXi Pharmaceuticals
DCR-MYC	Lipid nanoparticles	Myc	Hepatocellular carcinoma	I	Dicerna pharmaceuticals
Atu027	siRNA with 2′-O-Me and Cationic lipid	Protein Kinase N3 (PKN3)	Advanced solid tumors (metastatic pancreatic cancer)	I	Silence Therapeutics GmbH
AMG890	Cardiovascular disease	GALNac-siRNA conjugate	LP	I	Amgen
siG12D-LODER	LODER® (Polymer)	KRAS (mutation G12D in KRAS oncogene)	Solid tumors (advanced pancreatic cancer)	II	Silenseed Ltd
Onpattro	TTR mediated amyloidosis		GAL-siRNA conjugates	III	Alnylam Pharmaceuticals

vivo. siRNA conjugated to cholesterol in sense strand are able to slice multiple genes in mice, including endogenous apolipoprotein B gene expression in the liver and jejunum and p38 mitogen-activated protein (MAP) kinase in the lungs [12]. In vivo gene silencing to hepatocytes was observed by siRNA modified with Long chain fatty acids (> C_{18}) and bile-salt derivatives. A similar effect was also produced by siRNA based dynamic polyconjugate, consisting of acid responsive polymers that contain PEG and a NAG targeting ligand. When these nanoparticles interact with the endosome, PEG is released due to acid responsive polymer targeting the ligand, making a hydrophilic to hydrophobic transition, and resulting in endosomal disruption [13, 14, 15].

8.5 POLYETHYLENE AMINE AND CATIONIC BASED PEPTIDES AND PROTEINS (100 TO 300 NM IN SIZE)

The degradation of siRNA could be further prevented by encapsulating it into the cationic polymer, such as poly (ethylene imine) PEI. This type of carrier system enhances the circulation time and targets specific tissues or cells. The positive charge of the polymer strongly binds with the negatively charged phosphate groups of siRNA, leading to formation of nanoparticles. Studies show that chemically modified PEI (e.g., PEI-PEG) and other polymers, such as cyclodextrin-containing polycations, polylysine, and natural polymers such as chitosan, have proven the efficient delivery of siRNA to the targeted sites. However, cytotoxic effects of cationic polymers could be a major challenge to delivery of siRNA. Other large molecules, peptides, and proteins have drawn much attention to complexes with siRNA due to their less cytotoxic effect. Data reported that cationic peptide CADY and MPG-8 (a variant of MPG) can strongly bind with the siRNA. In addition, a CPP modified protein has strong affinity to the siRNA molecules, therefore are able to bind a strong complex, shield it from degradation, and efficiently deliver siRNA to various cell lines in vivo [14, 15].

8.6 CATIONIC BASED LIPID NANOPARTICLES (100 TO 300 NM IN SIZE)

Several lipid-based transfection agents are commercially available and are being used in vivo to study the mechanism and effects of siRNA delivery in mammalian cells. These include jetSi-ENDO, Lipofectamine RNAiMAX, siPORT NeoFX, DharmaFECT, X-tremeGENE, and TriFECTin. DOTAP, 1,2-dioleoyl-3-trimethylammonium-propane, and their in vivo siRNA delivery has been examined. Recently, a combinatorial library has been implemented for one by one screening and synthesis of cationic lipids for siRNA delivery. Lipidoid was developed for the efficient delivery of siRNA to the cells. It known as a combinatorial library of 1200 lipid like materials, which synthesize to primary or secondary amine with conjugation with alkyl-acrylates or alkyl-acrylamides. A number of materials were identified to substantially improve silencing over delivery of naked siRNA in vitro and in vivo in mice, rats, and nonhuman primates [15].

8.7 NEUTRAL LIPOSOMES (<200 NM IN SIZE)

siRNA based neutral liposomes are widely utilized to circumvent the potential toxicities of cationic polymers and lipids. They efficiently shield the siRNA from degradation, due to their hydrophobic surfaces and hydrophilic cores preventing the interaction with surrounding RNAses and facilitating internalization via membrane fusion or (receptor-mediated) endocytosis. Studies demonstrated that encapsulating siRNA between a cationic core composed of DOTAP and an outer lipid bilayer of 1,2-distearoyl-sn-glycero-3-phosphoethanolamine-N-(polyethylene glycol-2000) (PEG-DSPE) and egg phosphatidylcholine like sandwiches prolongs the circulation time after 20 hours of injection. Stable nucleic acid lipid particles, SNALPs, represented a major advancement in lipid-based siRNA delivery after injection of clinically relevant doses of ApoB-siRNA produced knockdown in the liver of non-human primates [13–15].

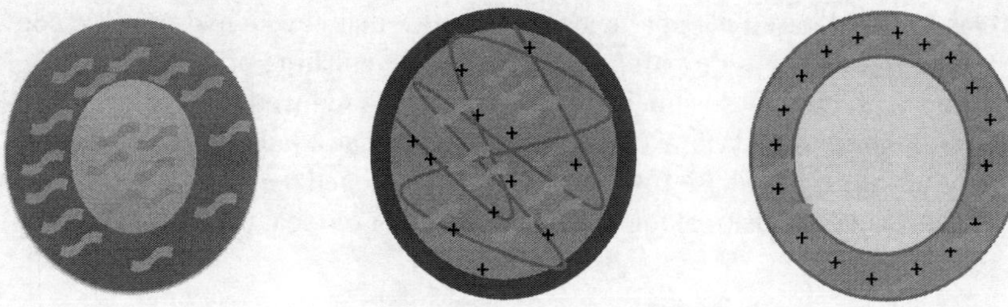

FIGURE 8.2 Different types of nanoparticles. (a) siRNA based liposomes (b) siRNA based polymeric nanoparticles (c) siRNA based metallic nanoparticles.

8.8 LIPOSOMES AND LIPOPLEXES

Liposomes are phospholipid bilayers consisting of a hydrophilic core to encapsulate hydrophilic agents surrounded by hydrophobic surfaces (Figure 8.2a). This surface possesses good permeation effect to the biological membrane. The size and shape of the liposomes can be adjusted to enhance their effect, prolong blood circulation and high loading of active molecules. Due to their non-toxic, biocompatible, and non-immunogenic properties, liposomal carriers have attained a major focus in siRNA delivery. Polyethylene glycol (PEG)-coated liposomes, also called "Stealth" liposomes, are currently widely used in the delivery of therapeutic molecules because the polymer coated surface of the molecules provides better stability with increasing residence time in the circulation and reduces the uptake by the reticuloendothelial system (RES) and phagocytosis, which can decrease the quantity of active agents delivered to the target site. Several PEGylated liposomes have been recently investigated, such as Doxil®, DaunoXome®, DepoCyt®, and ONCO-TCS®.

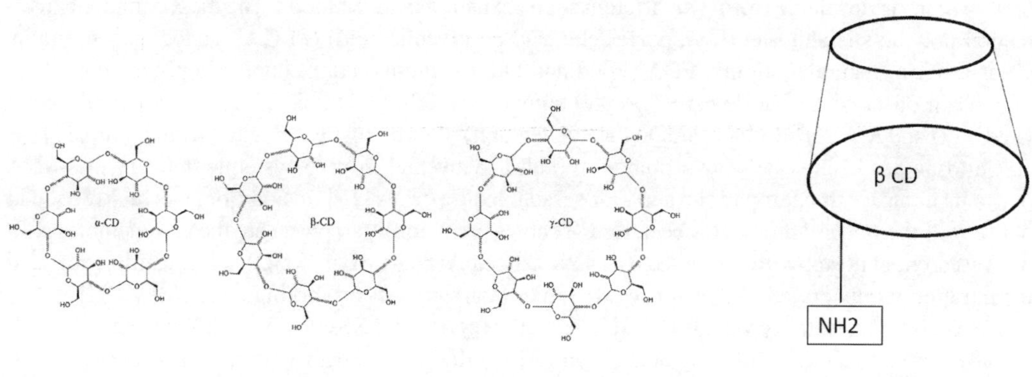

FIGURE 8.3 (a) Chemical structure of cyclodextrins (b) overview of β cyclodextrin.

Polycationic liposomes are considered as a major vehicle for siRNA delivery because of their ability to entrap negatively charged siRNA through electrostatic interactions and form lipid complexes systems called lipoplexes. They synthesize by coating the liposomes with oppositely charged polymers. So far, many liposomal formulations have received FDA approval, and many more are currently in development or in clinical trials.

Focusing on NSCLC, lipid-based nanoparticles play an important role in reducing the side effects of Cisplatin and improve the efficiency of the therapy. A randomized Phase III study on Lipoplatin in the treatment of NSCLC compared the results, responses, and toxicities of Lipoplatin + Paclitaxel versus Cisplatin + Paclitaxel used as a first-line treatment. Results of that study showed that there is an increase in tumor response rate in the Lipoplatin group (59.22%) compared to the Cisplatin group (42.42%), in addition to reducing the side effects of Cisplatin.

Another class of lipid-based nanoparticles are micelles that have a hydrophobic core filled with therapeutic agents and a PEG hydrophilic shell. Their distinguishing properties involve long bloodstream circulation, high binding specificity to target cells, and reduced side effects. Micelle formulations containing Doxorubicin, Paclitaxel, SN-38, Cisplatin, and Platinum II are undergoing clinical trials, with some advancing to Phase II studies. They have proved their effectiveness against various tumors and reduced side effects, making them promising for clinical use [13–15].

8.9 OTHER NANOPARTICLES

While cationic polymers, lipids, and liposomes have been shown to efficiently deliver siRNA, researchers are still looking to develop new types of nanoparticles to overcome the numerous limitations posed by the delivery of siRNA. Hyaluronic acid (HA) nanoparticles, also known as nanogels, with encapsulated siRNA were able to target HCT-116 cells in vitro, which over-express HA-specific CD44 receptors. Depending on the engineered nanogel characteristics, the siRNA release rate was also regulated. siRNA has also been incorporated in poly(D,L-lactic-co-glycolic) acid (PLGA) and calcium carbonate nanoparticles [15].

8.10 POLYPLEXES

The advantages of polymeric nanocarriers are that they are highly biocompatible, biodegradable, non-toxic in nature, and have outstanding controlled release character. Therefore, they offer a great platform for efficient delivery of siRNA. Currently, several natural and synthetic polymeric nanocarriers (PNCs) with siRNA that have been developed offer a great platform to deliver drugs and genes with improved efficacy (Figure 8.2b). The polycationic polymer has emerged as one of the most promising candidates for developing an efficient gene delivery vector [16]

Polymers are available from two different sources: natural and synthetic. The most commonly used natural polymers is chitosan (CS), poly(lactic acid-co-glycolic acid) (PLGA), atelocollagen, inulin. CS is a linear, natural, cationic, FDA (Food and Drug Administration) approved polysaccharide (a long chain of monosaccharide carbohydrate) composed of β-(1→4)-linked d-glucosamine (deacetylated unit) and *N*-acetyl-d-glucosamine (acetylated unit) after being deacetylated from chitin.[17].

Chitosan has been widely investigated in both in-vitro and in-vivo systemic delivery of siRNA due to its nontoxic, biocompatible, and biodegradable properties [17]. In addition, it is highly used in delivery of therapeutic molecules because CS shows great ability to permeate the cell membrane. In one research, chitosan with amino acid-functionalized Arg-Gly-Asp (RGD) significantly enhanced the intratumor delivery of siRNA for regulation of many growth-promoting genes (PLXDC1, FAK, and POSTN). This activity was observed in A2780, HeyA8, and SKOV3ip1 orthotopic animal studies of ovarian cancer. Besides natural polymers, synthetic polymers with nontoxic and biocompatible properties, such as PLL, PEI, PEG, dimethylamino ethyl methacrylate, polyfluorene, and cyclodextrin-based polycations, are incorporated in siRNA formulations. These polymers are linear and branch and efficient to shield and deliver the siRNA to the targeted sites. They basically transport the active agents by a passive route. For increasing the delivery concentration of siRNA on specific infected cells/tissues by escaping the normal cells and tissues, different actively targeted molecules are attached to the surfaces of PNCs [17–20]

8.11 NANOMICELLES

Micelles are formed by self-assembly of surfactants, lipids, and aqueous insoluble polymers, and they also term as colloidal suspension. Amphiphilic molecules with hydrophobic tail and hydrophilic head having self-assembly properties create micelles upon contact with the aqueous solutions. The structure of micelles consists of an inner side containing a hydrophobic core letting the polar side remain on the outer side, providing a hollow spherical or cylindrical shape in the aqueous medium [21–23]. Micelles accompanying anti-apoptotic Bcl-2 specific siRNA and an anti-cancer drug, docetaxel (DOX), showed efficient delivery of the drug and siRNA to yield a combined RNAi-chemotherapeutic benefit against hepatic cancer. Folic acid was conjugated with the surface of the micelles for a targeting ligand of hepatic cancer cells. PEG-pp-PEI-PE nanomicelles with siRNA and paclitaxel exhibits down regulation of GFP gene and survivin compared to free drug in a lung cancer cell line. In another study, siRNA loaded in PEGylated PEI-siRNA micellar nanoparticles shows high entrapment efficiency and long-term blood circulation of siRNA with good stability. The addition of polycationic PEI was used for high loading capacity of negatively charged siRNA inside the nanoparticles.[23]

8.12 CARBON-BASED NANOMATERIALS

These are synthesized from allotropes of carbon such as graphene, graphene oxide (GO), nanotubes, and fullerene. Studies show that carbon nanotubes complex with PEI and pyridinium exhibits 10–30% of silencing efficiency and 10–60% of cytotoxicity of siRNA. Studies have shown that carbon nanotube adjunct with PLL and Arg-Gly-Asp-Ser oligonucleotides actively target the tumors and control release of VEGF-siRNA, reducing the toxicity effect. These are hollow one-dimension nanostructures can be cylindrical, tubular, or needle in shape. Three types of Carbon nanotubes are discovered: single-walled carbon nanotubes with a thickness of 0.4–1 nm, double-walled carbon nanotubes with thickness of 1.4–20 nm, and multi-walled carbon nanotubes with a thickness of 20–100 nm [23].

C_{60} is the most commonly used fullerene, which contains 20 hexagonal rings and 12 pentagonal rings and is called as Buckyball. An internal view of the ball looks hollow, with an average diameter of about 100 nm [24, 25].

Pristine CNMs are less biocompatible and have less siRNA loading capacity due to electrostatic repulsion between the nanomaterials and drugs. However, conjugation or coating of Pristine CNMs with cationic polymers, such as PEG, PEI, chitosan, PAH (Poly (allylamine hydrochloride)), PLGA (Poly (lactic acid-co-glycolic acid)), is able to enhance the siRNA loading efficiency into carbon nanomembrane. Addition of ligands in pristine CNMs also improve the targeted delivery of siRNA to the cells [25–30].

8.13 DENDRIMERS

Dendrimers are highly branched, spherical shaped macromolecules with nanometer-scale dimensions. They mainly consist of three components: a central core, an interior dendritic structure (the branches), and an exterior surface with functional surface groups [31]. Because dendrimers contains numerous functional groups, various ligands can attach with it and facilitate the targeted delivery of therapeutic molecules [32, 33]. The main characteristics of dendrimers include extremely high stability, water solubility, and decreased antigenicity. Therefore, they can be used for many applications, including active and passive targeted drug delivery in modified chemotherapy treatments and gene delivery involving gene silencing techniques. Different types of dendrimers are available depending on its central core structure, such as PAMAM (Polyamidoamine), PPI (Poly (propylene imine), PLL, and poly(2,2-bis(hydroxymethyl)propionic acid. Large amount of drugs or genes can be loaded into this carrier, due to the presence of several voids/spaces deriving from branch chain polymers. Dendrimers with amine terminated cationic groups is considered as a good source of delivery of negatively charged siRNA therapeutics. Currently, the siRNA/paclitaxel based polypeptides-amphiphilic dendrimer has been developed, which is sensitive to the tumor microenvironment. This complex showed efficient regulation of VEGF mRNA synthesis in A375 xenograft mice and dual actions such as antitumor and antigenic effects in HT-1080 xenograft model [34].

8.14 METAL BASED NANOPARTICLES

Metal based nanoparticles are synthesized from inorganic materials that have hard, water insoluble, less biodegradable, and toxic properties, and hence have limitations in siRNA drug delivery (Figure 8.2c). Nanoparticles that come under this class are semiconductor nanoparticles, Carbon-based nanoparticles, silica nanoparticles, quantum dots, and fullerenes. They have drawn much attention from researchers and scientists due to their two-in-one role as both carriers for siRNA delivery to the targeted cells and use in the sophisticated bio imaging technology to accurately tracking the siRNA trajectory upon delivery and its activation at target sites. Although these nanocarriers have a toxicity effect, like other nanoparticles, they can be modified to reduce the toxicity levels. Moreover, coating the surface with several biocompatible, biodegradable, natural, or synthetic polymers and lipids enhances the biocompatibility and loading capacity of metal based nanoparticles. Moreover, their benefits, which include stability, noninvasive fluorescent nature, and controllability, might outweigh their disadvantages and drawbacks [35–39].

8.15 MESOPOROUS SILICA AND SILICON-BASED NANOPARTICLES

Mesoporous silica nanoparticles (MSNs) contain several ordered porous structural surfaces in which drugs and genes can be easily entrapped. Due to this specific characteristic, MSNs show some advantages over other nanoparticles in terms of high loading efficiency and controlled release of loaded therapeutic molecules. Moreover, their smaller size (50 to 200 nm) produce EPR effect by passively transporting across the cell membrane and accumulating in the tumor cells. MSNs coated with cationic polymers, PEG, PEI, and amine-terminated PAMAM dendrimers provide high loading and targeted delivery of siRNA [36–38].

8.16 HYBRID NANOPARTICLES

Both organic and inorganic NCS show many benefits and have been applied to remove various challenges in delivering therapeutic genes and drugs. However, they have some limitations due to toxicity, poor stability in vivo, and non-biodegradable substances. To overcome these challenges, using hybrid nanocarriers is the best solution. These nanoparticles are prepared by fusion of both organic and inorganic nanoparticles. To enhance their biocompatibility and loading efficiency, various polymers are grafted on the surface of the particles. Meanwhile the inorganic constituents provide diagnostic or therapeutic properties. Therefore, the hybrid NCs have benefits to offer the theragnostic application with high delivery efficiency [39, 40].

8.17 GOLD NANOPARTICLES

Gold nanoparticles have been studied extensively throughout the years as the ultimate candidates for siRNA delivery. Their special surface plasmon resonance (SPR) characteristic makes them beneficial for bio-imaging, and in addition to their stability and efficient delivery of siRNA, a lot of effort is being put to exploit the true potential of these particles and further improve their biocompatibility for prospective clinical applications [41–44].

8.18 IRON OXIDE NANOPARTICLES

Iron oxide nanocarriers have taken the spotlight due to their unique characteristics, such as thermal activation in the tumor microenvironment. They heat the tumors to lethal temperatures, inducing coagulative necrosis in tumor cells with the application of external magnetic fields. Furthermore, they have been reported to be potential candidate in siRNA therapy for many diseases, proving their MRI signal and magnetic properties allow them to be guided using an external magnetic field to accumulate in the tumor sites. Their magnetic properties have benefits to improve siRNA delivery to the targeted sites [45]. Studies prove that SPIO nanoparticles efficiently targeted the lung cancer cells upon application of external high-energy magnets positioned over a specific region of the lung. This approach was further elaborated in another study in which the use of high-energy magnets offered improved theragnostic effect of Doxorubicin-loaded iron-tagged nanocarriers, by magnetically targeting them towards metastatic tumor sites in the lungs [46, 47].

Many metal oxide particles, especially iron oxide particles, have taken the spotlight due to their unique characteristics. In addition to the fact that these nanoparticles can be used for bio-imaging, have excellent cellular absorption, and are stable and modifiable like other metal-based nanoparticles, they have the distinguishing feature of thermal activation that is also shared with gold nanoparticles. Iron oxide particles can be used to heat the tumors to lethal temperatures, causing the coagulative necrosis of tumor cells upon the application of alternating magnetic fields. Furthermore, SPIO nanoparticles were reported to be perfect candidates for future siRNA therapies for many diseases, including lung cancer, given their strong contrast (i.e., MRI signal) and their unique magnetic properties that allow them to be guided using an external magnetic field to accumulate in the tumor sites. Magnetic targeting can improve both the delivery of siRNA and/or therapeutic compounds (i.e., chemotherapeutic drugs) to improve cancer treatment. We have previously reported that targeting of intravenously injected SPIO nanoparticles to the lung was proved to be enhanced when using external high-energy magnets positioned over a specific region of the lung. This approach was further elaborated in another study in which the use of high-energy magnets offered improved theragnostic effect of Doxorubicin-loaded iron-tagged nanocarriers, by magnetically targeting them towards metastatic tumor sites in the lungs [44–48].

Taking all the information together, siRNA based nanoparticle formulations have drawn much more attention to the research and pharmaceutical industries. Within the past decade, the use of siRNA as a therapeutic agent has proven to be an effective nanomedicine for gene silencing. In recent years, the use of siRNA for gene silencing has proven to be an effective in nanomedicine. Although there are limitations in viral medicine, nanoparticle based delivery of siRNA overcomes those challenges. Moreover, siRNA delivery from tissue engineering matrices may also increase the number of cells exhibiting mRNA knockdown, while minimizing the quantity of the vector used, an important factor when dealing with doss-dependent toxicity. Yet despite all, the therapeutic activities of siRNA are promising and several clinical trials that are specific to tissue and diseases are currently progressing.

REFERENCES

1. Reischl D, and Zimmer A. Drug delivery of siRNA therapeutics: potentials and limits of nanosystems. *Nanomedicine* 2009;5(1):8–20.
2. Elbashir SM, Lendeckel W, and Tuschl T. RNA interference is mediated by 21- and 22-nucleotide RNAs. *Genes Dev* 2001;15(2):188–200.
3. Soutschek J, Akinc A, Bramlage B, et al. Therapeutic silencing of an endogenous gene by systemic administration of modified siRNAs. *Nature* 2004;432(7014):173–178.
4. Zimmermann TS, Lee ACH, Akinc A, et al. RNAi-mediated gene silencing in non-human primates. *Nature* 2006;441(7089):111–114.
5. Zhang H, Kolb FA, and Jaskiewicz L. Single processing center models for human Dicer and bacterial RNase III. *Cell* 2004;118:57–68.
6. Meister G, Landthaler M, Patkaniowska A, et al. Human Argonaute2 mediates RNA cleavage targeted by miRNAs and siRNAs. *Mol Cell* 2004;15:185.
7. Liu J, Carmell MA, Rivas FV, et al. Argonaute2 is the catalytic engine of mammalian RNAi. *Science* 2004;305:1437.
8. Bartel DP. MicroRNAs: genomics, biogenesis, mechanism, and function. *Cell* 2004;116:281–297.
9. Bartel DP, and Chen CZ. Micromanagers of gene expression: the potentially widespread influence of metazoan microRNAs. *Net Rev Genet* 2004;5(5):396–400.

10. Bagga SJ, Bracht S, Hunter K, et al. Regulation by let-7 and lin-4 miRNAs results in target mRNA degradation. *Cell* 2005;122:553–563.
11. Moschos SA, Jones SW, Perry MM, et al. Lung delivery studies using siRNA conjugated to TAT (48–60) and penetratin reveal peptide induced reduction in gene expression and induction of innate immunity. *Bioconjug Chem* 2007;18(5):1450–1459.
12. Golzio M, Mazzolini L, Ledoux A, et al. In vivo gene silencing in solid tumors by targeted electrically mediated siRNA delivery. *Gene Ther* 2007;14(9):752–759.
13. Oishi M, Nagasaki Y, Itaka K, et al. Lactosylated poly(ethylene glycol)siRNA conjugate through acid-labile β-thiopropionate linkage to construct pH-sensitive polyion complex micelles achieving enhanced gene silencing in hepatoma cells. *J Am Chem Soc* 2005;127(6):1624–1625.
14. De Paula D, Bentley MV, and Mahato RI. Hydrophobization and bioconjugation for enhanced siRNA delivery and targeting. *RNA* 2007;13(4):431–456.
15. Moschos SA, Jones SW, Perry MM, et al. Lung delivery studies using siRNA conjugated to TAT(48–60) and penetratin reveal peptide induced reduction in gene expression and induction of innate immunity. *Bioconjug Chem* 2007;18(5):1450–1459.
16. Egusquiaguirre SP, Igartua M, Hernandez RM, et al. Nanoparticle delivery systems for cancer therapy: advances in clinical and preclinical research. *Clin Transl Oncol* 2012;14(2):83–93.
17. Hu-Lieskovan S, Heidel JD, Bartlett DW, et al. Sequence-specific knockdown of EWS-FLI1 by targeted, nonviral delivery of small interfering RNA inhibits tumor growth in a murine model of metastatic Ewing's sarcoma. *Cancer Res* 2005;65(19):8984–8992.
18. Howard KA, Rahbek UL, Liu X. et al. RNA interference in vitro and in vivo using a novel chitosan/siRNA nanoparticle system. *Mol Ther* 2006;14(4):476–484.
19. Jean M, Smaoui F, Lavertu M, et al. Chitosan-plasmid nanoparticle formulations for IM and SC delivery of recombinant FGF-2 and PDGF-BB or generation of antibodies. *Gene Ther* 2009;16(9):1097–1110.
20. Alavi SE, Muflih Al Harthi S, Ebrahimi Shahmabadi H, et al. Cisplatin-loaded polybutylcyanoacrylate nanoparticles with improved properties as an anticancer agent. *Int J Mol Sci* 2019:20:1531.
21. Song W, Tang Z, Li M, et al. Polypeptide-based combination of paclitaxel and cisplatin for enhanced chemotherapy efficacy and reduced side-effects. *Acta Biomater* 2014;10:1392–1402.
22. Adeli M, Soleyman R, Beiranvanda Z, et al. Carbon nanotubes in cancer therapy: a more precise look at the role of carbon nanotube-polymer interactions. *Chem Soc Rev* 2013;42:5231–5256.
23. Aldieri E, Fenoglio I, Cesano F, et al. The role of iron impurities in the toxic effects exerted by short multiwalled carbon nanotubes (MWCNT) in murine alveolar macrophages. *J Toxicol Environ Health A* 2013;76:1056–1071.
24. Amenta V, and Aschberger K. Carbon nanotubes: potential medical applications and safety concerns. *Adv Rev* 2015;7:371–386.
25. Fenske DB, and Cullis PR. Liposomal nanomedicines. *Expert Opin Drug Deliv* 2008;5(1):25–44.
26. Elouahabi A, and Ruysschaert JM. Formation and intracellular trafficking of lipoplexes and polyplexes, *Mol Ther* 2005;11(3):336–347.
27. Taetz S, Bochot A, Surace C, et al. Hyaluronic acid-modified DOTAP/DOPE liposomes for the targeted delivery of anti-telomerase siRNA to CD44-expressing lung cancer cells. *Oligonucleotides* 2009;19(2):103–116.
28. Kostarelos K, Emfietzoglou D, Papakostas A, et al. Binding and interstitial penetration of liposomes within avascular tumor spheroids. *Int J Cancer* 2004;112(4):713–721.
29. Wu SY, and McMillan NAJ. Lipidic systems for in vivo siRNA delivery. *AAPS J* 2009;11(4):639–652.
30. Ozpolat B, Sood AK, and Lopez-Berestein G. Nanomedicine based approaches for the delivery of siRNA in cancer. *J Intern Med* 2010; 267(1):44–53
31. Kesharwani P, Tekade RK, Gajbhiye VK, et al. Cancer targeting potential of some ligand-anchored poly(propylene imine) dendrimers: a comparison. *Nanomedicine* 2011;7(3):295–304.
32. Lee JH, Cha EK, Kim MS, et al. Nanosized polyamidoamine (PAMAM) dendrimer-induced apoptosis mediated by mitochondrial dysfunction. *Toxicol Lett* 2009;190(2):202–207.
33. Patil ML, Zhang M, Taratula O, et al. Internally cationic polyamidoamine PAMAM-OH dendrimers for siRNA delivery: effect of the degree of Quaternization and cancer targeting. *Biomacromolecules* 2009;10(2):258–266.
34. Abbasi, E, Aval, SF, Akbarzadeh A, et al. Dendrimers: Synthesis, applications, and properties. *Nanoscale Res Lett* 2014;9:247.

35. Lai PS, Lou PJ, Peng CL, et al. Doxorubicindelivery by polyamidoamine dendrimer conjugation and photochemical internalization for cancer therapy. *J Control Release Off J Control Release Soc* 2007;122:39–46.
36. Zhang M, Zhu J, Zheng Y, et al. Doxorubicin-conjugated PAMAM dendrimers for pH-responsive drug release and folic acid-targeted cancer therapy. *Pharmaceutics* 2018; 10:162.
37. Zhu J, Wang G, Alves CS, et al. Multifunctional dendrimer-entrapped gold nanoparticles conjugated with doxorubicin for pH-responsive drug delivery and targeted computed tomography imaging. *Langmuir ACS J Surf Colloids* 2018;34:12428–12435.
38. Zhong Q, Merkel OM, Reineke JJ, et al. Effect of the route of administration and PEGylation of poly(amidoamine) dendrimers on their systemic and lung cellular biodistribution. *Mol Pharm* 2016;13:1866–1878.
39. Tan SJ, Kiatwuthinon P, Roh YH. et al. Engineering nanocarriers for siRNA delivery. *Small* 2011;7:841–856.
40. Ghosh P, Han G, De M, et al. Gold nanoparticles in delivery applications *Adv Drug Deliv Rev* 2008;60(11):1307–1315.
41. Rosi NL, Giljohann DA, Thaxton CS, et al. Oligonucleotide-modified gold nanoparticles for infracellular gene regulation, *Science* 2006;312(5776):1027–1030.
42. Kong WH, Bae KH, Jo SD, et al. Cationic lipid-coated gold nanoparticles as efficient and non-cytotoxic intracellular siRNA delivery vehicles. *Pharm Res* 2012;29(2):362–374.
43. Bonoiu AC, Mahajan SD, Ding H, et al. Nanotechnology approach for drug addiction therapy: gene silencing using delivery of gold nanorod-siRNA nanoplex in dopaminergic neurons. *Proc Natl Acad Sci U S A* 2009;106(14):5546–5550.
44. Hilger I, Hiergeist R, Hergt R, et al. Thermal ablation of tumors using magneticnanoparticles: an in vivo feasibility study. *Investig Radiol* 2002;37:580–586.
45. Price PM, Mahmoud WE, Al-Ghamdi A. et al. Magnetic drug delivery: where the field is going. *Front Chem* 2018;6:619.
46. Schleich N, Po C, Jacobs D, et al. Comparison of active, passive and magnetic targeting to tumors of multifunctional paclitaxel/SPIO-loaded nanoparticles for tumor imaging and therapy. *J Control Release Off J Control Release Soc* 2014;194:82–91.
47. Price PM, Mahmoud WE, Al-Ghamdi AA, et al. Magnetic drug delivery: where the field is going. *Front Chem* 2018;6:619.

9 Clinical Perspectives on Gene Therapy for Retinal and Eye Diseases

Devika S. Joshi, Gaurav M. Karve and Shrikant D. Joshi
Deenanath Mangeshkar Hospital & Research Centre, Pune, India

CONTENTS

9.1 INTRODUCTION

Retinal diseases are an important cause of vision impairment and blindness globally. Among the various reasons of visual impairment in adults aged 50 years or more, diseases like glaucoma (11%), age related macular degeneration (5.6%) and diabetic retinopathy (2.5%) are leading causes.[1] In the younger age group, prevalence of retinal diseases like retinitis pigmentosa (RP) (1 in 4,000, non-syndromic RP), inherited retinal diseases like cone rod dystrophies (1 in 40,000), Leber's hereditary optic neuropathy (LHON) (1 in 30,000 to 50,000) and Leber's congenital amaurosis (LCA) (1 in

DOI: 10.1201/9781003186083-9

33,000 to 50,000) contribute to the burden of visual impairment. With the increasing number of the elderly population, it is projected that 111.8 million people will have glaucoma in 2040.[2]

9.1.1 Age Related Macular Degeneration (AMD or ARMD)

The two major subtypes of the disease include dry form and wet (exudative or neovascular) form. Presently there is no definitive treatment for the dry form. Treatment for the exudative form includes various modalities, such as intravitreal anti-VEGF injections, laser, and photodynamic therapy. However, the aim of all these treatments is to delay disease progression rather than being able to halt it completely. In this premise, gene therapy appears to be a promising new approach aimed at halting the disease process itself.

9.1.2 Retinitis Pigmentosa (RP)

Retinitis pigmentosa, also known as hereditary retinal dystrophy is a group of retinal disorders causing progressive loss of vision due to pigmentary changes in the retinal pigment epithelial layer. It exists in syndromic (occurring in association with systemic disorders, e.g., Usher's syndrome) and non-syndromic forms. Historically, it has been an untreatable disease, but newer scientific advances like gene therapy, retinal implants, etc., have allowed good improvement in vision.

9.1.3 Leber's congenital amaurosis (LCA)

Leber's congenital amaurosis is an inherited retinal disorder which causes severe vision loss in infancy. Patients usually present with nystagmus, sluggish or near-absent pupillary responses, severely decreased visual acuity, photophobia, and high hypermetropia. There is no definitive treatment or cure for this condition. LCA occurs due to mutation in the RPE65 gene, which encodes a key enzyme that is required for regeneration of visual chromophore in the retinal cycle. The deficiency of all-trans retinyl ester isomerase leads to LCA.

In 2017, the USFDA has approved voretigene neparvovec (Luxturna Spark Therapeutics, Inc., Philadelphia, PA), for the treatment of biallelic RPE65 mutation-associated- LCA2(The first USFDA approved gene therapy for the eye) offering some hope for patients suffering from this condition.

Gene therapy for a few other diseases, like Stargardt's disease, Leber's hereditary optic neuropathy (LHON), and Chorideremia, is also under investigation.

A recent review of the number of ongoing gene therapy clinical trials has shown that about 1.8% are targeting eye diseases, of which a single one was targeting glaucoma.

The use of gene therapy has expanded from the original concept of only replacing the mutated gene causing the disease. It is now also used to control non physiological levels of expression or to modify pathways known to affect the disease. Genes offer numerous advantages over conventional drugs. They have longer duration of action and are more specific.[3] A direct repair of gene or compensation of defective genes is possible via gene therapy. The therapeutic benefits of gene therapy are maintained over long periods of time.

9.2 CATEGORIES OF GENE THERAPY

a) Gene Augmentation – to provide functional copies of defective genes, to boost the dosage of the gene. This approach is commonly used in Autosomal Recessive (AR) and X linked diseases(e.g., Duchenne Muscular Dystrophy, Spinal Muscular Atrophy).
b) Gene targeting – to introduce genetic material to repair or regenerate gene function. Mainly used in autosomal dominant (AD) and non-genetic diseases.
c) Genome editing – directly corrects and transforms the mutated gene to normal. This mode is beneficial in all modalities of inherited diseases.[4]

9.3 EYE AS A LUCRATIVE TARGET FOR GENE THERAPY

There are several advantages for the eye as a target for gene therapy.

- The eye is a unique organ, with a relatively small anatomical size. It is further subdivided into well-defined compartments, making it a very organized anatomical organ.
- The other unique aspect is the easy accessibility to all these sub compartments, using relatively safe surgical techniques. This facilitates effective delivery and concentration of the gene therapy vector.
- Genetic disorders usually manifest bilaterally. Thus, since the eye is a paired organ, one eye can serve as a control for other, allowing comparative analysis.
- The eye is an immunologically privileged site because of its less immunogenic microenvironment compared to other organs of the body, in addition to the physical and blood retina barrier.
- Retinal cells do not proliferate after birth, as they are post mitotic cells. Thus, vectors are capable of producing long lasting gene expression regardless of transduction.
- The optical transparency of ocular media permits accessibility for microsurgical delivery of vector suspension to the retina under direct visualization and superior, yet simple, in vivo imaging techniques make it possible to effectively study and monitor the effects of gene therapy.[3]

9.3.1 Ocular Gene Delivery Vectors

Ocular gene therapy can be administered via various types of vectors, which are broadly divided into viral [adenoviral (Ad), adeno associated viral vector (AAV), lentiviral] and non-viral (naked plasmid DNA, oligonucleotides, and RNA).

9.3.2 Ocular Routes of Administration

Various routes of delivery of vectors in the eye can be as follows:

- Intravitreal injection (IVT):
  This presents a relatively safe and the least invasive mode of delivery, offering effective and widespread circulation of the vector mediated by the vitreous humor to the adjacent layers of the retina. This mode, however, offers multiple physical barriers, like the internal limiting membrane (ILM), and therefore, less delivery to RPE and photoreceptors (outer retina). Hence, currently, IVT delivery of AAV is used in a limited number of clinical trials, particularly for Leber hereditary optic neuropathy (LHON), for which photoreceptors are not the primary target.

- Subretinal injections:
  The subretinal route of injection of viral vector suspension has been used in several trials. This necessitates a rather invasive surgery via the pars plana approach by highly trained surgeons. It can also be done under intraoperative Optical Coherence Tomography (OCT) guidance. It involves raising a bleb in the subretinal space, causing a transient localized retinal detachment in an already fragile and degenerating retina. Known complications seen in previous such trials of LCA2 include macular holes, unresolved retinal detachment requiring surgical repair, choroidal effusions, hypotonia, and retinal tears. Also the benefit of treatment is limited to bleb area.[5]

- Suprachoroidal space:
  Vector suspension can be injected in the potential suprachoroidal space, useful for choroidopathies. This therapy route is also currently being used to deliver stem cells into the subretinal space.

Other common and less invasive routes, like topical instillation, periocular routes like retrobulbar, subtenon, or subconjunctival injections are not found to be much effective.

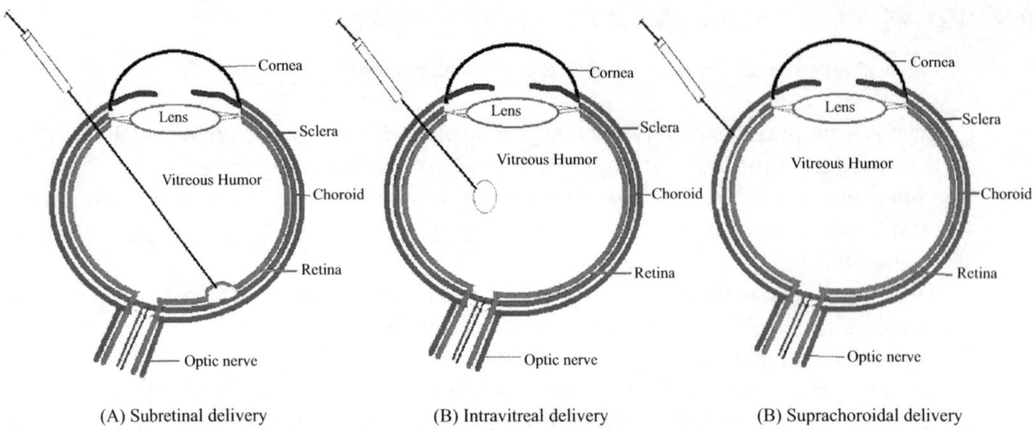

FIGURE 9.1 Delivery routes for gene therapy. Schematic figure of the eye, depicting the possible routes of delivery to introduce viral vectors into the eye.

Image adapted from: de Guimaraes, T. A. C., Georgiou, M., Bainbridge, J. W., & Michaelides, M. (2021). Gene therapy for neovascular age-related macular degeneration: rationale, clinical trials and future directions. *British Journal of Ophthalmology, 105*(2), 151–157.

9.3.3 Outcome Measures in Clinical Trials for Study of Effectivity of Gene Therapy in Eye Diseases

Clinically meaningful endpoints: These include the mean change or mean rate of change in the best corrected distance visual acuity, visual field sensitivity (including analysis of hill of vision volumes), retinal sensitivity measured by full-field stimulus testing (FST), and multiluminance mobility tests. Mobility course performance is being used to assess functional vision in a fixed environment as part of performance-based endpoints.

The eye being a unique organ with clear media, various latest technologies are put to use to assess imaging-based endpoints. With the help of artificial intelligence and analysis of in vivo retinal structure by high resolution images obtained from spectral domain OCT, treatment potential for IRDs can be assessed. Also, it is helpful in evaluating therapeutic outcomes in RP and Stargardt's disease. In Stargardt's disease, fundus autofluorescence imaging is being used to describe the leading disease front and evaluate therapeutic outcomes. Some surrogate endpoints used in various clinical trials, especially those for IRDs, include the mean change or mean rate of change in electrophysiological measures of retinal function, such as electroretinogram (ERG), optical coherence tomography (OCT) documenting the rate of loss of photoreceptors, and changes in lesion size on fundus autofluorescence (hypo or hyper fluorescence).[6]

There are several concerns about the induction of inflammation during AAV mediated gene therapy and efficacy of therapy found in various animal models. Inflammation during ocular gene therapy is widely variable in severity in both animal and clinical studies. In cases where inflammation was more severe or prolonged, it was associated with reduced efficacy of the administered gene therapy. There is no clear consensus on the approaches, effectiveness, or the standard treatment protocol for treatment of such inflammation during animal and clinical studies.[7]

In the above context, designing clinical trials, especially in retinal diseases like inherited retinal disorders, can be quite challenging because of the prohibitive costs, participation of industry, fewer number of participants, more patients from the pediatric age group, ethical principles, and safety concerns.

9.4 CURRENT THERAPIES FOR RETINAL DISEASES UNDER INVESTIGATION

9.4.1 Leber's Congenital Amaurosis (LCA)

RPE65 gene therapy for LCA has been one of the most successful examples of gene therapy for ocular diseases till date.

The first trial of ocular gene therapy for humans with RPE65-LCA was reported in 2008.[8]

In 2017, the US FDA approved voretigene neparvovec (Luxturna TM, Spark Therapeutics Inc.) for the treatment of RPE65-associated LCA. The long-term sustained effects on safety and efficacy of AAV2-RPE65 gene delivery in biallelic RPE65-mediated LCA were assessed over a period of three years, with the initial improvements in visual function peaking at 6 to 12 months after injection. [9]Longer term findings have shown that, despite improved function of surviving retina, the durability of benefit can be limited by progressive retinal degeneration.[10]

9.4.2 Achromatopsia

This disorder leads to loss of colour vision, nystagmus, photophobia, and poor visual acuity. About 70 to75% of mutations for this autosomal recessive disease are seen in the CNGA3 (25%) and CNGB3 genes (40 to 50%).

In rodent models of both CNGA3-ACHM31 and CNGB3-ACHM32, and in canine models of CNGB3-ACHM, AAV-mediated gene replacement therapy can improve cone function. Early phase clinical trials in humans of the above genes will further help to elucidate whether it is effective in humans as well.[4]

9.4.3 Retinitis Pigmentosa (RP)

Apart from usual gene therapy modalities, an optogenetic therapy relies on delivering a gene encoding a light sensitive channel protein (such as channelrhodospin [ChR] or halorhodospin) to create artificial photoreceptors by either reactivating dormant cones or activating other retinal neurons during the late stages of degeneration. Optogenetic therapy involves genetically modifying neurons to express light-sensitive ion channels. It has been shown to successfully restore light responses and visual function in multiple mouse models of retinal degeneration.[3] This type of gene therapy may be able to offer a wider treatment for a variety of retinal degenerations, with a possibly larger area of opportunity for useful intervention in progressive disease.

9.4.4 Leber's Hereditary Optic Neuropathy (LHON)

Leber's hereditary optic neuropathy (LHON) is a common inherited mitochondrial disorder. It is characterized by the degeneration of the optic nerves, leading to vision loss. The major mutations in the mitochondrial genes *ND1*, *ND4*, and *ND6* of LHON subjects are found to increase the oxidative stress experienced by the optic nerve cell, thereby leading to nerve cell damage and vision loss.

The missense, m.3460G > A m.11778G.A and m.14484T > C mutation constitutes 90% of the mutation spectrum in this disorder.

A unique approach in this disease, involves inducing expression of these mitochondrial genes, in the nucleus.[4]

Currently, various drugs like Idebenone, EPI 743 and Bendavia are used to treat vision loss due to oxidative stress in LHON largely due to antioxidant properties.

Gene therapy for LHON is still under clinical trials. One such trial in Phase I, involving adeno-associated virus as a vector of different doses has shown slight improvement in the visual loss of LHON subjects with G11778A mutation of *ND4*.[11]

Other Clinical trials for gene therapy are currently underway for several diseases.

9.4.5 AMD

Campochiaro et al. performed a phase 1 clinical trial to assess the safety of intravitreal injections of AdPEDF. OPTIC is a phase 1, 2-year, multicenter, open-label study of ADVM-022 in neovascular AMD. RGX-314 trial (RGX-314) has been showing some encouraging results because only 25% of the patients in cohort five needed rescues, meaning gene therapy was making sufficient anti-VEGF protein in 75% of patients to suppress exudation from CNV.[12]

9.4.6 Retinoblastoma

Chévez-Barrios et al, reported the results of a phase 1 clinical trial investigating the safety and efficacy of intravitreal injections of a herpes simplex thymidine kinase gene (AdV-TK) followed by systemic ganciclovir for the treatment of bilateral retinoblastoma with vitreous tumor seeding. They found that AdV-TK followed by ganciclovir can be administered safely to children with retinoblastoma. Their results may suggest that in children with vitreous seeds that are complicating retinoblastoma, gene therapy may be a useful adjuvant to other standard therapies.[13]

9.4.7 Other Retinal Diseases

In addition to Luxturna TM, three clinical trials advanced to Phase III collectively, including antisense oligonucleotides-based QR-110 for LCA10, AAV5-RPRG gene-drug for XLRP, and AAV2-REP1 for choroideremia.[14] A phase III clinical trial comparing high and low titers of recombinant AAV2.REP1 to choroideremia patients is now ongoing (NCT03496012).

Active clinical trials in gene therapy for eye diseases are summarized in Table 9.1.

9.5 GENE THERAPY IN OTHER EYE DISEASES

In situations like high-risk corneal grafts with poor survival, especially those in which disease recurs, such as full thickness keratoplasty in MPS4, Reis-Bücklers, and other TGFBI corneal dystrophies, gene therapy as a "single shot treatment" is a very good possible treatment option.[15]

Various targets for gene therapy for glaucoma include the following:

- transferring genes to Retinal Ganglion Cells (RGCs) via AAV,
- transferring genes to the trabecular meshwork (T M) allowing the transgene modification in outflow facility,
- Using adenoviral vectors to identify and modify the small GTP-binding protein RhoA involved in regulation of intraocular pressure (IOP).
- Adenoviral vector carrying the inhibitor of calcification matrix Gla gene (Adh.MGP) inhibits bone morphogenetic 2 (BMP2)–induced mineralization in TM cells and the involvement of the Wnt signaling pathway.
- Insertion of regulatory elements for expressing therapeutic genes only when needed, such as vectors developed to deliver matrix metallopeptidase 1 (*MMP1*) gene regulated by glucocorticoid response elements (Adh.GRE.MMP1) in the case for the treatment of steroid glaucoma is also a promising approach.[16]

Genes selected to protect RGCs have traditionally been genes encoding neurotrophins, antiapoptotic, and defense genes, which are also being explored for potential use in gene therapy.

TABLE 9.1

Sr. No.	NCT Number	Title	Acronym	Conditions	Interventions	Sponsor/Collaborators	Phases
1	NCT03748784	ADVM-022 Intravitreal Gene Therapy for Wet AMD	OPTIC	Wet Age-related Macular Degeneration|Neovascular Age-related Macular Degeneration	Biological: ADVM-022	Adverum Biotechnologies, Inc.	Phase 1
2	NCT03066258	Safety and Tolerability of RGX-314 Gene Therapy for Neovascular AMD Trial		Neovascular Age-related Macular Degeneration|Wet Age-related Macular Degeneration	Genetic: RGX-314	Regenxbio Inc.	Phase 1|Phase 2
3	NCT03406104	RESCUE and REVERSE Long-term Follow-up	RESCUE/ REVERSE	Leber Hereditary Optic Neuropathy	Genetic: GS010|Other: Sham	GenSight Biologics	Phase 3
4	NCT03597399	A Patient Registry Study for Patients Treated With Voretigene Neparvovec in US		Confirmed Biallelic RPE65 Mutation-associated Retinal Dystrophy	Biological: AAV2-hRPE65v2,voretigene neparvovec-rzyl	Spark Therapeutics	
5	NCT04418427	ADVM-022 Intravitreal Gene Therapy for DME	INFINITY	Diabetic Macular Edema|Diabetic Retinopathy	Biological: 6E11 vg/eye of ADVM-022|Biological: 2E11 vg/eye of ADVM-022|Biological: Aflibercept	Adverum Biotechnologies, Inc.	Phase 2
6	NCT03507686	A Safety Study of Retinal Gene Therapy for Choroideremia With Administration of BIIB111	GEMINI	Choroideremia	Drug: BIIB111	NightstaRx Ltd, a Biogen Company|Biogen	Phase 2

(continued)

TABLE 9.1 (Continued)

<table>
<tr><th>Sr. No.</th><th>NCT Number</th><th>Title</th><th>Acronym</th><th>Conditions</th><th>Interventions</th><th>Sponsor/Collaborators</th><th>Phases</th></tr>
<tr><td>7</td><td>NCT02077361</td><td>An Open Label Clinical Trial of Retinal Gene Therapy for Choroideremia</td><td></td><td>Choroideremia</td><td>Genetic: rAAV2.REP1 vector</td><td>Ian M. MacDonald|Alberta Innovates Health Solutions|Canada Foundation for Innovation|Canadian Institutes of Health Research (CIHR)|Choroideremia Research Foundation Canada|Foundation Fighting Blindness|Imperial College London|University of Oxford|University of Alberta</td><td>Phase 1|Phase 2</td></tr>
<tr><td>8</td><td>NCT03428178</td><td>Efficacy Study of Gene Therapy for The Treatment of Acute LHON Onset Within Three Months</td><td>LHON</td><td>Acute LHON|Onset Within Three Months|Onset Between 3 to 6 Months|Onset Between 6 to 12 Months|Onset Between 12 to 24 Months|Onset Between 24 to 60 Months|Onset Over 60 Months</td><td>Drug: rAAV2-ND4</td><td>Bin Li|Huazhong University of Science and Technology</td><td>Not Applicable</td></tr>
<tr><td>9</td><td>NCT03252847</td><td>Gene Therapy for X-linked Retinitis Pigmentosa (XLRP) Retinitis Pigmentosa GTPase Regulator (RPGR)</td><td></td><td>X-Linked Retinitis Pigmentosa</td><td>Genetic: AAV2/5-RPGR</td><td>MeiraGTx UK II Ltd|Syne Qua Non Limited|Bionical Emas</td><td>Phase 1|Phase 2</td></tr>
</table>

<table>
<tr><td>10</td><td>NCT02161380</td><td>Safety Study of an Adeno-associated Virus Vector for Gene Therapy of Leber's Hereditary Optic Neuropathy</td><td>LHON</td><td>Leber's Hereditary Optic Neuropathy</td><td>Drug: injection of scAAV2-P1ND4v2 1.18x10e9 vg (Low),|Drug: injection of scAAV2-P1ND4v2 5.81 X10e9 vg (Med)|Drug: injection of scAAV2-P1ND4v2 2.4 X10e10vg (High)|Drug: injection of scAAV2-P1ND4v2 1.0 X10e11vg (Higher)</td><td>Byron Lam|National Eye Institute (NEI)|University of Miami</td><td>Phase 1</td></tr>
<tr><td>11</td><td>NCT00481546</td><td>Phase I Trial of Gene Vector to Patients With Retinal Disease Due to RPE65 Mutations</td><td>LCA</td><td>Amaurosis of Leber|Retinal Diseases</td><td>Genetic: rAAV2-CBSB-hRPE65</td><td>University of Pennsylvania|National Eye Institute (NEI)</td><td>Phase 1</td></tr>
<tr><td>12</td><td>NCT03293524</td><td>Efficacy & Safety Study of Bilateral IVT Injection of GS010 in LHON Subjects Due to the ND4 Mutation for up to 1 Year</td><td>REFLECT</td><td>Leber Hereditary Optic Neuropathy</td><td>Genetic: GS010|Drug: Placebo</td><td>GenSight Biologics</td><td>Phase 3</td></tr>
<tr><td>13</td><td>NCT03602820</td><td>Long-term Follow-up Study in Subjects Who Received Voretigene Neparvovec-rzyl (AAV2-hRPE65v2)</td><td></td><td>Inherited Retinal Dystrophy Due to RPE65 Mutations</td><td>Biological: AAV2-hRPE65v2</td><td>Spark Therapeutics</td><td></td></tr>
<tr><td>14</td><td>NCT00999609</td><td>Safety and Efficacy Study in Subjects With Leber Congenital Amaurosis</td><td></td><td>Inherited Retinal Dystrophy Due to RPE65 Mutations|Leber Congenital Amaurosis</td><td>Biological: AAV2-hRPE65v2,voretigene neparvovec-rzyl</td><td>Spark Therapeutics|Children's Hospital of Philadelphia|University of Iowa</td><td>Phase 3</td></tr>
</table>

(*continued*)

TABLE 9.1 (Continued)

Sr. No.	NCT Number	Title	Acronym	Conditions	Interventions	Sponsor/Collaborators	Phases
15	NCT02407678	REP1 Gene Replacement Therapy for Choroideremia	REGENERATE	Choroideremia	Genetic: AAV-mediated REP1 gene replacement	University of Oxford|Moorfields Eye Hospital NHS Foundation Trust|University College, London	Phase 2
16	NCT04516369	Study of Efficacy and Safety of Voretigene Neparvovec in Japanese Patients With Biallelic RPE65 Mutation-associated Retinal Dystrophy		Biallelic RPE65 Mutation-associated Retinal Dystrophy	Genetic: voretigene neparvovec	Novartis Pharmaceuticals|Novartis	Phase 3
17	NCT03585556	AAVCAGsCD59 for the Treatment of Wet AMD		Wet Age-related Macular Degeneration	Drug: Intravitreal anti-VEGF|Biological: Intravitreal AAVCAGsCD59|Drug: Oral prednisolone	Janssen Research & Development, LLC	Phase 1
18	NCT01678872	A Follow-up Study to Evaluate the Safety of RetinoStatÂ® in Patients With Age-Related Macular Degeneration		Age-Related Macular Degeneration	Drug: RetinoStat	Oxford BioMedica	Phase 1
19	NCT03153293	A Single Intravitreal Injection of rAAV2-ND4 for the Treatment of Leber's Hereditary Optic Neuropathy		Leber Hereditary Optic Neuropathy	Drug: rAAV2-ND4	Huazhong University of Science and Technology|Shiyan Taihe Hospital	Phase 2|Phase 3
20	NCT02416622	Safety and Efficacy of rAAV-hRS1 in Patients With X-linked Retinoschisis (XLRS)		X-linked Retinoschisis	Biological: rAAV2tYF-CB-hRS1	Applied Genetic Technologies Corp	Phase 1|Phase 2
21	NCT02556736	RST-001 Phase I/II Trial for Advanced Retinitis Pigmentosa		Advanced Retinitis Pigmentosa	Drug: RST-001	Allergan	Phase 1|Phase 2

<table>
<tr><td>22</td><td>NCT02341807</td><td>Safety and Dose Escalation Study of AAV2-hCHM in Subjects With CHM (Choroideremia) Gene Mutations</td><td></td><td>Choroideremia|CHM (Choroideremia) Gene Mutations</td><td>Biological: AAV2-hCHM</td><td>Spark Therapeutics|Children's Hospital of Philadelphia|University of Pennsylvania|Massachusetts Eye and Ear Infirmary</td><td>Phase 1|Phase 2</td></tr>
<tr><td>23</td><td>NCT03913143</td><td>A Study to Evaluate Efficacy, Safety, Tolerability and Exposure After a Repeat-dose of Sepofarsen (QR-110) in LCA10 (ILLUMINATE)</td><td>ILLUMINATE</td><td>Leber Congenital Amaurosis 10|Blindness|Leber Congenital Amaurosis|Vision Disorders|Sensation Disorders|Neurologic Manifestations|Eye Diseases|Eye Diseases, Hereditary|Eye Disorders Congenital|Retinal Disease</td><td>Drug: sepofarsen|Other: Sham</td><td>ProQR Therapeutics</td><td>Phase 2|Phase 3</td></tr>
<tr><td>24</td><td>NCT03913130</td><td>Extension Study to Study PQ-110-001 (NCT03140969)</td><td>INSIGHT</td><td>Leber Congenital Amaurosis 10|Blindness|Leber Congenital Amaurosis|Vision Disorders|Sensation Disorders|Neurologic Manifestations|Eye Diseases|Eye Diseases, Hereditary|Eye Disorders Congenital|Retinal Disease</td><td>Drug: QR-110</td><td>ProQR Therapeutics</td><td>Phase 1|Phase 2</td></tr>
<tr><td>25</td><td>NCT04560790</td><td>Safety and Efficacy of CRISPR/Cas9 mRNA Instantaneous Gene Editing Therapy to Treat Refractory Viral Keratitis</td><td></td><td>Viral Keratitis|Blindness Eye|Herpes Simplex Virus Infection|Cornea</td><td>Drug: BD111 Adult single group Dose</td><td>Shanghai BDgene Co., Ltd.|Eye & ENT Hospital of Fudan University</td><td>Phase 1|Phase 2</td></tr>
<tr><td>26</td><td>NCT01208389</td><td>Phase 1 Follow-on Study of AAV2-hRPE65v2 Vector in Subjects With Leber Congenital Amaurosis (LCA) 2</td><td></td><td>Leber Congenital Amaurosis</td><td>Biological: voretigene neparvovec-rzyl</td><td>Spark Therapeutics</td><td>Phase 1|Phase 2</td></tr>
</table>

(continued)

TABLE 9.1 (Continued)

<table>
<tr><th>Sr. No.</th><th>NCT Number</th><th>Title</th><th>Acronym</th><th>Conditions</th><th>Interventions</th><th>Sponsor/Collaborators</th><th>Phases</th></tr>
<tr><td>27</td><td>NCT02223819</td><td>Crizotinib in High-Risk Uveal Melanoma Following Definitive Therapy</td><td></td><td>Uveal Melanoma</td><td>Drug: Crizotinib</td><td>Columbia University|Pfizer</td><td>Phase 2</td></tr>
<tr><td>28</td><td>NCT03173144</td><td>Chronic Inflammatory Disease, Lifestyle and Treatment Response</td><td>BELIEVE</td><td>Autoimmune Diseases|Inflammatory Bowel Diseases|Crohn Disease (CD)|Colitis, Ulcerative (UC)|Arthritis, Rheumatoid (RA)|Spondylarthropat hies|Arthritis, Psoriatic (PsA)|Psoriasis|Hidradenitis Suppurativa (HS)|Uveitis</td><td>Other: Primary exposure variable|Other: Other (exploratory) exposure variables</td><td>University of Southern Denmark|Hospital of Southern Jutland|University Hospital Bispebjerg and Frederiksberg|University of Copenhagen|Odense University Hospital|University of Aarhus|Aalborg University Hospital|Herning Hospital|Hospital of South West Jutland|Sygehus Lillebaelt|Aalborg University|University of Kiel|Colitis-Crohn Foreningen|Herlev Hospital|Regionshospitalet Silkeborg|Aarhus University Hospital|The Danish Psoriasis Association</td><td></td></tr>
<tr><td>29</td><td>NCT00342797</td><td>Retinoblastoma Biomarker Study</td><td></td><td>Retinoblastoma|-Melanoma|Sarcoma</td><td></td><td>National Cancer Institute (NCI)|National Institutes of Health Clinical Center (CC)</td><td></td></tr>
</table>

Trial Data as on June 2021
Tables adopted from:
*Rodrigues GA, Shalaev E, Karami TK, Cunningham J, Slater NKH, Rivers HM. Pharmaceutical Development of AAV-Based Gene Therapy Products for the Eye. Pharm Res. 2018;36(2):29. Published 2018 Dec 27. doi:10.1007/s11095-018-2554-7
**Lee JH, Wang JH, Chen J, et al. Gene therapy for visual loss: Opportunities and concerns. Prog Retin Eye Res. 2019;68:31-53. doi:10.1016/j.preteyeres.2018.08.003 (reference number 4 mentioned in 'References')

9.6 FUTURE DIRECTIONS

The initial gene-therapy trials have allowed us to improve our approaches with viral vectors, transgene selection, and delivery. There are still several unanswered questions. The success of gene therapy may help us consider newer avenues, like the fusion of gene therapy along with stem cell therapy.

Silencing genes with siRNA is a novel method that is beginning to be explored in gene therapy. The proof of concept was established by delivering the naked glucocorticoid receptor siRNA to the human TM in an organ culture perfusion system. The transferred siRNA ablated the dexamethasone induction of genes, such as myocilin (*MYOC*) and *ANGPTL7*.[3]

The clustered regularly interspaced short palindromic repeat (CRISPR)/Cas system of gene editing, also called as 'molecular surgery' is a promising new technique for future treatment of inherited disease because it is capable of correcting gene defects by manipulating the genome at the particular site. The advantages of the CRISPR/Cas9 system are its ease of use and the speed with which desired genomic modifications can be generated compared with other systems.

9.7 CONCLUSION

Gene therapy has the potential for changing the practice of classical clinical medicine to that of molecular medicine. Newer techniques in gene therapy are likely to bring forth new and better therapies, which may, in the near future, replace conventional pharmacological therapies and pave the way for complete disease cure.

ACKNOWLEDGEMENT

We thank Dr Kalpana S. Joshi for her help with fine tuning the manuscript and constant encouragement. We thank Ms. Mrunmayee Joshi for her unwaivering patience.

REFERENCES

1. GBD 2019 Blindness and Vision Impairment Collaborators; Vision Loss Expert Group of the Global Burden of Disease Study. Causes of blindness and vision impairment in 2020 and trends over 30 years, and prevalence of avoidable blindness in relation to VISION 2020: the right to sight: an analysis for the global burden of disease study. *Lancet Glob Health.* 2021 Feb; 9(2): e144–e160. doi: 10.1016/S2214-109X(20)30489-7. Epub 2020 Dec 1. Erratum in: Lancet Glob Health. 2021 Apr;9(4):e408. PMID: 33275949; PMCID: PMC7820391.
2. Kang JM, and Tanna AP. Glaucoma. *Med Clin North Am.* 2021 May; 105(3): 493–510. doi: 10.1016/j.mcna.2021.01.004. Epub 2021 Apr 2. PMID: 33926643.
3. Borrás T. The pathway from genes to gene therapy in glaucoma: a review of possibilities for using genes as glaucoma drugs. *Asia Pac J Ophthalmol (Phila).* 2017; 6(1): 80–93. doi:10.22608/APO.2016126
4. Lee JH, Wang JH, Chen J, Li F, Edwards TL, Hewitt AW, and Liu GS. Gene therapy for visual loss: Opportunities and concerns. *Prog Retin Eye Res.* 2019 Jan; 68: 31–53. doi: 10.1016/j.preteyeres.2018.08.003. Epub 2018 Aug 29. PMID: 30170104.
5. Bordet T, and Behar-Cohen F. Ocular gene therapies in clinical practice: viral vectors and nonviral alternatives. *Drug Discov Today.* 2019 Aug; 24(8): 1685–1693. doi: 10.1016/j.drudis.2019.05.038. Epub 2019 Jun 5. PMID: 31173914.
6. Thompson DA, Iannaccone A, Ali RR, Arshavsky VY, Audo I, Bainbridge JW, Zacks DN, et al. Advancing clinical trials for inherited retinal diseases: recommendations from the second monaciano symposium. *Transl Vis Sci Technol.* 2020; 9(7): 2–2.
7. Chan YK, Dick AD, Hall SM, Langmann T, Scribner CL, and Mansfield BC, Ocular Gene Therapy Inflammation Working Group. Inflammation in viral vector-mediated ocular gene therapy: a review and report from a workshop hosted by the foundation fighting blindness, 9/2020. *Transl Vis Sci Technol.* 2021; *10*(4): 3–3.

8. Bainbridge JW et al. Effect of gene therapy on visual function in Leber's congenital amaurosis. *N Engl J Med.* 2008; 358: 2231–2239. doi:10.1056/NEJMoa0802268
9. Ziccardi L, Cordeddu V, Gaddini L, Matteucci A, Parravano M, Malchiodi-Albedi F, and Varano M. Gene therapy in retinal dystrophies. *Int J Mol Sci.* 2019; 20(22): 5722.
10. Prado, D A, Acosta-Acero, M, and Maldonado, R S, Gene therapy beyond luxturna: a new horizon of the treatment for inherited retinal disease. *Curr Opin Ophthalmol.* 2020 May; 31(3): 147–154. doi: 10.1097/ICU.0000000000000660
11. Manickam AH et al. Mitochondrial genetics and therapeutic overview of Leber's hereditary optic neuropathy. *Indian journal of ophthalmology* 65,11 (2017): 1087–1092. doi:10.4103/ijo.IJO_358_17
12. Gene Therapy: An Emerging Frontier in the Treatment of Neovascular Diseases SUPPLEMENT TO RETINA TODAY | APRIL 2020
13. Chévez-Barrios P, Chintagumpala M, Mieler W, et al. Response of retinoblastoma with vitreous tumor seeding to adenovirus-mediated delivery of thymidine kinase followed by ganciclovir. *J Clin Oncol.* 2005; 23: 7927–7935. [PubMed: 16258092]
14. Chiu W, Lin TY, Chang YC, et al. An update on gene therapy for inherited retinal dystrophy: experience in Leber congenital amaurosis clinical trials. *Int J Mol Sci.* 2021; 22(9): 4534. Published 2021 Apr 26. doi:10.3390/ijms22094534
15. Williams Keryn A., and Irani Yazad D. Gene therapy and gene editing for the corneal dystrophies, *Asian-Pacific J. Ophthalmol.* 2016 July/August; 5(4): 312–316. doi: 10.1097/APO.0000000000000215
16. Borrás Terete, Advances in glaucoma treatment and management: gene therapy. *Invest. Ophthalmol. Vis. Sci.* 2012; 53(5): 2506–2510. doi: 10.1167/iovs.12-9483o

10 Herpesvirus microRNAs for Use in Gene Therapy Immune-Evasion Strategies

Vineet Mahajan
University of Pittsburgh, Pittsburgh, PA, USA

Shruti Saptarshi
University of South Florida, Tampa, FL, USA

Yashwant Pathak
University of South Florida, Tampa, FL, USA

CONTENTS

10.1 INTRODUCTION

10.1.1 What is Gene Therapy?

Modern molecular biology has made it possible to treat human disease through gene therapy. Thus, one can integrate a specific gene into a viral vector and use it to infect a patient whose gene is defective in order to rectify that defective gene. Currently, this approach is being tested in clinical trials for several diseases and has been successful in aiding patients with lethal immune disorders caused by genes encoding adenosine deaminase. Gene therapy uses specific genes to treat human diseases, and efficient gene delivery systems are important to its success.

10.1.1.1 Importance of Gene Therapy in Transplantation Biology

A paradigm shift is taking place in transplantation medicine, away from traditional chemical immunosuppression regimens that dominate clinics today and leaning toward modalities that are tissue and cell specific. As a result, xenotransplantation is sure to become a less debated issue in the coming years, and a greater challenge for clinicians, transplant immunologists, and biologists. The ability to deliver exogenous nucleic acids to cells of various tissues is the basis for a successful use of gene therapy in transplantation.

DOI: 10.1201/9781003186083-10

10.1.1.2 Transplantation and Graft Rejection

Genetically corrected autologous cells as well as allogeneic cells are potent approaches to restoring cellular functions in patients suffering from genetic diseases.[1] The recipient's immune response against non-self-antigens may compromise the survival of these cells.[2]

Antibodies produced by non-self-antigens may compromise the recipient's immune system cell survival after grafting. Immunosuppressive drugs may be required for the rest of the recipient's life when receiving the graft.[3] In addition to immune-evasion molecules, another approach that could reduce graft rejection is using DNA mutations. It is possible to subvert an immune response by expressing such molecules in the tissues of a graft. A virus in particular is extremely adept at manipulating the immune response of its hosts.[1] The Herpes Simplex Virus (HSV) genome size and ability to be latent in neuronal cells make it unique for such gene delivery in the designated system, such as the nervous system, regarding its large genome size and for delivery of genes.[1] The Herpesviridae family provides a reasonable proof of concept for this, since they have the ability to establish latency in the infected host and to persist for life. Many viral proteins involved in immune evasion have been characterized; however, the Herpesviridae also encode a large and diverse range number of viral microRNAs (miRNAs). Immune evasion is linked to some of them. Researchers have shown that several miRNAs inhibit the host's immune system from destroying infected cells.[4, 5] In this chapter, miRNAs from some common herpesviruses that modulate immune responses will be discussed, culminating with a discussion of their potential application in non-immune generating cell therapy.

RNA molecules smaller than 21 nucleotides are known as microRNAs (miRNAs). A pair of complementary strands of mRNA can interact when they are of complementary lengths. This reduces by obstructing or degrading translation. In the world of microRNA research, microRNAs are often called the "fine tuners" of the gene expression. Since most of the posttranslational regulation occurs after this process, it produces results that are rather subtle. Cellular miRNAs are abundantly present and are involved in almost all biological processes. The diversity of their DNA makes them perfect targets for viruses. Indeed, Herpesviruses use cellular miRNAs to benefit from replication and survival. In addition to being encoded by herpesviruses, miRNAs may also be effective in rendering cellular grafts as part of gene therapy techniques. RNA-mediated immune evasion by herpesvirus is known to involve herpesvirus miRNAs. These have only recently gained attention, and to our knowledge, there has been no evaluation of miRNAs as a tool for gene editing therapy. Specifically, we have defined and discussed miRNAs derived from herpesviruses that are associated with immune evasion..

10.2 WHAT ARE miRNAs

miRNAs are small non-coding RNAs that are involved in post-transcriptional gene silencing and regulation of gene expression via physiological and pathological mechanisms. miRNAs were first identified in the nematode, *C.elegans*, almost a decade ago.[6] High-throughput sequencing technologies and computational analysis tools have been key to the discovery of miRNAs. miRNAs are small and single-stranded molecules, approximately 22 nucleotides in length (occurring as a hairpin like duplex) and exert their effect by binding to the 3′ untranslated region (3′UTR) of target messenger RNAs (mRNAs). This interaction is responsible for the degradation or translational inhibition of the target mRNA.

miRNA sequences appear to be conserved across species. A huge variety of miRNAs have been identified in plants and animals, including humans and viruses. Currently, over 2,000 human miRNAs are recognized in the comprehensive miRNA database miRbase 22 (http://www.mirbase.org). It is estimated that about 60% of human genes may be subjected to miRNA regulation.[4] As miRNAs and their targets are not paired, a single miRNA can potentially affect more than 300 different mRNA transcripts, having a profound impact on gene expression patterns.

Cellular miRNAs are known to be involved in biological processes such as cell differentiation, metabolism, homeostasis, and apoptosis. Some studies have also indicated that miRNAs may have hormone-like function, as release of miRNAs into the extracellular fluids may modulate cellular activities in an autocrine, endocrine, or paracrine fashion.[7]

Viruses have evolved to regulate cellular (host) and viral genes via miRNAs, including mechanisms to avoid being targeted by cellular miRNAs.[6] Moreover, viruses may upregulate host miRNAs for immunological escape. For example, human cytomegalovirus replication was shown to suppress suppressing type I IFN response following host miR-146a upregulation, and miR-23a was found to facilitate HSV-1 replication by targeting the interferon regulatory factor 1 (IRF1) and inhibiting the interferon pathway.[8] Most DNA viruses apply gene regulation via miRNAs transcribed from their DNA genomes.[9] Whether RNA viruses express miRNAs needs further elucidation, although some research indicates that RNA viruses can produce functional miRNAs or some non-canonical miRNA-like RNA fragments that have been detected in RNA virus infections.[10] It has now become increasingly apparent that dysregulation of miRNAs may have a prominent role in the development of disease and cancer.[11]

The canonical pathway for miRNA biogenesis involves the following steps (See Figure 10.1). RNA Polymerase II (Pol II) initially transcribes the miRNA from the protein-coding/noncoding region of the genome as a long, capped poly A precursor called primary-miRNA (pri-miRNA). Next, the nuclear RNAse III Drosha trims the pri-miRNAs into precursor-miRNA hairpin intermediates (pri-miRNAs 60-70 nucleotides long), which are subsequently transported to the cytoplasm and cleaved by RNAse II Dicer into 20 nucleotide-long double-stranded miRNAs. The resulting miRNA/miRNA* duplex is then loaded into a member of the Argonaute family of proteins. The argonaut protein unwinds and selects a guide strand from this duplex to become the mature miRNA, whereas the complementary strand is degraded. The strand selection is determined by the duplex stability at the 5′ end of each miRNA arm, and highly abundant miRNAs typically originate more frequently from the 5p strand than the 3p strand. The mature miRNA is then loaded into an effector complex: the RNA-induced silencing complex (RISC). The RISC-bound miRNA complex thus

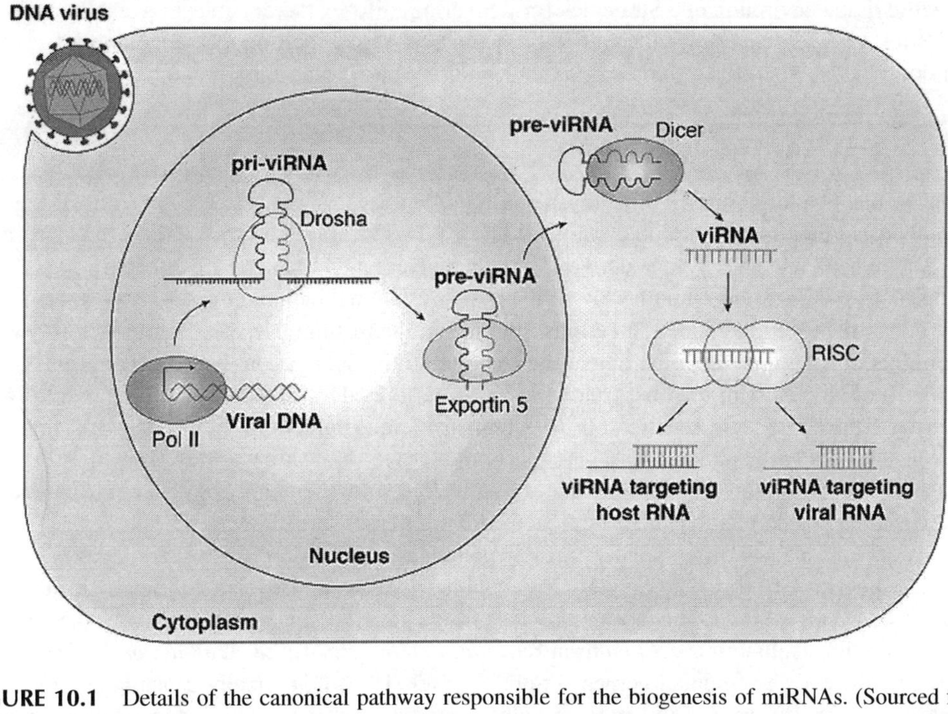

FIGURE 10.1 Details of the canonical pathway responsible for the biogenesis of miRNAs. (Sourced from Mishra et.al 2020.)

guides the recognition of a sequence in the 3′-untranslated region (3′-UTR) of a target mRNA, which has either extensive or partial complementarity to the miRNA, to induce mRNA cleavage and degradation or inhibition of mRNA translation.[7] Pairing interactions between miRNAs and their target RNAs are referred to as "seed sequence" which involve 2–8 nucleotides of the miRNA. Unlike DNA viruses, most RNA viruses replicate in in the cytoplasm and so are unable to access the nuclear enzyme Drosha necessary for the miRNA processing.[12]

The degree of sequence complementarity between the miRNA and its target mRNA and the associated Argonaute protein determine the stability of an miRNA.[4] Short, single-stranded RNA/DNA molecules called as antisense oligonucleotides (ASOs), antagomirs or blockmirs have been shown to inhibit specific miRNAs in mammalian cells.[13] These are chemically engineered antagonists of miRNAs and have therapeutic implications. miRNAs can be cleaved by small RNA degrading nucleases, exoribonuclease, and other endo and exonuclease. miRNA sponges are plasmid or viral vectors containing tandemly arrayed miRNA binding sites; separated with a small nucleotide spacer, they can inhibit an entire family of miRNA by using the common seed sequence, and can therefore inhibit multiple miRNAs at once.[14]

Targeted transfer of a therapeutic transgene is central to safe and effective gene therapeutic procedures.[15] Currently, microRNA-dependent post-transcriptional suppression of transgene expression is an emerging new technology. Identification of viral miRNA as targets for treating viral diseases and associated cancers is also being explored. Several miRNA-targeted therapeutics are now in clinical development, prominent examples being a mimic of the tumor suppressor miRNA miR-34, which reached phase I clinical trials for treating cancer, and hepatitis therapeutic, antimiRs targeted at miR-122, in phase II trials.[16] A significant barrier to miRNA-based therapy, however, is the development of targeted delivery paradigms with minimum toxicity. Recently, an engineered miRNA-based regulatory element was shown to control deleterious overexpression of a target gene in a Rett syndrome murine model.[17] Another study reported miRNA-mediated transgene de-targeting to promote immune tolerance of a transgene-encoded antigen.[4] Combination of the pleiotropic regulatory potential of miRNAs with gene therapy can allow targeted and potent expression of transgenes in specific tissue environments. Strategies for delivering miRNA therapeutics in a stable manner to target sites involve delivering miRNAs with synthetically modified oligoribonucleotides (ORNs) that mimic the native miRNA duplex or nanotechnology-based platforms.[18]

10.3 WHAT ARE HERPES VIRUSES

Herpesviridae is a large family of encapsulated DNA viruses. A typical herpes virion structure is made up of an inner core of double stranded DNA 124-235 kilobase (kbp) pair in size, a protein capsid, tegument, and an envelope derived from the nuclear membrane of the infected cell and long glycoprotein spikes. A large number of clinical manifestations, including malignancies, have been recorded for herpes virus related infections in humans. Interestingly, herpes viruses not only cause acute infections that may even be transmitted via organ transplantation, but also latent infections. The tendency to persist in the host following an initial infection, without producing overt disease is latency. Thus, herpesviruses are able to establish lifelong infections in the host and transition between the lytic (productive) and latent (non-productive) replication cycles. The microenvironment, host immunity, cell autonomous factors, as well as viral elements, may all contribute to this latency establishment, maintenance, and reactivation of herpes virus infection.[1]

Herpes virus species differ greatly in their genomic composition, with genome sizes ranging from 125 kbp to 230 kbp. The herpes virus genome is large and encodes at least 100 different proteins. Based on genome sequence and biology, there are eight human herpesviruses that are subdivided into three subfamilies; *Alphaherpesvirinae*; *Betaherpesvirinae*; and *Gammaherpesvirinae*. α-herpesviruses, such as human herpes simplex virus 1 (HSV-1), typically cause latent infections of neurons of the sensory ganglia and, cause lesions on adjacent mucosal surfaces following reactivation. Human cytomegalovirus (HCMV), a β-herpesvirus, is responsible for birth defects and establishes latency in hematopoietic progenitor cells. Human γ-herpesviruses establish latent infections primarily in B cells and are linked to several cancers.

The role of miRNAs in herpes virus infections has been studied extensively. Several herpes viruses exploit the most miRNA networks to facilitate their replication and survival. Following sections describe the details of viral miRNA-host interactions in some prominent herpesviruses. Herpesviridae themselves encode a range of miRNAs. Exploring their potential as tools in gene therapy has garnered significant research interest. When compared to each other, Herpesviruses seem to have co-evolved miRNAs as host immune regulators to escape immune cell recognition.[13]

10.4 EPSTEIN BARR VIRUS AND ASSOCIATED MICRORNAS

Epstein-Barr virus (EBV) is a DNA virus from the gamma herpesvirus family. Although mostly asymptomatic, EBV infection can be found in the majority of the population worldwide. Diseases caused by EBV include, acute infection such as mononucleosis and severe diseases such as Burkitt lymphoma, diffuse large B cell lymphoma, nasopharyngeal carcinoma, and post-transplant lymphoproliferative disease.[5] EBV exhibit latency after the primary infection, similar to other herpesviruses. In a latent infection, the EBV exists as an episome within the nucleus of memory B-lymphocytes. Depending on the expression pattern of EBV latent genes, four types of latency (0, I, II, III) have been described.[19] EBV are able to establish a persistent infection in host cells with the help of a vast array of viral proteins and miRNAs. The EBV double stranded (~170kbp) genome encodes 80 genes and 44 mature miRNAs derived from 25 pre-miRNAs.[1] These miRNAs are distributed into three clusters: the *Bam*H I fragment A rightward transcript (BART)-cluster 1, BART-cluster 2, and the *Bam*H I fragment H rightward open reading frame 1 (BHRF1)-cluster. The BART-1 family of miRNAs is the most diverse and involved in viral replication and apoptosis. BART-2 miRNAs are linked to host-cell transformation and in the prevention of apoptosis, thereby contributing to oncogenesis. The third BHRF1-cluster, miRNAs typically influence genes related to immune evasion.[20]

Expression of miRNA from each cluster is not co-regulated, but depending on the stage of latency, may vary (Wang et al., 2018). Some of the EBV miRNAs may work in a co-operative fashion to sustain viral latency.[21]

During the lytic replication cycle, EBV encoded proteins help in immune evasion, while miRNAs primarily serve this purpose during the latent cycle. EVB miRNAs target both the host mRNAs and the viral mRNAs, thus "fine tuning" gene regulation.[22] Examples of EBV miRNAs-viral gene regulation include suppression of viral membrane antigens LMP1 and LMP2 by the EBV miRNAs from BART-cluster 1 (ebv-miR-BART16, ebv-miR-BART17-5p, and evb-miR-BART1-5p) to help escape immune surveillance. Alternatively, EBV miRNAs-cellular mRNA interaction can affect protein synthesis of key cytokines to reduce immune response. For example, EBV-miR-BHRF1-3 and the T-cell chemokine CXCL11; ebv-miR-BART2-5p and the NK ligand MCIB; and ebv-miR-BART15 and NLRP3, a cellular protein that contributes to the production of pro-inflammatory cytokines such as IL-1β and IL-18. miR-BART1 and miR-BART2h miRNAs were also found to degrade the mRNA of the IL-12p40 subunit, thereby reducing IL-12B and IL-23 protein levels and leading to impaired T-cell differentiation from naive to Th1 cells.[1]

EBV genome encodes several miRNAs, most of which have been implicated in oncogenesis.[23] This may be due to the large array of pro-apoptotic proteins or host tumor suppressor genes targeted by EBV miRNAs. A prominent cellular target of EBV miRNAs is the pro-apoptotic protein PUMA (p53-upregulated modulator of apoptosis), which is regulated by miR-BART5-5p.[24] Zheng and colleagues demonstrated that EBV miRNA miR-BART5-3p can inhibit the expression of the critical tumor suppressor gene p53 by directly targeting its 3′-UTR, contributing to tumorigenesis.[25] The pro-apoptotic protein TOMM22 (translocase of outer mitochondrial membrane 22 homolog), part of the mitochondrial pore receptor complex for the pro-apoptotic protein BAX (BCL2-associated X protein), was identified as a potential target for ebv-miR-BART16. So are the BCL2-associated death promoter (BAD) protein, another target of the ebv-miR-BART20-5p, and the pro-apoptotic protein caspase 3 identified as being a target of ebv-miR-BART16 and ebv-miR-BART1-3p. Transport of EBV miRNAs from tumors into the blood stream has been implicated in the progression of EBV associated malignancies. This was demonstrated when EBV BART miRNAs were detected as extracellular vesicles secreted by LCLs and nasopharyngeal carcinoma (NPC) cells. BART1-5p, BART5, BART7-3p, BART12, and BART13 were also found in the circulating extracellular vesicles from NPC patients. Clinically, EBV miRNAs my prove to be promising biomarkers for detection of EBV related malignancies. EBV miR-BART3, miR-BART7, and miR-BART13 were highly expressed in EBV-positive NPC. A total of 12 BART miRNAs (miR-BART1-3p, 2-5p, 5, 6-3p, 6-5p, 7, 8, 9, 14, 17-5p, 18-5p, and 19-3p) were validated to be upregulated in NPC tissues versus their non-tumorous biopsies.[26] A negative regulatory role several BART-2 miRNAs was demonstrated for the DNA repair gene BRCA1 expression, which also increased the cell sensitivity to DNA damaging chemotherapeutic drugs, cisplatin and doxorubicin, thus facilitating novel NPC therapies.[27]

Interestingly, EBV-positive lymphomas frequently possess large deletions in viral microRNA (miRNA) clusters, suggesting that some miRNAs negatively regulate tumorigenesis.[28]

10.5 HUMAN CYTOMEGALOVIRUS AND ASSOCIATED MICRORNAS

The human cytomegalovirus (HCMV) is a beta herpes virus with a 230 kbp double stranded DNA, implicated in congenital infections that can be transmitted across the placental barrier, infected body fluids, and organ transplantation.[29] HCMV infect a broad range of cell types via varied mechanisms. Among all Herpesviridae, HCMV has the largest genome. HCMV infection in the immunocompromised and individuals undergoing immunosuppressive treatments can be life threatening.

Like other herpes viruses, HCMV miRNAs have been demonstrated to play a crucial role in host immune evasion for viral proliferation/maintaining latency. However, unlike EBV, HCMV miRNAs are not clustered in the viral genome.[1] The HCMV miRNAs are designated as UL or US miRNAs based on their location within the HCMV genome. A total of 26 such miRNAs have been identified along with potential targets, scattered throughout the genome. There is no clear-cut distinction between the function of the UL and US-encoded miRNAs, although it may be possible that the expression and function of each isolated HCMV miRNA may be regulated by its own regulatory sequence.[30]

This section highlights the regulatory strategies and targets of HCMV encoded miRNAs. Cytokine regulation by HCMV miRNA (miR-UL148D-1) was demonstrated for ACVR1B gene to limit IL-6 and RANTES/CCL5 secretion.[31] Similarly, the HCMV miRs UL112-3p, US5-1, UL112-1, US25-1-5p, and UL148D target multiple host inflammatory genes, with the result of mitigation of inflammatory response. Furthermore, miR-UL112-3p was shown to target the TLR2 transcript to limit signaling through the TLR2-IRAK1 axis and block NFκB activation.[29] Although some HCMV miRNAs and their targets have been reported, many more miRNAs, their target antiviral mechanisms, and their biological functions still remain unclear. A recent study reported for the first time the role of HCMV miRNA hcmv-miR-US33as-5p in targeting the IFN signaling pathway during both lytic and latent infection.[32]

Herpes viruses are able to maintain latency within the host by targeting viral transcripts. Several HCMV miRNAs, such as miR-UL148D and miR-US22, have been reported to play a key role in latency as well as reactivation. Interestingly, unlike other herpes viruses, very few HCMV miRNAs have been reported to target viral transcripts, although HCMV miR-UL112-1 was shown to downregulate the major immediate early trans activator, IE72, resulting in a reduction of viral replication and promotion of latency.[22] HCMV miR-US5-1 miRNA was shown to regulate the kinetics of expression of the DNA replication inhibitor Geminin (GMNN) in order to sustain latency.[33]

miRNAs also influence the viral ability to either switch between or maintain lytic/latent replication. HCMV miRNAs involved in the lytic phase of HCMV infection include miR-112-1, miR-US25-5p and miR-US29-5p; whereas, miR-UL36, miR-112-3p, miR-US5-2, miR-US22-5p, miRUS25-1-5p, miR-US25-2-5p, and miR-US29-3p are involved in the latent phase of the infection.[1]

HCMV have been shown to antagonize the innate antiviral immune response via the HCMV-encoded miR-US25-1-5p miRNA that targets CD147/EMMPRIN, a type-I transmembrane glycoprotein of the immunoglobulin superfamily.[34]

10.6 KAPOSI'S SARCOMA ASSOCIATED HERPESVIRUS AND ASSOCIATED MICRORNAS

Kaposi's sarcoma-associated herpesvirus (KSHV) is a double-stranded gamma herpes virus. This DNA oncovirus is the causative agent for Kaposi's sarcoma (KS), primary effusion lymphoma, and multicentric Castleman disease, commonly seen in immunocompromised individuals.[35] Similar to other herpesviruses, KSHV also exhibits latent and lytic life cycle phases. KSHV express 13 precursor miRNAs that generate 25 mature miRNAs, the majority of which are implicated in tumor genesis and immune evasion. Individual miRNAs may be expressed at up to 2,200 copies per cell.[36] Across the KSHV genome, the miRNAs appear to be clustered in the K12 intron and K12 ORF, so are designated as miR-K12-n. KSHV miRNAs are key to the establishment and maintenance of the latent infection/oncogenesis *in vivo*.[36]

During a latent infection, KSHV expresses a number of viral genes that confer anti-apoptotic, inflammatory, and angiogenic benefits to the infected cell. The major latent genes expressed by KHSV are latency-associated nuclear antigen (LANA), vCyclin; ORF72, vFLIP, kaposin (K12), and a cluster of viral microRNAs, all of which are located within the KSHV latency associated region (KLAR). KSHV miRNAs that are typically enhanced during the viral-lytic phase are ones originating from pre-miR-K10 and pre-miR-K12 pre-miRNAS.[12] The delicate balance between these phases in KSHV is controlled by miRNAs. It was shown that the viral miRNA miR-K3 can suppress viral lytic replication and gene expression. This KSHV miRNA can further maintain latency by targeting cellular transcriptions factor in infected cells. (Lu et al., 2010) KSHV miRNAs were shown to repress multiple mRNA targets in the STAT3 signaling cascade, deregulate cytokine-mediated gene activation, and suppress an interferon response, facilitating the lytic replication phase of the virus in an *in vitro* test system.[37]

A recent study demonstrated the role of KSHV miRNA in conferring antiapoptotic advantages to infected cells,[38] thus promoting oncogenesis. In addition, KSHV miRNAs have been shown to share seed sequences with cellular miRNAs, making them functional orthologs that are able to tap into established cellular miRNA target networks of the host. Three examples of this miRNA mimicry strategy include the miR-K12-10 (miR-142-3p ortholog) that inhibits the cellular TGF-β pathway to promote cell survival; miR-K12-3 (miR-23 ortholog) that inhibits caspases 3 and 7 to inhibit apoptosis; and miR-K12-11 (miR-155 ortholog) that suppresses several of the interferon response signaling pathway components to and promote cell survival.[22]

KSHV miRNAs have evolved to successfully target host cellular genes to their advantage and are often key proteins with broad influence on cell function. For example, KSHV miRNAs miR-K12-1 and miR-K12-7, have been shown to target MICB mRNA and subsequently reduce the protein expression by inhibition of translation in order to evade natural killer (NK) cell detection.[22]

miR-K12-5 was demonstrated to mitigate both mRNA and protein expression of MYD88, a cytosolic protein involved in toll like receptor/IL-1R signaling that is central to the innate and adaptive immune response arms. KSHV miR-K1 enhances the NF-κB pathway activity by directly targeting the key component IκB to inhibit viral lytic replication. Furthermore, KSHV v-miRNAs reduce expression of C/EBPβ p20 (LIP), a known negative regulator of IL6 and IL10 cytokines, to regulate the cytokine signaling in infected cells.[1] KSHV miRNAs (miR-K12-1, miR-K12-3-3p, miR-K12-6-3p, and miR-K12-11) have also been shown to influence the cell cycle regulation of the host in order to indirectly promote viral pathogenesis by acting on key genes such as p21, p53, and Thrombospondin 1 (THBS1) involved in cell cycle progression.[13]

10.7 HERPES SIMPLEX VIRUS AND ASSOCIATED MICRORNAS

Herpes simplex virus (HSV) is a widely disseminated human pathogen. Once infected, HSV infections remain dormant in the host. Occasionally, the virus can reactivate and cause recurrent disease, thus establishing a lifelong infection. The most frequent form of transmission is by physical contact with infected lesions or body fluids. HSV-1 and HSV-2 are the two closely related members of this group of viruses responsible for the development of oral and genital lesions, eczema, conjunctivitis, etc. HSV are spherical, measuring 120-200 nm in diameter, with a double stranded linear DNA core. HSV are successfully able to employ a range of strategies to maintain latency within the infected host. These include latency-associated transcripts (LATs), less characterized non-coding RNAs (ncRNAs), coding RNAs, and more recently, miRNAs. Another strategy adopted by HSV seem to involve modulation of host miRNAs to promote a favourable environment facilitating viral replication.[39]

HSV-1 and HSV-2 are known to encode 27 and 24 functional miRNAs, respectively. These are also known to share functional miRNA homologs.[40] These miRNAs are not clustered nor designated in groups, but are situated in a small segment of the genome. Some of the HSV-1 and HSV-2 miRNAs share the same seed sequence.[1] Most HSV miRNAs have been described to function during either the reactivation or in the latency phase of viral infection, and not so much to evade host immune response. This may be attributed to the fact that most of the miRNA targets of HSV seem to be of viral origin.[1] HSV-1-encoded miR-H2-3p was shown to attenuate cytosolic DNA-stimulated antiviral immune response by manipulating host DNA sensor molecule DDX41 to enhance virus replication in cultured cells.[41] HSV-1 microRNAs (miR-H28 and miR-H29) may be expressed late during an infection cycle and exported in exosomes.[42] HSV-1 miRNA miR-H2 bears a complete sequence complementarity to ICP0 (viral gene activator). Although inhibition of ICP0 expression by miR-H2 has been hypothesized to explain HSV latency, a recent study reported otherwise in cultured cells. Currently, there is very little knowledge on the functions of HSV miRNAs, and the area needs further elucidation.[43]

REFERENCES

1. Bots, S. T. F. & Hoeben, R. C. Herpesvirus microRNAs for use in gene therapy immune-evasion strategies. *Gene Ther* **24**, 385–391, doi:10.1038/gt.2017.37 (2017).
2. Joudeh, A., Saliba, K. A., Topping, K. A. & Sis, B. Pathologic basis of antibody-mediated organ transplant rejection: from pathogenesis to diagnosis. *Curr Opin Organ Transplant* **18**, 478–485, doi:10.1097/MOT.0b013e3283636ce6 (2013).
3. Dobbels, F., Hames, A., Aujoulat, I., Heaton, N. & Samyn, M. Should we retransplant a patient who is non-adherent? A literature review and critical reflection. *Pediatr Transplant* **16**, 4–11, doi:10.1111/j.1399-3046.2011.01633.x (2012).
4. Broderick, J. A. & Zamore, P. D. MicroRNA therapeutics. *Gene Ther* **18**, 1104–1110, doi:10.1038/gt.2011.50 (2011).
5. Murer, A. et al. MicroRNAs of epstein-barr virus attenuate T-cell-mediated immune control in vivo. *mBio* **10**, doi:10.1128/mBio.01941-18 (2019).
6. Sorel, O. & Dewals, B. G. MicroRNAs in large herpesvirus DNA genomes: recent advances. *Biomol Concepts* **7**, 229–239, doi:10.1515/bmc-2016-0017 (2016).
7. O'Brien, J., Hayder, H., Zayed, Y. & Peng, C. Overview of microRNA biogenesis, mechanisms of actions, and circulation. *Front Endocrinol (Lausanne)* **9**, 402, doi:10.3389/fendo.2018.00402 (2018).
8. Venuti, A. et al. HSV-1\EGFP stimulates miR-146a expression in a NF-kappaB-dependent manner in monocytic THP-1 cells. *Sci Rep* **9**, 5157, doi:10.1038/s41598-019-41530-5 (2019).
9. Zou, X. L. X. An overview of RNA virus-encoded microRNAs. *ExRNA* **1**, 1–5 (2019).
10. Varble, A. et al. Engineered RNA viral synthesis of microRNAs. *Proc Natl Acad Sci U S A* **107**, 11519–11524, doi:10.1073/pnas.1003115107 (2010).

11. Peng, Y. & Croce, C. M. The role of microRNAs in human cancer. *Signal Transduct Target Ther* **1**, 15004, doi:10.1038/sigtrans.2015.4 (2016).
12. Mishra, R., Kumar, A., Ingle, H. & Kumar, H. The interplay between viral-derived miRNAs and host immunity during infection. *Front Immunol* **10**, 3079, doi:10.3389/fimmu.2019.03079 (2019).
13. Boss, I. W. & Renne, R. Viral miRNAs: tools for immune evasion. *Curr Opin Microbiol* **13**, 540–545, doi:10.1016/j.mib.2010.05.017 (2010).
14. Simonson, B. & Das, S. MicroRNA therapeutics: the next magic bullet? *Mini Rev Med Chem* **15**, 467–474, doi:10.2174/1389557515666150324123208 (2015).
15. Geisler, A. & Fechner, H. MicroRNA-regulated viral vectors for gene therapy. *World J Exp Med* **6**, 37–54, doi:10.5493/wjem.v6.i2.37 (2016).
16. Rupaimoole, R. & Slack, F. J. MicroRNA therapeutics: towards a new era for the management of cancer and other diseases. *Nat Rev Drug Discov* **16**, 203–222, doi:10.1038/nrd.2016.246 (2017).
17. Sinnett, S. E., Boyle, E., Lyons, C. & Gray, S. J. Engineered microRNA-based regulatory element permits safe high-dose miniMECP2 gene therapy in Rett mice. *Brain*, doi:10.1093/brain/awab182 (2021).
18. Adams, B. D., Parsons, C., Walker, L., Zhang, W. C. & Slack, F. J. Targeting noncoding RNAs in disease. *J Clin Invest* **127**, 761–771, doi:10.1172/JCI84424 (2017).
19. Munz, C. Immune escape by non-coding RNAs of the epstein barr virus. *Front Microbiol* **12**, 657387, doi:10.3389/fmicb.2021.657387 (2021).
20. Zhang, J. et al. The oncogenic role of epstein-barr virus-encoded microRNAs in epstein-barr virus-associated gastric carcinoma. *J Cell Mol Med* **22**, 38–45, doi:10.1111/jcmm.13354 (2018).
21. Lung, R. W. et al. EBV-encoded miRNAs target ATM-mediated response in nasopharyngeal carcinoma. *J Pathol* **244**, 394–407, doi:10.1002/path.5018 (2018).
22. Bernier, A. & Sagan, S. M. The diverse roles of microRNAs at the host(-)virus interface. *Viruses* **10**, doi:10.3390/v10080440 (2018).
23. Jing, J. J., Wang, Z. Y., Li, H., Sun, L. P. & Yuan, Y. Key elements involved in epstein-barr virus-associated gastric cancer and their network regulation. *Cancer Cell Int* **18**, 146, doi:10.1186/s12935-018-0637-5 (2018).
24. Marquitz, A. R., Mathur, A., Nam, C. S. & Raab-Traub, N. The epstein-barr virus BART microRNAs target the pro-apoptotic protein bim. *Virology* **412**, 392–400, doi:10.1016/j.virol.2011.01.028 (2011).
25. Zheng, X. et al. Epstein-barr virus microRNA miR-BART5-3p inhibits p53 expression. *J Virol* **92**, doi:10.1128/JVI.01022-18 (2018).
26. Wang, M. et al. The function and therapeutic potential of epstein-barr virus-encoded microRNAs in cancer. *Mol Ther Nucleic Acids* **17**, 657–668, doi:10.1016/j.omtn.2019.07.002 (2019).
27. Lung, R. W. et al. EBV-encoded miRNAs can sensitize nasopharyngeal carcinoma to chemotherapeutic drugs by targeting BRCA1. *J Cell Mol Med* **24**, 13523–13535, doi:10.1111/jcmm.16007 (2020).
28. Kimura, H., Okuno, Y., Sato, Y., Watanabe, T. & Murata, T. Deletion of viral microRNAs in the oncogenesis of epstein-barr virus-associated lymphoma. *Front Microbiol* **12**, 667968, doi:10.3389/fmicb.2021.667968 (2021).
29. Landais, I. et al. Human cytomegalovirus miR-UL112-3p targets TLR2 and modulates the TLR2/IRAK1/NFkappaB signaling pathway. *PLoS Pathog* **11**, e1004881, doi:10.1371/journal.ppat.1004881 (2015).
30. Zhang, L., Yu, J. & Liu, Z. MicroRNAs expressed by human cytomegalovirus. *Virol J* **17**, 34, doi:10.1186/s12985-020-1296-4 (2020).
31. Lau, B. et al. The expression of human cytomegalovirus microRNA MiR-UL148D during latent infection in primary myeloid cells inhibits activin A—triggered secretion of IL-6. *Sci Rep* **6**, 31205, doi:10.1038/srep31205 (2016).
32. Zhang, Q. et al. Human cytomegalovirus miR-US33as-5p targets IFNAR1 to achieve immune evasion during both lytic and latent infection. *Front Immunol* **12**, 628364, doi:10.3389/fimmu.2021.628364 (2021).
33. Jiang, S. et al. Human cytomegalovirus miR-US5-1 inhibits viral replication by targeting Geminin mRNA. *Virol Sin* **32**, 431–439, doi:10.1007/s12250-017-4064-x (2017).
34. Chen, J. et al. Human cytomegalovirus encoded miR-US25-1-5p attenuates CD147/EMMPRIN-mediated early antiviral response. *Viruses* **9**, doi:10.3390/v9120365 (2017).
35. Liu, X., Happel, C. & Ziegelbauer, J. M. Kaposi's sarcoma-associated herpesvirus microRNAs target GADD45B to protect infected cells from cell cycle arrest and apoptosis. *J Virol* **91**, doi:10.1128/JVI.02045-16 (2017).

36. Cai, X. et al. Kaposi's sarcoma-associated herpesvirus expresses an array of viral microRNAs in latently infected cells. *Proc Natl Acad Sci U S A* **102**, 5570–5575, doi:10.1073/pnas.0408192102 (2005).
37. Ramalingam, D. & Ziegelbauer, J. M. Viral microRNAs target a gene network, inhibit STAT activation, and suppress interferon responses. *Sci Rep* **7**, 40813, doi:10.1038/srep40813 (2017).
38. Lu, C. C. et al. MicroRNAs encoded by Kaposi's sarcoma-associated herpesvirus regulate viral life cycle. *EMBO Rep* **11**, 784–790, doi:10.1038/embor.2010.132 (2010).
39. Wang, X. et al. ICP4-induced miR-101 attenuates HSV-1 replication. *Sci Rep* **6**, 23205, doi:10.1038/srep23205 (2016).
40. Cokaric Brdovcak, M., Zubkovic, A. & Jurak, I. Herpes simplex virus 1 deregulation of host microRNAs. *Noncoding RNA* **4**, doi:10.3390/ncrna4040036 (2018).
41. Duan, Y. et al. Herpes simplex virus type 1-encoded miR-H2-3p manipulates cytosolic DNA-stimulated antiviral innate immune response by targeting DDX41. *Viruses* **11**, doi:10.3390/v11080756 (2019).
42. Han, Z. et al. miR-H28 and miR-H29 expressed late in productive infection are exported and restrict HSV-1 replication and spread in recipient cells. *Proc Natl Acad Sci U S A* **113**, E894–901, doi:10.1073/pnas.1525674113 (2016).
43. Pan, D., Pesola, J. M., Li, G., McCarron, S. & Coen, D. M. Mutations inactivating herpes simplex virus 1 microRNA miR-H2 do not detectably increase ICP0 gene expression in infected cultured cells or mouse trigeminal ganglia. *J Virol* **91**, doi:10.1128/JVI.02001-16 (2017).

11 Gene Therapy and Small Molecules Used in the Treatment of Cystic Fibrosis

Manish P. Patel, Uma G. Daryai, Mansi N. Athalye and Praful D. Bharadia
L. M. College of Pharmacy, Ahmedabad, Gujarat, India.

Jayvadan Patel
Sankalchand Patel University, Gujarat, India

CONTENTS

DOI: 10.1201/9781003186083-11

11.1 INTRODUCTION

Cystic fibrosis is an autosomal recessive genetic disease occurring due to mutations in the cystic fibrosis transmembrane conductance regulator (CFTR) gene (Yang et al., 2019) which is comprised of 27 exons and is situated on the long arm of chromosome 7 (Burney and Davies, 2012). The disease is known to have a prevalence of roughly 1:2,500 and a carrier rate of around 1:25 within the Caucasian population, along with estimated worldwide numbers of 60,000 to 70,000 affected patients (Thursfield and Davies, 2012). Almost 2,000 distinct CFTR variants have been identified. However, more research is needed to understand whether many of these variants cause disease or not; the complete information of the same is provided on the following link (http://www.genet.sickkids.on.ca/app) (Yan et al., 2019). CFTR gene mutation leads to dysfunction affecting the functions of many organs; however, lung disease is responsible for the vast majority of mortality in patients with cystic fibrosis. The cystic fibrosis phenotype is designated by progressive lung disease that, over time, limits the ability to breath properly, exocrine pancreatic insufficiency that results in intestinal abnormalities, and gastrointestinal malabsorption leading to malnutrition, impaired growth, and also a variety of other manifestations that include sinusitis and diabetes (Ratjen et al., 2015). The disease is also characterized by malabsorption of fat and protein, fatty stools, growth failure, and pulmonary infection (Davis, 2006). The male reproductive tract is another targeted area that leads to bilateral absence of the vas deferens and resultant obstructive azoospermia in 98% of male patients (Thursfield and Davies, 2013). Abnormally high rates of sodium and water absorption are found in cystic fibrosis airway epithelia that are responsible for dehydration of the airway surface liquid (Burney and Davies, 2012). The mechanism behind dehydration is that the cystic fibrosis lung disease exemplifies how the defective functioning of a single ion channel, (i.e., CFTR ion channel) results in serious disturbances in airway surface liquid (ASL) physiology. With the loss of CFTR function in cystic fibrosis epithelial cells, Cl^- is retained within the epithelial cells, while Na^+ absorption by epithelial sodium channel (ENaC) will increase, resulting in augmented epithelial cytosolic NaCl contents. This augmented cytosolic contents of NaCl in epithelial cells produce an osmotic drive promoting net movement of water from the airway surface liquid (ASL) layer into the epithelial cells, thus leading to ASL layer dehydration (Lewis, Patial, and Saini, 2019); once the airway surface gets dehydrated, mucociliary clearance fails to remove any of the inhaled bacteria, which is responsible for infecting lower airways, finally causing inflammation. Also the appearance of various inflammatory cell contents, such as DNA and elastase in the airway, leads to increased mucus viscosity and contributes to tissue breakdown. (Burney and Davies, 2012.)

11.2 HISTORY OF CYSTIC FIBROSIS TRANSMEMBRANE CONDUCTANCE REGULATOR (CFTR) GENE

The study on the CFTR gene started in 1981 when Michael Knowles and colleagues revealed that there is abnormal potential difference in the nasal mucosa of patients with cystic fibrosis, which provides the direct evidence of primary epithelial dysfunction (Knowles et al., 1981). In 1983, Quinton observed that the chloride impermeability in sweat glands was the basis for raised sweat electrolytes in patients with cystic fibrosis (Quinton, 1983). The important advances were observed for the basic defect as a cell membrane transport problem, which provided the first description of a basic cellular defect that has since then been seen in all cystic fibrosis affected cells. It was observed that the problems with the mucus were not due to abnormalities with its synthesis or any change in composition; rather with the fluid environment into which it was secreted. From the early 1980s, various groups tried 'reverse genetics' to identify the Cystic Fibrosis Gene, as the protein was unknown. The studies were carried out on families with more than one affected child. In 1985, using 'reverse genetics,' Eiberg in Copenhagen exhibited a linkage to the enzyme paraoxinase, which exists in two forms, but is present in the same form in 90% of CF siblings (Eiberg et al., 1985). In the same year, experiments were carried out by Lap-Chee Tsui in Toronto on a mouse, revealing a marker on chromosome 7 that was linked to both paraoxinase and cystic fibrosis (Tsui et al., 1985) Finally in 1989, the CF gene was identified by Lap-Chee Tsui, Francis Collins, and Jack Riordan and was termed as the 'cystic fibrosis transmembrane conductance regulator' (Littlewood, 2007).

11.3 PATHOPHYSIOLOGY OF CFTR GENE

The disease is caused by mutations in the CFTR gene. This gene is a unique member of the ATP-binding cassette (ABC) or traffic ATPase family of genes, which carries a regulatory domain that is actively phosphorylated. The gene encodes for a 1,480-amino-acid protein known as the cystic fibrosis transmembrane conductance regulator (CFTR) (Lane and Doe, 2014); this regulator functions primarily as an apical anion channel of chloride and bicarbonate, instead of acting as an active pump. The gene includes two nucleotide-binding domains (NBDs), encoding sites capable of binding and hydrolyzing ATP (Walker A and B motifs) and membrane-spanning domains, which function as the ion channel pore through the plasma membrane (Ratjen et al., 2015). The domain layout is illustrated in Figure 11.1 (Turnbull, Rosser, and Cyr, 2007). In both the respiratory and

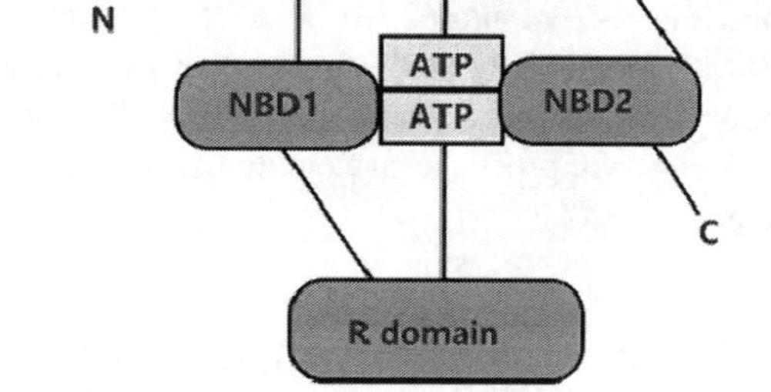

FIGURE 11.1 CFTR domain layout in the ER membrane lipid bilayer during ATP hydrolysis, depicting the membrane spanning domains (MSD), nucleotide binding domains (NBD) and regulatory domain (R).

(Turnbull, Rosser, and Cyr, 2007.)

gastrointestinal tracts, bicarbonate is responsible for driving the ionic content. On the epithelial surfaces, bicarbonate allows mucins to unfold and become slippery, contributing to innate immunity, whereas in the GI tract it neutralizes gastric acid to support digestion and absorption. When CFTR is dysfunctional, there is less bicarbonate secretion, hence, this disfunction is responsible for disruption of these normal processes that directly lead to the clinical symptoms and signs of Cystic Fibrosis (Borowitz, 2015). As the person ages, the mucus becomes increasingly concentrated, leading to reduced mucociliary clearance, which results in the development of chronic polymicrobial infection with microbes like staphylococcus aureus, H-Influenzae, Pseudomonas aeruginosa, etc. (Mall and Stuart Elborn, 2014).

11.4 IONIC CONTENT AND FLUID ON EPITHELIAL SURFACES

As the CFTR gene is responsible for secretion of bicarbonate and chloride, it works as a chloride-bicarbonate exchanger and is known to have an exchange of one chloride molecule for two bicarbonate molecules by the apical chloride-bicarbonate exchanger in the SLC26 family. In addition to direct conductance of ions through CFTR, local pH affects the proteins that act on different ion channels, and most specifically in the airways (Borowitz, 2015). Normal airway surface liquid (ASL) has a pH of around 7.0, with the bicarbonate concentration of 10 to 20 mMol. The literature describes the evidence that airway surface liquid pH is controlled by bicarbonate secretion via CFTR and other secreted proteins. Airways clearances are also affected by changes in pH, based on the observation that transient acidification of normal airway surface liquid increases the rate of airway surface liquid absorption by the epithelia, while alkalinization of airway surface liquid leads to a slowing of airway surface liquid absorption and transiently restores airway surface liquid height into the normal range. Thus an epithelial sodium channel (ENaC) plays a major role in ionic content and fluid on the airway epithelial surface (Borowitz, 2015). Prostasin, a membrane-anchored serine protease with a trypsin-like substrate specificity, is required to activate ENaC but is inactive below pH 7.0. Another regulatory protein is a short palate lung and nasal epithelial clone 1 (SPLUNC1, also known as BPIFA1) that has been shown to be a pH-sensitive regulator of ENaC that is not able to inhibit ENaC within the acidic cystic fibrosis airway environment (Garland et al., 2013). Hence, CFTR is responsible for controlling ENaC by regulating the local acid–base balance. Hence, both the gastrointestinal tract and airways are responsible for bicarbonate secretion, which regulates ionic content and fluid secretion on the epithelial surfaces (Borowitz, 2015). Although the 'low volume' hypothesis, (Boucher, 2007) suggests that decrease in the transepithelial chloride transport due to mutated CFTR and increase in transepithelial sodium absorption due to lack of CFTR-dependent inhibition of the epithelial sodium channel leads to an increase in water absorption into the tissue and, therefore, decreases airway surface liquid and mucociliary clearance (Uta Griesenbach, Pytel, and Alton, 2015). The work of chloride secretion is not only limited to CFTR—calcium-dependent chloride channels but it may also act as alternative channels if CFTR is non-functioning. There are at minimum two agents that stimulate this secretion, which are in clinical trials. Denufosol has been shown to stimulate chloride secretion (Accurso et al., 2008). The other drug is Lancovutide or Moli1901, which has been shown to improve Nasal Potential Difference (NPD) when applied to the nasal epithelium (Grasemann et al., 2007).

11.5 ROLE OF CYSTIC FIBROSIS TRANSMEMBRANE CONDUCTANCE REGULATOR (CFTR)

The role of CFTR in the human lung is not clearly understood, but during animal studies mostly in lambs it is found that there is active chloride secretion in the lung early in the gestation period, which leads to the production of the fetal lung liquid. But few days before birth, the amount of fluid secretion falls, and after birth the lungs absorb rather than secreting the fluid through the sodium absorption; whereas in the humans, the gene encoding CFTR is expressed throughout the gestation, which becomes localized to the airway in the third trimester of pregnancy. After birth, most CFTR expression occurs in the serous cells of the submucosal gland; however after a period of two years from birth, the percentage of serous cells increases, and a few new glands are formed that also increases the gland area. The gland is not able to approach the form characteristic of an adult until 13 years of age; this is relevant if the submucosal glands are to be targeted by gene therapy in youngsters. At present, epithelial cells of the airway are targeted; also, it is estimated that as little as 6% of cells need to be transfected to correct the chloride defect (Jaffe et al., 1999). CFTR also has a crucial role in the conduction of anions across epithelia to manage the pH, mucus viscosity of liquid secretions, and volume (Pezzulo et al., 2012; Wine et al., 2018). The gene interacts with other ion channels, such as the epithelial sodium channel (ENaC), to regulate the fluid movement across epithelia, and thus can have cell-intrinsic functions based on the composition of other channels co-expressed with CFTR (Shei et al., 2018). The epithelial sodium channel is multimeric transmembrane protein consisting of three subunits (2α, 1β, and 1γ), all contributing to the central channel pore. Following the cloning of epithelial sodium channel subunits and co-expression of CFTR and ENaC in heterologous cells manifested that cAMP-dependent activation of wild type but not mutant CFTR results in inhibition of epithelial sodium channel activity. Findings lead to the hypothesis that CFTR dysfunction results in the basic defect of epithelial ion transport, which is characterized by deficient cAMP dependent anion secretion and abnormal ENaC-mediated sodium absorption across cystic fibrosis surfaces(Mall and Stuart Elborn, 2014); also in case of diabetes, abnormal chloride channel function due to cystic fibrosis results in thick viscous secretions, causing obstructive damage to the exocrine pancreas, with progressive fibrosis and fatty infiltration. This leads to disruption and destruction of islet architecture, resulting in loss of endocrine beta, alpha, and pancreatic polypeptide cells (O'Riordan et al., 2009). Also, CFTR is known to act as a receptor for the binding, endocytosing, and clearing of the pseudomonas aeruginosa, but due to dysfunction in the CFTR gene, this important function is absent in patients with cystic fibrosis. So, defected genes lead to an increase in the binding sites for bacteria, which causes inflammation (Jaffe et al., 1999).

11.6 CYSTIC FIBROSIS RELATED LUNG DISEASE

As cystic fibrosis widely affects lungs due to accumulation of mucus in the airway epithelium, therefore, the study of airway epithelium provides the ways to overcome the mucus accumulation and progression of further lung destruction. Airway epithelium act as a barrier to numerous pathogens and foreign particles, thereby preventing infection and tissue injury by secreting the mucus and through the action of mucociliary clearance. There are four main types of respiratory epithelial cells, such as ciliated cells, goblet cells, club cells, and airway basal cells. The ciliated cells serve as the main component in the mucociliary clearance mechanism each epithelial cell is known to have around 200 cilia that beat constantly at a rate of between 10 and 20 times/second. Goblet cells are columnar epithelial cells that secrete mucus as part of the airway surface liquid; the composition of goblet cells is tightly regulated by mucus that helps to maintain epithelial moisture, and also traps particulate material and pathogens moving through the airway and determines how well mucociliary clearance works. The basal cells are small, nearly cuboidal cells that responds to injury of the airway epithelium, migrating to cover a site and finally differentiating to restore a healthy epithelial cell layer. Club cells carry out the same functions in the more distal airways (Wikipedia The free encyclopedia, 2021). Topical delivery of gene transfer agents to the lung is currently the most preferred method for airway gene transfer. But before the gene transfer agents reach the surface of the epithelial cells, they have to overcome a huge number of extracellular physical and immunological barriers (Ferrari et al., 2003; Weiss, 2002). It has been observed that mucus extensively reduces the transfection efficiency of most viral and non-viral gene transfer agents. However, transfection efficiency can be increased by prior treatment with mucolytics or anticholinergic drug glycopyrrolate, both in vitro and in vivo (Ferrari, Geddes, and Alton, 2002).

11.7 SIGNS AND SYMPTOMS OF CYSTIC FIBROSIS

11.7.1 Respiratory Symptoms

Various respiratory symptoms are observed that involve recurrent pneumonia, bronchiolitis, staphylococcal pneumonia, persistent pneumonia, and pansinusitis (Stern, 1997).

11.7.2 Gastro-enterologic Symptoms

This includes voluminous, foul-smelling stools, failure of proper growth during infancy and childhood, distal intestinal obstruction in children and young adults, rectal prolapse, intussusception in childhood or adolescence, hepatic cirrhosis, cholelithiasis, or pancreatitis before the age of 30, and deficiency of vitamin A, D, E, or K (Stern, 1997).

11.7.3 Miscellaneous Symptoms

Other symptoms observed are salty-tasting skin, hyponatremic dehydration or metabolic alkalosis in infancy, and recurrent hyponatremic heat prostration in adults, clubbing in children and young adults, azoospermia, congenital absence of the vas deferens, hypoproteinemia, or anasarca (in infancy) (Stern, 1997), lack of appetite, loss of taste along with abdominal pain, and flatulence (Tabori et al., 2017).

11.7.4 Symptoms that may Indicate the Presence of CFRD (Cystic Fibrosis Related Diabetes)

Patients with cystic fibrosis related diabetes are observed to have symptoms as unexplained polyuria, failure to maintain weight, delayed progression of puberty, and unexplained chronic decline in pulmonary function (O'Riordan et al., 2009).

11.8 DIAGNOSIS

Various diagnostic tests are performed to identify the persons with cystic fibrosis. The following section summarizes all of them.

11.8.1 Sweat Testing

The identification of cystic fibrosis can be made on the concept of a sweat chloride concentration of 60 mEq/L or bigger, plus either a sibling or first cousin with cystic fibrosis or a person having the respiratory illness of an applicable character or patient with pancreatic insufficiency (Davis, 2006). Sweat chloride concentration is known to be the standard screening test. Tests that are frequently strongly positive (chloride concentration, ≥ 80 mmol/liter) along with a characteristic clinical manifestation or a family history confirms the diagnosis. Although most of the patients have sweat chloride concentrations (60 to 79 mmol/liter), some have minimally abnormal concentrations (40 to 59 mmol/liter in childhood and slightly higher concentrations in adults), but some patients having normal sweat chloride concentrations are known to have cystic fibrosis. In some cases, sweat chloride measurements can be incorrectly low if the sweat sample is inadequate or if the patient has edema (Stern, 1997). However, in most cases the diagnosis of cystic fibrosis can be confirmed by measurement of chloride concentration in sweat after iontophoresis of pilocarpine. Sweat testing should be carried out in accordance with the rules of the National Committee for Clinical Laboratory Standards. The only acceptable procedure is the quantitative pilocarpine iontophoresis sweat test. A minimum acceptable volume (15 μl for the Wescor Macroduct coil system) or weight (75 mg for the Gibson-Cooke procedure) of sweat must be collected during a 30 minute period to ensure an average sweat rate of more than 1gm/m²/min (LeGrys et al., 1994). Alternative sweat test procedures are direct-reading conductivity measurements (Kopito and Shwachman, 1969) or a paper-patch indicator system which involves the chloride concentration test; the patch contains an inner white circular area surrounded by an outer brown circular area, with a small rectangular tab extending from it. This patch is designed in such a way that the border between the inner and outer circular areas breaks, and thus, the original brown area turns white whenever the chloride concentration reaches approximately 45 mEq/l (Yeung et al., 1984). However, the described tests show an incidence of false-positive and false-negative results. Hence, they should never be used as the basis of a definitive CF diagnosis (Rosenstein and Cutting, 1998).

11.8.1.1 Interpretation of Sweat Test

For the interpretation of results, values of sweat chloride have been applied to all patients regardless of age [i.e., normal (≤39 mmol/liter), intermediate (40 to 59 mmol/liter), and abnormal (≥60 mmol/liter)]. Sweat chloride values in infancy were found by carrying out the study on 103 infants without CF. The value was found as a mean sweat chloride value (±1 standard deviation) of 23.3 ± 5.7 mmol/L at age 3 to 7 days, which decreases further to 17.6 ± 5.6 mmol/L by age 8 to 14 days, and finally to 13.1 ± 7.4 mmol/L after age 6 weeks.(Farrell et al., 2008).

11.8.2 Genotyping

Although 500 cystic fibrosis transmembrane conductance regulator (CFTR) mutations related to cystic fibrosis are known (Stern, 1997), the commercially available probes tests are only for 70 CFTR mutations. Even though these 70 mutations can be used to identify more than 90% of all cystic fibrosis genes, failure to seek out two abnormal genes does not rule out the disease. It was found that in approximately 1% of patients with the disease no abnormal gene can be found, and in about 18% more, only one abnormal gene will be identified. Even if both genes are abnormal, the patient might have an ameliorating or neutralizing second mutation somewhere else. For example, patients homozygous for delta F508 (the most common CFTR mutation) have normal sweat electrolyte concentrations if a second mutation R553Q is also present (Dörk et al., 1991). So, the fusion of two cystic fibrosis mutations plus abnormal concentration of sweat electrolytes is normally accepted as sufficient for diagnosis. Occasionally, family testing (genotyping or analysis of restriction-fragment–length polymorphisms) can provide additional information (Stern, 1997).

11.8.3 Semen Analysis

Obstructive azoospermia holds to be strong evidence of cystic fibrosis and should be confirmed by testicular biopsy (Stern, 1997).

11.8.4 Sinus Radiographs

Pansinusitis is almost common in cystic fibrosis; hence, its presence strongly suggests a diagnosis of cystic fibrosis (Stern, 1997).

11.8.5 Tests of Exocrine Pancreatic Function

Most clinicians accept the signs and symptoms of malabsorption and add on clear response to pancreatic enzyme treatment as sufficient evidence of exocrine insufficiency (Stern, 1997). This test supports the diagnosis, but the gold standard (intubation, pancreatic-duct isolation, and analysis of secretions before and after stimulation with secretin and cholecystokinin) may be necessary to detect minute abnormalities (Lankisch, 1982).

11.8.6 Nasal Potential Difference Measurements

Nasal potential difference measurements are measurements of the potential difference (voltage) that use as the reference electrode a saline-filled butterfly needle inserted subcutaneously within the forearm, and as the mucosal electrode, a saline-filled polyethylene tube that lightly touches the mucosa close to the anterior turbinate. The test should never be performed in patients that are recovering from an acute respiratory tract infection. The voltage measured (normal mean [±SD], −24.7 ± 0.9 mV; abnormal, −53 ± 1.8 mV) correlates with the movement of sodium across cell membranes – a physiologic function rendered abnormal by mutant CFTR. As amiloride blocks the epithelium sodium channel, it leads to a large drop in potential difference, which is usually greater in patients with cystic fibrosis (73%) compared to those in normal subjects (53%). Hence, for a clearer idea, measurements are repeated with amiloride. Also, subsequent perfusion with a chloride-free solution and isoproterenol normally produces sharp increases in potential difference (of about 30 mV), but has little effect when CFTR function is abnormal. This test may demonstrate abnormal CFTR function more reliably than the sweat test (Knowles, Paradiso, and Boucher, 1995).

11.8.7 Bronchoalveolar Lavage

The bronchoalveolar lavage procedure is basically done to collect a sample from the lungs for diagnostic purpose. The results of diagnosis shows a high percentage of neutrophils (≥50% in patients with cystic fibrosis as compared with normal subjects having 3%); the presence of large numbers of neutrophils in lavage fluid, even in the absence of pathogens, is strong evidence of cystic fibrosis (Stern, 1997).

11.8.8 Respiratory Tract Microbiology

Characterization of the respiratory microbial flora can be diagnostically helpful in the evaluation of patients with atypical features of cystic fibrosis. The predilection of Pseudomonas aeruginosa to colonize the respiratory tract in cystic fibrosis is documented. The presence of the mucoid phenotype of P. aeruginosa within the respiratory tract, particularly if persistent, is highly suggestive of cystic fibrosis. Persistent colonization with other organisms also support diagnosis of cystic fibrosis, though many of the pathogens are also found in other conditions (Rosenstein and Cutting, 1998).

11.8.9 Newborn Screening

Patients can also be identified by newborn screening. Newborn screening is the nationwide program that helps to spot the babies born with cystic fibrosis and assist health care suppliers to take the mandatory steps to retain the babies as healthy as attainable. The immunoreactive trypsin test was done, in which results showed evidence that immunoreactive trypsin levels measured in blood spots collected at birth were increased in most patients with cystic fibrosis (Davis, 2006).

11.9 NATIONAL REGISTRIES: THE CRUCIAL PROVISION TO MONITOR PROGRESS

Data collection was becoming increasingly important for both the national and local Cystic Fibrosis centers. Hence, the Cystic Fibrosis Foundation's Patient Registry was developed by Warren Warwick in 1964. In 1995, the UK Cystic Fibrosis (CF) Database replaced the original CF Survey started by the working party in 1982; the UK CF Database was replaced by CF Trust's CF Registry in 2007. Currently; CF Trust's CF Registry in 2019 is used, which gives complete information on data collection and currently available treatment strategies (Littlewood, 2007).

11.10 DIFFERENT CLASS MUTATIONS OF CYSTIC FIBROSIS

Cystic fibrosis is caused due to class mutations. Some characteristics of class mutations are as follows.

Class I defects account for up to 10% of all cystic fibrosis mutations and are found in more than 60% of selected population groups, e.g., Israeli patients (Lane and Doe, 2014),

The class II mutation i.e.,F508del accounts for two-third of all cystic fibrosis mutations worldwide, and about 50% of affected patients are homozygous for this mutation (Lane and Doe, 2014).

Class III mutations works upon channel activation by preventing binding and hydrolysis of ATP at one of the two NBDs (NBD1 or NBD2) (Wilschanski, 2010). This class III mutation G551D accounts for about 5% of all CF mutations (Van Goor et al., 2009)

Class IV mutations create a protein with impaired function due to abnormal anion conduction (Wilschanski, 2010)

Class V mutations reduce the number of normally functioning CFTR molecules on the apical surface (Wilschanski, 2010).

Class VI mutations result from truncation of the C terminus of CFTR, and produce a functional protein that is unstable at the apical membrane surface (Wilschanski, 2010).

Details of various class mutations are given in Figure 11.2 and Table 11.1 (Cystic Fibrosis Trust, 2014; Thursfield and Davies, 2013; Brodlie et al., 2015).

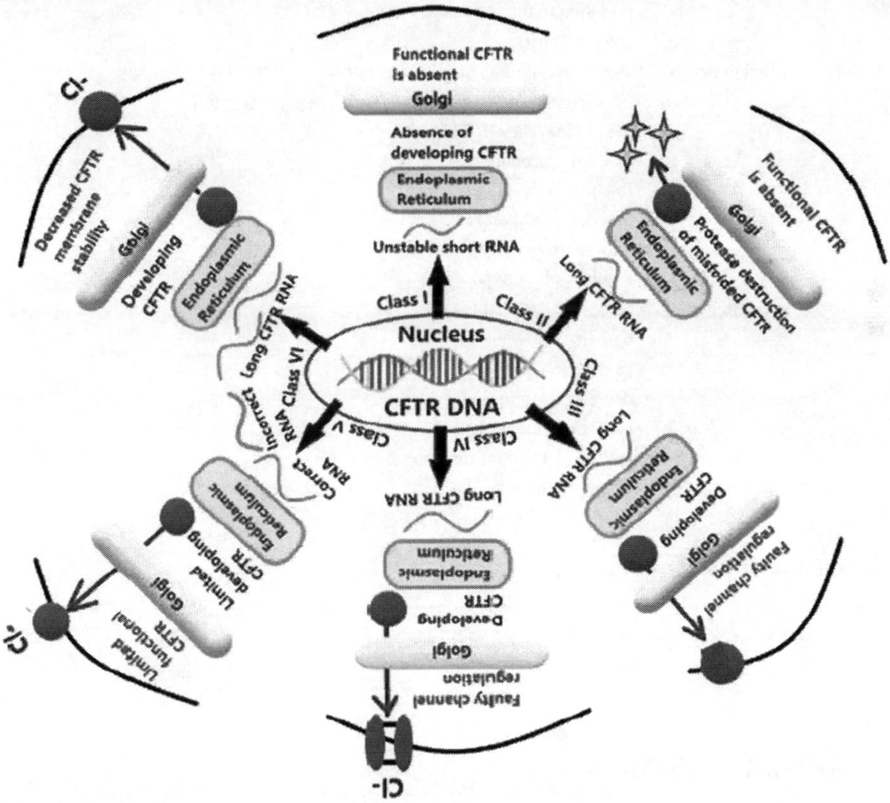

FIGURE 11.2 CFTR different class mutations.

(Cystic Fibrosis Trust, 2014.)

TABLE 11.1
Different Classes of CFTR Gene Mutation and Therapeutic Strategy to Overcome Mutation

Mutation Class	Basic Defect	CFTR Protein Abnormality	Old Name	New Name	Therapeutic Strategy
I	Nonsense mutations	Premature termination codon leads to a truncated protein with no function	G542X W1282X	Gly542X Trp1282X	correctors (ataluren)
II	Trafficking defect	Misfolding causes the protein to fail to be trafficked to the apical surface of the cell	DeltaF508 N1303K	Phe508del Asn1303Lys	Corrector plus potentiator (lumacaftor plus ivacaftor, VX-661 plus ivacaftor)
III	Gating defect	Normal quantities of CFTR protein reach the cell membrane but the channel remains closed, preventing normal function	G551D	Gly551Asp	Potentiator (ivacaftor)
IV	Decreased conductivity	Protein reaches the apical surface but conductance across the channel is poor due to the abnormal conformation	R117H	Arg117His	Potentiator (ivacaftor)
V	Splicing defect	Decreased amounts of CFTR reach the apical membrane The CFTR that does, probably functions normally; hence, many patients with these mutations have a milder phenotype	38949 + 10kbC.T	c.3717 + 12191C.T	NA
VI	Decreased half-life	CFTR is functional but half-life is decreased, leading to overall decreased amount of CFTR present at the cell surface, and it may impair regulation of other channels	Q1412X	Gln1412X	NA

Source: Thursfield and Davies (2013); Brodlie et al. (2015).

11.11 TREATMENT OPTIONS

Treatments Available for cystic fibrosis are Gene therapy, Small molecules used in the treatment of cystic fibrosis, Managing a good nutritional state in older children, Gastrointestinal therapies, Pulmonary therapies, Organ transplantations.

11.11.1 Gene Therapy

Gene therapy is outlined as a transfer of genetic material to cure a sickness or at least to enhance the clinical status of a patient. One of the basic ideas of gene therapy treatment is it transforms viruses into genetic shuttles, which could deliver the gene of interest into the target cells to treat the illness (Akhtar et al., 2011). Currently, two chiefly used approaches for genetic corrections of the cells are named as virus mediated and another via physical mechanisms from preparations obtained by advanced nanotechnology techniques. Target cell types of gene therapy are subdivided into two massive groups: gene therapy of the germline and gene therapy of somatic cells (Gonçalves and de Melo Alves Paiva, 2017). Gene therapy for treatment of the cystic fibrosis lung has been an energetic space of analysis. Advances created with delivery vehicles for gene therapy can inform development of delivery vehicles for gene editing. It's vital to notice that the cargoes for gene therapy and gene editing likely will differ. Gene therapy delivers DNA with the goal of the introduced DNA persisting, whereas in gene editing, the gene editor need only be present long enough to accomplish the correction (Hodges and Conlon, 2019).

11.11.1.1 Importance of Gene Therapy

As gene therapy involves insertion of the correct gene in place of a mutated gene, this helps in the treatment of disease at the targeted site. Also, the success of providing CF patients with a gene therapy treatment can substantially improve their quality of life by preventing irreversible lung damage (Carlon, Vidović, and Birket, 2017).

11.11.1.2 Challenges for Gene Therapy

A respiratory organ is known to be a complex and tough organ due to the presence of various potent intracellular and extracellular barriers that have evolved to guard the lungs from foreign bodies, including bacteria, viruses, and gene transfer agents (Griesenbach, Pytel, and Alton, 2015). The most important challenges for gene therapy relate to technical difficulties caused by the mucus barrier and the innate and adaptive immune systems in the lung. From a structural perspective, the lung is well equipped to defend against particle uptake because of a mucus layer serving as a primary barrier. Thus, the physical barrier is intensified in patients with cystic fibrosis, who have a thicker and more viscous mucus layer that can impede the uptake of the vector and prevent its delivery to distal airways (Prickett and Jain, 2015).

11.11.1.3 Gene Editing

In recent years, gene editing has been promoted as a possible treatment strategy for many genetic diseases along with cystic fibrosis. This therapeutic approach permits mediation at the chromosomal level of a specific gene, resulting in the correction of all disease-associated mutations. CFTR function can be saved by either editing the endogenous gene or inserting a wild kind CFTR at a safe harbor locus, such as AAVS1(Adeno-associated virus). The gene editing techniques that have been investigated as potential cystic fibrosis treatment approaches are zinc finger nucleases (ZFNs), transcription activator-like effector nucleases (TALENs), and clustered frequently interspaced palindromic repeats (CRISPR)/CRISPR-associated systems (Cas). It was found that ZFN-based gene editing successfully corrected the F508del mutation in CFTR29 cells (Lee et al., 2012) so employing a similar approach, evoked pluripotent stem cells (iPSC) from CF patients were corrected and differentiated into airway-like epithelial cells by the TALEN gene editing technique (Suzuki et al., 2016). The CRISPR/Cas technique that is capable of accurate correction without a mutagenic double-strand break is now available. Geurts et al. (2020) revealed the ability of base editor systems to recover the genetic modifications and functions of cystic fibrosis intestinal organoids. Consequently, the CRISPR/Cas gene editing system could function as a possible tool for cystic fibrosis treatment in the near future (Almughem et al., 2020).

11.11.1.4 Gene Transfer Involving Both Viral and Non-viral Gene Therapy

11.11.1.4.1 Viral Vectors

11.11.1.4.1.1 DNA Vectors

11.11.1.4.1.1.1 Adeno-Associated Virus The adeno-associated virus (AAV) is a small single-stranded DNA virus belonging to the Parvovirus family; this virus lacks pathogenicity, hence is known to be good vector (Burney and Davies, 2012). The AAV genome consists of a 4.68-kb, linear single strand of deoxyribonucleic acid consisting of inverted terminal repeats and rep and cap proteins. These inverted terminal repeats represent 145 nucleotide of DNA and are absolutely required for integration, replication (ori), excision, and packaging. Single-stranded adeno-associated virus must be transformed into a double stranded form for gene expression (Guggino and Cebotaru, 2020). As it is a small virus, large genes such as CFTR may be difficult to package, hence some researchers have tried to create a functional CFTR "mini gene," using techniques such as cutting the CFTR gene in half and using the two complementary AAVs to solve the problem (Burney and Davies, 2012). AAV undergoes site specific stable integration into human chromosome 19 (Guggino and Cebotaru, 2020); and the virus gets entry into cells by binding to cell surface proteoglycans (sialic acid, galactose, or heparan sulfate) (Mietzsch et al., 2014) and also to cell surface receptors such as fibroblast growth factor receptor or integrins (Nonnenmacher and Weber, 2012). In addition, Faust and colleagues have revealed that depletion of CpG dinucleotides from the AAV vector reduces activation of Toll-like receptor 9 (TLR9)-mediated adaptive immune responses after the intramuscular injection of the vector, resulting in reduced inflammation and prolonged gene expression (Faust et al., 2013). Several other AAV isoforms, for example, either naturally occurring (e.g., AAV1, AAV5, and AAV6) or generated through rational design (e.g., AAV6.2) have been assessed in the lung, and some have shown that they have even higher transduction efficiency than the initial AAV2 vector (Griesenbach, Pytel, and Alton, 2015). In further studies, Yan and colleagues reported the generations of a chimeric virus that consist of a human bocavirus-1 (HBoV1) capsid, which comfortably accommodated an AAV2 genome carrying CFTR and patented transcriptional regulatory elements (Yan et al., 2013). The chimeric rAAV2/HBoV1 vector transduced polarized airway epithelial cells more efficiently than AAV1 and AAV2 in vitro and corrected CFTR-dependent chloride transport in vitro (Griesenbach, Pytel, and Alton, 2015). Adeno-associated virus is advantageous for gene therapy as it has only two sets of genes involved in replication and capsid formation, and both these sets of genes are removed to produce a recombinant virus. To produce recombinant adeno-associated virus for cystic fibrosis gene therapy, basically three components are needed: CFTR cDNA with inverted terminal repeats (ITRs), plasmid containing the rep and cap genes, and a method for providing the helper function associated with the adenovirus (Guggino and Cebotaru, 2020).

11.11.1.4.1.1.2 Adenovirus Adenoviruses belong to the family Adenoviridae. These are nonenveloped, double-stranded DNA viruses composed of a complex icosahedral capsid (Volpers and Kochanek, 2004) and (Saban et al., 2006). Studies realized that the coxsackie and adenovirus receptor (CAR) receptors are located on the basolateral rather than the apical membrane of airway epithelial cells that led to assessment of tight junction openers, such as sodium caprate and lysophosphatidylcholines, which showed moderate success (Griesenbach, Pytel, and Alton, 2015). The knob domain of the fiber binds to the coxsackie and adenovirus receptor (CAR) on the cell surface and facilitates virus entry, (Volpers and Kochanek, 2004) although integrin receptors, such as $\alpha v\beta 3$ and $\alpha v\beta 5$ and the major histocompatibility complex (MHC) class I, may also play a role in virion–cell interaction. After arrival, the virus remains in an episomal state within the nucleus. In one research study, a serotype 2 adenoviral (Ad2) vector carrying the human CFTR cDNA was administrated nasally (Griesenbach, Pytel, and Alton, 2015).

11.11.1.4.1.2 RNA Vectors

11.11.1.4.1.2.1 Sendai virus Sendai virus is a single-stranded RNA virus which belongs to the family of Paramyxoviridae. This virus is effective at transfecting airway epithelial cells, as sialic acid and cholesterol receptors are present on their apical surface (Burney and Davies, 2012). The key factors that are important for entry of virus into cells are the fusion (F) and hemagglutinin neuraminidase (HN) envelope proteins. The HN protein mediates virus–host cell attachment via sialic acid receptors whereas the F protein is essential in fusing virus and cell membranes, allowing the viral nucleocapsid to be released into the host cell (Griesenbach, Pytel, and Alton, 2015).

11.11.1.4.1.2.2 Lentivirus Lentiviruses are enveloped RNA viruses belonging to the family Retroviridae (Segura et al., 2013). Lentivirus has the capacity to transfect both dividing and non-dividing cells (Burney and Davies, 2012). This virus is known to transduce into dividing and nondividing cells, which makes them a suitable vector for transduction of terminally differentiated cells in the airways (Segura et al., 2013). Lentiviral vectors are more effective in escaping adaptive immune response mechanisms than adenoviral and adeno-associated viral vectors (Griesenbach, Pytel, and Alton, 2015).

11.11.1.4.2 Other Viral Vectors

Other viruses used are Human (HIV), simian (SIV) and feline (FIV) immunodeficiency virus, as well as equine infectious anemia virus (EIAV). Vector platforms are being developed for cystic fibrosis gene therapy. As these vectors do not have a natural tropism for the lung, pseudotyping with appropriate envelope proteins is required to achieve efficient transduction. The vesicular stomatitis virus G (VSVG) protein, which is widely used, does not transduce airway epithelium efficiently via the apical membrane, but the transduction efficiency of vesicular stomatitis virus G can be increased by pre-administration of the tight junction openers that allow access of the virus to the basolateral membrane (Griesenbach, Pytel, and Alton, 2015).

11.11.1.4.3 Non-Viral Gene Therapy

Nonviral gene transfer formulations consist of two components: one is nucleic acid, that is, the therapeutic cDNA and appropriate regulatory elements, and another is a carrier molecule that binds to the DNA. A large number of carrier molecules have been developed that may be characterized as either cationic lipids that form lipoplexes or cationic polymers that form polyplexes and combination of both lipopolyplexes; but in preclinical studies these are found less economical than microbial vectors (Griesenbach, Pytel, and Alton, 2015) and (Montier et al., 2004).

11.11.1.4.3.1 Cationic lipids The lipoplex approach is derived from the neutral liposomal delivery method that was first applied to nucleic acids in the 1970s (Montier et al., 2004). As neutral liposomes were known to have a poor encapsulation rate for nucleic acids, natural and artificial cationic lipids were subsequently enclosed within the complexes, yielding a 'lipoplex.' The addition of positive charges to the complex accelerated the interactions between the cationic lipids and the negatively charged nucleic acids, along with generated electrostatic interactions between the complex and the plasma membrane that enhanced gene transfer. In addition, this sort of structure permits the association and or transportation of larger DNA structures, as compared with viral vectors (Zhdanov et al., 2002). It was found that interaction between the lipoplex and the target cells is nonspecific to the cationic lipid in the lipoplex, hence, numerous targeting approaches have been investigated, along with incorporating peptides into the lipidic vesicles to hasten tissue targeting, particularly by antibodies, and incorporating glycosylated lipids for glycol-targeting (Montier et al., 2004). Alton and colleagues, for the first time, administered the cationic lipid formulation GL67A complexed with pCFTR to the lungs of U.K. patients with cystic fibrosis. The study conjointly showed that in contrast to the previous nose studies, administration of nonviral vectors to the lung caused gentle transient inflammation (Griesenbach, Pytel, and Alton, 2015). When cationic lipososmes are combined with DNA, they form particles of 100–500 nm in diameter that can easily penetrate cell membranes and enter cells. As the complex of the two is usually resistant to nuclease degradation, this process improves the gene therapy success rate (Burney and Davies, 2012). Hence, GL67 ([Cholest-5-en-3-ol(3b)- 3-[(3-aminopropyl)[4-[(3-aminopropyl)amino]butyl] carbamate]) emerged as a promising lipid for efficient lung transfection, based on structure-function studies (Conese et al., 2011).

11.11.1.4.3.2 Cationic Polymers A second method of non-viral gene delivery consists of complexes of DNA and cationic polymers that induce DNA condensation and a considerable size reduction of the complex within the plasma. Various cationic polymers have been used to form polyplexes with DNA, like histidylated polylysine, polylysine, polyethyleneimine, polyamidoamine dendrimer, chitosan, and polyallylamine (Montier et al., 2004). The high molecular weight, large extent of polymerization, and high degree of polydispersity of cationic polymers renders their characterization difficult (Midoux and Pichon, 2002). Therefore, low molecular weight cationic peptides have been also developed to provide controlled synthesis and defined purity of the polymers (Montier et al., 2004).

11.11.1.4.3.3 DNA Nanoparticles Nanoparticle formulation consisting of a single plasmid molecule compacted with polyethyleneglycol (PEG)-substituted polylysine (polymer of 30 lysines) has been developed. These particles have a rod-like structure with 12–15 nm diameter and 100– 150 nm length. A single-dose escalation study to judge the safety of nasal administration into cystic fibrosis patients was allotted in 12 subjects. In addition of assessing safety, secondary end points included assessment of electrical correction of the ion transport defects and molecular analysis for the presence of vector-specific DNA and mRNA. Administration of the nanoparticles was considered safe. Partial correction of the chloride transport defect was manifested in seven out of 12 patients, which persisted for up to 15 days (U. Griesenbach, Geddes, and Alton, 2004). Another study involves use of muco-inert nanoparticles in cystic fibrosis sputum pretreated with N-acetyl cysteine; it was found that if particles were densely coated with polyethylene glycol, they were able to penetrate the sputum. It was concluded that N-acetyl cysteine facilitated rapid diffusion of PEG-coated nanoparticles in CF sputum (Suk et al., 2011).

11.11.1.4.3.4 Naked DNA The use of uncomplexed DNA is appealing since complexing agents themselves can increase toxicity. However without complexing agents, DNA is vulnerable to degradation (Ziady, Davis and Konstan, 2003). Naked DNA is known to have poor mobility in cytoplasm, although some tissues such as muscle or liver have excellent uptake, and expression of naked DNA has been demonstrated in airways, results of gene delivery with naked DNA applied to luminal surface have been mixed. Reports from a number of clinical trials have demonstrated that following luminal application, DNA alone is as effective or even more effective than DNA complexed to various carriers (Ziady, Davis and Konstan, 2003). Zabner and colleagues reported that similar levels of correction of nasal potential difference in cystic fibrosis patients was achieved with naked DNA as was obtained with DNA complexed to Genzyme's lipid 67(N^4-Cholesteryl-Spermine HCl salt), most efficient lipid (Zabner et al., 1997).

11.11.1.4.3.5 Using an mRNA The use of the new strategy can be used to express CFTR in target cells by use of mRNA-based nonviral gene transfer (Conese et al., 2011). The advantage of this approach is that the nuclear envelope, which is one of the major obstacles to nonviral gene therapy, doesn't have to be overcome. In the past, mRNA-based gene therapy strategies were limited by the cost of producing large amounts of translatable mRNA. However currently, mRNA generated by in vitro transcription from plasmid templates can be obtained from commercial sources at reasonable costs. There have conjointly been issues regarding the stability of mRNA and handling issues. However, many papers have demonstrated that mRNA can be successfully used to express transgenes in mammalian cells and tissues. Moreover, this methodology doesn't modify the genome, avoiding the danger of insertional mutagenesis (Conese et al., 2011).

11.11.1.4.3.6 Using Antisense Oligonucleotide Since the CFTR gene mutation causes hyperactivation of the epithelial sodium channel (ENaC), the absorption of Na^+ ions is enhanced and the lung airway mucus becomes dehydrated. Hence, the inhibition of ENaC expression could serve as a promising therapeutic approach for the treatment of cystic fibrosis. One method for the inhibiting the expression of ENaC-encoding genes (SCNN1A, SCNN1B, SCNN1G, and SCNN1D encoding α, β, γ, and δ ENaC subunits, respectively) involves the use of a single strand nucleic acid known as antisense oligonucleotide (ASO) (Almughem et al., 2020). When this oligonucleotide is hybridized to mRNA, RNase H is triggered to slice the hybridized mRNA. Targeting the α-subunit of ENaC in the respiratory organ, using ASO might inhibit the cationic channel activity (Almughem et al., 2020). Another study showed the possibility of using aerosolized ENaC antisense oligonucleotide containing wing modifications to inhibit ENaC mRNA in CF-like mice models (Hajj and Whitehead, 2017). This aerosolized ENaC antisense oligonucleotide helped to cure various cystic fibrosis symptoms, like airway hyper-responsiveness and inflammation (Crosby et al., 2017).

Milestones in the development of gene therapy for treatment of cystic fibrosis and data of various viral and non-viral vectors used in clinical trials are illustrated in Tables 11.2 and 11.3 (Cooney, McCray, and Sinn, 2018; Prickett and Jain, 2015).

TABLE 11.2
Milestones in the Development of Gene Therapy for the Treatment of Cystic Fibrosis

1989	CFTR gene was discovered
1990	It was concluded that gene therapy is feasible for the treatment of cystic fibrosis
1991	It was found that CFTR is Cl– channel
1992–93	First cystic fibrosis mouse was generated
1993–94	First adenovirus-CFTR clinical trial was conducted
1994–96	Gene transfer studies with the help of adenovirus was completed in rats and non-human primates.
1995–96	It was concluded that immune system is barrier to the treatment by adenovirus. Also first non-viral clinical trial was done.
1996–97	Hepler dependent adenovirus was developed
1997–98	It was found that EGTA lead to increased lentivirus, adeno-associated virus, and adenovirus transduction.
1996–99	Efforts were made to block the immune responses of body to adenovirus
1999–2000	First adeno-associated virus clinical trial was conducted
2001	Last adenovirus-CFTR clinical trial was done
2004–05	It was determined that proteasome inhibitor increases adeno-associated virus expression.
2007–08	Last clinical trial for adeno-associated virus was completed

(continued)

TABLE 11.2(Continued)
Milestones in the Development of Gene Therapy for the Treatment of Cystic Fibrosis

2008	Cystic fibrosis pig was generated
2002–09	Study was conducted, and it was concluded lentivirus pseudotyping improves airway delivery.
2013–14	Helper dependent adenovirus was readministered in pigs
2016–17	Lentivirus and Adeno-associated virus were used for correction of cystic fibrosis in pigs
2015–18	Gene editing of CFTR

Source: Cooney, McCray, and Sinn (2018).

Note: CFTR – cystic fibrosis transmembrane conductance regulator, EGTA – ethylene glycol-bis(β-aminoethyl ether)-N,N,N′,N′-tetraacetic acid.

TABLE 11.3
Data of Various Viral and Non-Viral Vectors used in Clinical Trials

Work Performed by Scientists	Vector	Route	Safety Concerns	Results
Crystal et al.	Adenovirus	Nose–lung	Yes, shows transient inflammation at highest dose	NPD:variable CFTRmRNA: detectable in nose but not in lung
Zuckerman et al.	Adenovirus	Lung	Flu-like symptoms	Vector DNA: detectable
Wagner et al.	Adeno-associated virus	Nose	No	NPD: partial correction CFTR mRNA: none detected
Flotte et al.	Adeno-associated virus	Nose–lung	Well tolerated, possible side effects reported	NPD: unchanged
Moss et al.	Adeno-associated virus	Lung	No	FEV1, sputum markers, antibiotic days: no improvement
Caplen et al.	DC-Chol/DOPE	Nose	No	NPD: partial correction CFTR mRNA: not detected
Porteous et al.	DOTAP	Nose	No	NPD: partial correction CFTR mRNA: detected in some
Zabner et al.	GL67 vs naked pDNA	Nose	No	NPD: significant response with both vectors CFTR mRNA: non detected
Alton et al.	GL67	Nose–lung	Flu-like symptoms	NPD: significant correction CFTR mRNA: undetectable CFTR DNA: detectable
Ruiz et al.	GL67	Lung	Flu-like symptoms	CFTR mRNA: partially detectable
Konstan et al.	DNA nanoparticles	Nose	No	NPD: partial correction CFTR DNA: detectable in treatment and some controls

Source: Prickett and Jain (2015).

Note: CFTR – cystic fibrosis transmembrane conductance regulator, FEV1 – forced expiratory volume, NPD – nasal potential difference, DC-Chol – 3b-[N-(N9, N9-dimethylaminoethane)-carbamoyl]-cholesterol, DOPE – 1,2-dioleoyl-sn-glycerol-phosphoethanolamine, DOTAP – 1,2-dioleoyl-3-trimethylammoniopropane, pDNA – plasmid DNA.

11.11.1.5 In-utero Gene Therapy for Cystic Fibrosis

This technique allows targeting of tissues/organs which are inaccessible later in life and before early onset tissue damage occurs. The small size of the fetus also provides an advantageous vector to cell ratio, which is applicable for considerations of both efficiency and vector production costs. Successful fetal gene therapy could therefore offer prenatal prevention of disease (Waddington et al., 2005). In utero gene therapy requires accurate prenatal diagnosis (Conese et al., 2011).

11.11.2 Small Molecules Used in the Treatment of Cystic Fibrosis

Although, the conventional therapy currently used for cystic fibrosis primarily controls only the disease symptoms without treating the underlying cause, small molecules or CFTR modulators might reduce disease progression and enhance the patient's quality of life. The recent discovery of CFTR modulators has provided a progress in CF treatment, since they can be used to improve and or probably correct the CFTR function and other related complications. Fast developments in the field of drug discovery, owing to high-throughput screening technologies, have led to the screening of the activity of thousands of small molecules simultaneously, thus aiding the identification of molecules that enhance the function of defective CFTR proteins (Almughem et al., 2020).

11.11.2.1 Potentiator

The potentiator targets the CFTR protein on cell membranes, leading to the opening of CFTR ion channels and improving CFTR function (Almughem et al., 2020). This increases the function of CFTR channels, which are expressed at the apical surface of epithelial cells, and also are used in class III or IV CFTR mutations in which CFTR reaches the surface of cells but is dysfunctional (Brodlie et al., 2015). Ivacaftor (VX-770) is a small-molecule drug that is identified via high-throughput screening (HTS) involving a library of nearly 230,000 potential therapeutic compounds (Brodlie et al., 2015) Ivacaftor significantly increases chloride transport, increased airway surface liquid height, and cilia beat frequency in airway epithelial cells expressing Gly551Asp-CFTR mutation (Van Goor et al., 2009). It is the first FDA-approved potentiator drug and is used especially in cases of the G551D class III mutation that produces a defect in CFTR channel gating. Meanwhile, several other potentiators, like QBW251 and PTI-808, are in the process of undergoing clinical trials (Almughem et al., 2020).

11.11.2.2 Stabilizer

Stabilizers, like N91115 (Cavosonstat), are another class of modulators. They act by reducing macromolecule degradation within the endoplasmic reticulum and increasing macromolecule residence time on the cell membrane (Almughem et al., 2020).

11.11.2.3 Corrector

Correctors improve CFTR folding either by direct binding or by adapting protein homeostasis (Mijnders, Kleizen, and Braakman, 2017). This modulator corrects the misfolded CFTR protein, thereby restoring it to its original three-dimensional form. Thus, it accelerates the functioning of the CFTR protein through its movement toward the cell surface (Almughem et al., 2020). These also increase gating and conductance. The first two correctors identified by high-throughput screening of a small-molecule library were bis-aminomethylbithiazole C4 (Corr-4a) and quinazolinole C3 (VRT-325) (Mijnders, Kleizen, and Braakman, 2017). VX-809 (Lumacaftor) is the first corrector to undergo extensive trials (Shanthikumar and Massie, 2017). It is used to correct the F508del mutation, the most common type of CF mutation. This helps avoid the ER-mediated degradation of the CFTR macromolecule by enhancing the interaction between the NBD1, MSD1, and MSD2 domains (Almughem et al., 2020). VX-661 is considered to be an improved VX-809 analog. The correctors, VX-440, VX-152, and VX-659, are also studied in clinical trials (Mijnders, Kleizen, and Braakman, 2017). Correctors GLPG2222 and GLPG2851 (C1) are additive to GLPG2737 and GLPG3221 (C2) and may be combined in therapy. GLPG2222 have structural similarities with VX-809 and VX-661 but is known to be more potent. Combinations of C1, C2, and a potentiator developed by Galapagos/ AbbVie significantly increase chloride transport over Orkambi (VX-809 and VX-770) in vitro (Mijnders, Kleizen, and Braakman, 2017). Corrector modulators are usually used in combination with potentiator modulators in CF treatment. Examples are the FDA-approved Orkambi® (lumacaftor combined with ivacaftor) and Symdeko™ (tezacaftor combined with ivacaftor) (Almughem et al., 2020). The tezacaftor/ivacaftor combination has been shown to have fewer side effects, such as chest tightness and drug interactions, than lumacaftor/ivacaftor (Cystic Fibrosis Foundation) Tezacaftor, in combination with Lumacaftor and Ivacaftor, have shown a promising moderate improvement (Fiore et al., 2019). On October 21 2019, the US Food and Drug Administration approved a new triple-combination therapy (elexacaftor/tezacaftor/ivacaftor), including two correctors and one potentiator combination, for individuals with CF who are aged 12 years and older and have at least one allele with the F508del mutation (2019 PATIENT REGISTRY ANNUAL DATA REPORT, 2021). The efficacy of Trikafta in patients with cystic fibrosis aged 12 years and older was manifested in two trials. The first trial consisted of a 24-weeks randomized, double-blind, placebo-controlled trial done on 403 patients who had an F508del mutation and a mutation on the second allele that results in either no CFTR protein or a CFTR protein which is not responsive to ivacaftor or tezacaftor/ivacaftor alone. The second trial was a 4-weeks randomized, double-blind, active-controlled trial with 107 patients who had two identical F508del mutations (U.S. Food & Drug Administration, 2019)

In the United States (Trikafta™) has been recently approved for the treatment of CF patients 12 years and older, carrying a minimum of one F508del mutation (Almughem et al., 2020).

11.11.2.4 Amplifier

Amplifiers are modulators that increase CFTR expression (Mijnders, Kleizen, and Braakman, 2017). PTI-428 is an amplifier that increases the amount of CFTR mRNA and protein loaded on to the endoplasmic reticulum. The amplifier is usually combined, either with a corrector or potentiator modulator. These modulators play an important role in the treatment of most CFTR class mutations except class I mutation, owing to a premature termination codon mutation that leads to defective protein production. To overcome this drawback, a combination of an aminoglycoside antibiotic, like gentamicin or tobramycin, and PTC124 (Ataluren) is employed (Almughem et al., 2020). A small molecule HDAC7 inhibitor SAHA1 not only amplifies F508del-CFTR expression, but also promotes significant transport of F508del-CFTR to the cell surface by reshaping CFTR's proteostasis network (Mijnders, Kleizen, and Braakman, 2017).

Details of various small molecules, their development stage and targeted applications are given in Table 11.4 (Almughem et al., 2020).

11.11.3 Managing a Good Nutritional State in Older Children

A good nutritional state can help manage the disease as a study showed that children affected with cystic fibrosis have a poor nutritional state (Littlewood, 2007).

11.11.4 Gastrointestinal Therapies

Various GI therapies are prescribed as proton pump inhibitors that are prescribed more often than H2 blockers, along with vitamins A, D, E and K. Also, individuals with abnormal liver function were prescribed ursodeoxycholic acid (2019 patient registry).

TABLE 11.4
Small Molecules: Their Development Stage and Targeted Applications

Name	Therapeutic Approach	Target CFTR Mutations	Development Stage
Ivacaftor	Potentiator	Class III (G551D mutation)	FDA-Approved, 2012
N91115 (Cavosonstat)	Stabilizer	Class II (F508del homozygous)	Phase II/Discontinued
Orkambi (lumacaftor + ivacaftor)	Corrector + potentiator	Class II (F508del homozygous)	FDA-Approved, 2015
Symdeko™ (tezacaftor + ivacaftor)	Corrector + potentiator	Class II (F508del homozygous)	FDA-Approved, 2018
Trikafta™ (ivacaftor + tezacaftor + elexacaftor)	Potentiator + corrector + corrector	Class II (F508del heterozygous)	FDA-Approved, 2019
PTI-801/PTI-808 /PTI-428	Corrector + potentiator + amplifier	Class II (F508del homozygous)	Phase II
PTC124 (Ataluren)	Read-through	Class I (PTC mutation)	Phase III/Discontinued
Duramycin	Cl– stimulator through CaCCs	—	Phase II/Discontinued
Denufosol	Cl– stimulator through CaCCs	—	Phase III/Discontinued
SPX-101	ENaC inhibitor	—	Phase II

Source: Almughem et al. (2020).

11.11.5 Pulmonary Therapies

Various pulmonary therapies are prescribed, e.g., bronchodilators, azithromycin, dornase alfa, and inhaled antibiotics like tobramycin, aztreonam, and colistin (2019 patient registry).

11.11.6 Organ Transplantations

The last option that is used to prevent disease is use of an organ transplantation procedure. In 2019, there were around 1,957 individuals in the registry who were reported to have ever received a lung, kidney, heart, or liver transplant (2019 Patient Registry).

11.12 COMMON ISSUES COMPLICATING CYSTIC FIBROSIS AND ITS TREATMENT

Issues complicating cystic fibrosis and its treatment are depression, anxiety, risk of thrombosis, antibiotic hypersensitivity reactions and intolerance, vestibulo-auditory disturbance, including tinnitus, obesity, and airway complications (Bell et al., 2020). Although most people tolerate antibiotics well, side effects are common and include disturbance of the natural host flora, with diarrhea, allergy, and in some cases antimicrobial resistance. Some evidence also links the use of antibacterial drugs with an increased incidence of fungal and mycobacterial infection in patients with cystic fibrosis (Zemanick et al., 2017) Bell et al., 2020). Risk of cancer increases during transplantation, like colorectal cancer (Safaeian et al., 2016), esophageal cancers, and non-Hodgkin lymphoma (Fink et al., 2017). Chronic kidney disease, sometimes leading to chronic renal failure requiring hemodialysis and renal transplantation, is an emerging complication in older adults with cystic fibrosis (Berg et al., 2018) and (Quon et al., 2011). Also, cost is an issue, i.e., treatments with small molecules increase medication cost (Bell et al., 2020).

11.13 CONCLUSION

Cystic Fibrosis is a monogenic disorder caused by mutations in the cystic fibrosis transmembrane conductance regulator gene. To treat the disorder, various treatments are available, like pulmonary therapy, gastro-intestinal therapies which give symptomatic relief. To get the effect at the desired location, gene therapy and small molecules treatments hold the best. Hence, Gene therapy and Small molecule treatments will definitely bring overall benefit into the lives of people suffering with cystic fibrosis by targeting the mutated genes and curing the class defects through modulation of the ion channels.

REFERENCES

"2019 PATIENT REGISTRY ANNUAL DATA REPORT." Cystic Fibrosis Foundation. (accessed on March 24, 2021).

Accurso, Frank. "The Tiger-1 phase 3 clinical trial of denufosol in cysticfibrosis: S18. 4." *Pediatric Pulmonology* 43 (2008):191.

Akhtar, Naveed, M. Akram, H. M. Asif, Khan Usmanghani, SM Ali Shah, Saeed Ahmad Rao, M. Uzair, Ghazala Shaheen, and Khalil Ahmad. "Gene therapy: a review article." *Journal of Medicinal Plant Research* 5, no. 10 (2011): 1812–1817.

Almughem, Fahad A., Ahmad M. Aldossary, Essam A. Tawfik, Mohammad N. Alomary, Waleed S. Alharbi, Mohammad Y. Alshahrani, and Abdullah A. Alshehri. "Cystic fibrosis: overview of the current development trends and innovative therapeutic strategies." *Pharmaceutics* 12, no. 7 (2020): 616.

Bell, Scott C., Marcus A. Mall, Hector Gutierrez, Milan Macek, Susan Madge, Jane C. Davies, Pierre-Régis Burgel et al. "The future of cystic fibrosis care: a global perspective." *The Lancet Respiratory Medicine* 8, no. 1 (2020): 65–124.

Berg, Kristina H., Lene Ryom, Daniel Faurholt-Jepsen, Tania Pressler, and Terese L. Katzenstein. "Prevalence and characteristics of chronic kidney disease among Danish adults with cystic fibrosis." *Journal of Cystic Fibrosis* 17, no. 4 (2018): 478–483.

Borowitz, Drucy. "CFTR, bicarbonate, and the pathophysiology of cystic fibrosis." *Pediatric Pulmonology* 50, no. S40 (2015): S24–S30.

Boucher, Richard C. "Airway surface dehydration in cystic fibrosis: pathogenesis and therapy." *Annual Review of Medicine* 58 (2007): 157–170.

Brodlie, Malcolm, Iram J. Haq, Katie Roberts, and J. Stuart Elborn. "Targeted therapies to improve CFTR function in cystic fibrosis." *Genome Medicine* 7, no. 1 (2015): 1–16.

Burney, Tabinda J., and Jane C. Davies. "Gene therapy for the treatment of cystic fibrosis." *The Application of Clinical Genetics* 5 (2012): 29.

Carlon, Marianne S., Dragana Vidović, and Susan Birket. "Roadmap for an early gene therapy for cystic fibrosis airway disease." *Prenatal Diagnosis* 37, no. 12 (2017): 1181–1190.

Conese, Massimo, Fiorentina Ascenzioni, A. Christopher Boyd, Charles Coutelle, Ida De Fino, Stefaan De Smedt, Joanna Rejman et al. "Gene and cell therapy for cystic fibrosis: from bench to bedside." *Journal of Cystic Fibrosis* 10 (2011): S114–S128.

Cooney, Ashley L., Paul B. McCray, and Patrick L. Sinn. "Cystic fibrosis gene therapy: looking back, looking forward." *Genes* 9, no. 11 (2018): 538.

Crosby, Jeff R., Chenguang Zhao, Chong Jiang, Dong Bai, Melanie Katz, Sarah Greenlee, Hiroshi Kawabe et al. "Inhaled ENaC antisense oligonucleotide ameliorates cystic fibrosis-like lung disease in mice." *Journal of Cystic Fibrosis* 16, no. 6 (2017): 671–680.

Cystic Fibrosis Trust. 2014. "UK Cystic Fibrosis Registry Annual Data Report 2013." In Registry. Cystic Fibrosis Foundation. "About Cystic Fibrosis." Accessed March 20, 2021. https://www.cff.org/intro-cf/about-cystic-fibrosis

Davis, Pamela B. "Cystic fibrosis since 1938." *American Journal of Respiratory and Critical Care Medicine* 173, no. 5 (2006): 475–482.

Dörk, Thilo, Ulrich Wulbrand, Thomas Richter, Thomas Neumann, Heiner Wolfes, Brigitte Wulf, Günter Maass et al. "Cystic fibrosis with three mutations in the cystic fibrosis transmembrane conductance regulator gene." *Human Genetics* 87, no. 4 (1991): 441–446.

Eiberg, H., J. Mohr, K. Schmiegelow, L. S. Nielsen, and R. Williamson. "Linkage relationships of paraoxonase (PON) with other markers: indication of PON-cystic fibrosis synteny." *Clinical Genetics* 28, no. 4 (1985): 265–271.

Farrell, Philip M., Beryl J. Rosenstein, Terry B. White, Frank J. Accurso, Carlo Castellani, Garry R. Cutting, Peter R. Durie et al. "Guidelines for diagnosis of cystic fibrosis in newborns through older adults: cystic fibrosis foundation consensus report." *The Journal of Pediatrics* 153, no. 2 (2008): S4–S14.

Faust, Susan M., Peter Bell, Benjamin J. Cutler, Scott N. Ashley, Yanqing Zhu, Joseph E. Rabinowitz, and James M. Wilson. "CpG-depleted adeno-associated virus vectors evade immune detection." *The Journal of Clinical Investigation* 123, no. 7 (2013): 2994–3001.

Ferrari, S., U. Griesenbach, D. M. Geddes, and E. Alton. "Immunological hurdles to lung gene therapy." *Clinical and Experimental Immunology* 132, no. 1 (2003): 1–8.

Ferrari, Stefano, Duncan M. Geddes, and Eric WFW Alton. "Barriers to and new approaches for gene therapy and gene delivery in cystic fibrosis." *Advanced Drug Delivery Reviews* 54, no. 11 (2002): 1373–1393.

Fink, Aliza K., Elizabeth L. Yanik, Bruce C. Marshall, Michael Wilschanski, Charles F. Lynch, April A. Austin, Glenn Copeland et al. "Cancer risk among lung transplant recipients with cystic fibrosis." *Journal of Cystic Fibrosis* 16, no. 1 (2017): 91–97.

Fiore, Michele, Claudia Cossu, Valeria Capurro, Cristiana Picco, Alessandra Ludovico, Marcin Mielczarek, Israel Carreira-Barral et al. "Small molecule-facilitated anion transporters in cells for a novel therapeutic approach to cystic fibrosis." *British Journal of Pharmacology* 176, no. 11 (2019): 1764–1779.

Garland, Alaina L., William G. Walton, Raymond D. Coakley, Chong D. Tan, Rodney C. Gilmore, Carey A. Hobbs, Ashutosh Tripathy et al. "Molecular basis for pH-dependent mucosal dehydration in cystic fibrosis airways." *Proceedings of the National Academy of Sciences* 110, no. 40 (2013): 15973–15978.

Geurts, Maarten H., Eyleen de Poel, Gimano D. Amatngalim, Rurika Oka, Fleur M. Meijers, Evelien Kruisselbrink, Peter van Mourik et al. "CRISPR-based adenine editors correct nonsense mutations in a cystic fibrosis organoid biobank." *Cell Stem Cell* 26, no. 4 (2020): 503–510.

Gonçalves, Giulliana Augusta Rangel, and Raquel de Melo Alves Paiva. "Gene therapy: advances, challenges and perspectives." *Einstein (Sao Paulo)* 15, no. 3 (2017): 369–375.

Van Goor, Fredrick, Sabine Hadida, Peter DJ Grootenhuis, Bill Burton, Dong Cao, Tim Neuberger, Amanda Turnbull et al. "Rescue of CF airway epithelial cell function in vitro by a CFTR potentiator, VX-770." *Proceedings of the National Academy of Sciences* 106, no. 44 (2009): 18825–18830.

Grasemann, Hartmut, Florian Stehling, Helmut Brunar, Rudolf Widmann, Terry W. Laliberte, Luis Molina, Gerd Döring et al. "Inhalation of Moli1901 in patients with cystic fibrosis." *Chest* 131, no. 5 (2007): 1461–1466.

Griesenbach, U., D. M. Geddes, and E. W. F. W. Alton. "Gene therapy for cystic fibrosis: an example for lung gene therapy." *Gene Therapy* 11, no. 1 (2004): S43–S50.

Griesenbach, Uta, Kamila M Pytel, and Eric WFW Alton. "Cystic fibrosis gene therapy in the UK and elsewhere." *Human Gene Therapy* 26, no. 5 (2015): 266–275.

Guggino, William B., and Liudmila Cebotaru. "Gene therapy for cystic fibrosis paved the way for the use of adeno-associated virus in gene therapy." *Human Gene Therapy* 31, no. 9–10 (2020): 538–541.

Hajj, Khalid A., and Kathryn A. Whitehead. "Tools for translation: non-viral materials for therapeutic mRNA delivery." *Nature Reviews Materials* 2, no. 10 (2017): 1–17.

Hodges, Craig A., and Ronald A. Conlon. "Delivering on the promise of gene editing for cystic fibrosis." *Genes & Diseases* 6, no. 2 (2019): 97–108.

Jaffe, A., A. Bush, D. M. Geddes, and E. W. F. W. Alton. "Prospects for gene therapy in cystic fibrosis." *Archives of Disease in Childhood* 80, no. 3 (1999): 286–289.

Knowles, M. R., A. M. Paradiso, and R. C. Boucher. "In vivo nasal potential difference: techniques and protocols for assessing efficacy of gene transfer in cystic fibrosis." *Human Gene Therapy* 6, no. 4 (1995): 445–455.

Knowles, Michael, John Gatzy, and Richard Boucher. "Increased bioelectric potential difference across respiratory epithelia in cystic fibrosis." *New England Journal of Medicine* 305, no. 25 (1981): 1489–1495.

Kopito, Louis, and Harry Shwachman. "Studies in cystic fibrosis: determination of sweat electrolytes in situ with direct reading electrodes." *Pediatrics* 43, no. 5 (1969): 794–798.

Lane, Matthew A., and Simon J. Doe. "A new era in the treatment of cystic fibrosis." *Clinical Medicine* 14, no. 1 (2014): 76.

Lankisch, P. G. "Exocrine pancreatic function tests." *Gut* 23, no. 9 (1982): 777.

Lee, Ciaran M., Rowan Flynn, Jennifer A. Hollywood, Martina F. Scallan, and Patrick T. Harrison. "Correction of the ΔF508 mutation in the cystic fibrosis transmembrane conductance regulator gene by zinc-finger nuclease homology-directed repair." *BioResearch Open Access* 1, no. 3 (2012): 99–108.

LeGrys, Vicky Allen. Sweat testing: sample collection and quantitative analysis: approved guideline. NCCLS, 1994.

Lewis, Brandon W., Sonika Patial, and Yogesh Saini. "Immunopathology of airway surface liquid dehydration disease." *Journal of Immunology Research* 2019 (2019):1–16.

Littlewood, J. "The history of cystic fibrosis." *Cystic Fibrosis* 1 (2007): 19.

Mall, Marcus A., and J. Stuart Elborn, eds. *Cystic Fibrosis: ERS Monograph*. Vol. 64. European Respiratory Society, 2014.

Midoux, Patrick, and Chantal Pichon. "Peptide-based gene delivery systems." In *Pharmaceutical Perspectives of Nucleic Acid-Based Therapeutics*, pp. 334–366. Taylor & Francis, (2002). edited by- Ram I. Mahato, Sung Wan Kim.

Mietzsch, Mario, Felix Broecker, Anika Reinhardt, Peter H. Seeberger, and Regine Heilbronn. "Differential adeno-associated virus serotype-specific interaction patterns with synthetic heparins and other glycans." *Journal of Virology* 88, no. 5 (2014): 2991–3003.

Mijnders, Marjolein, Bertrand Kleizen, and Ineke Braakman. "Correcting CFTR folding defects by small-molecule correctors to cure cystic fibrosis." *Current Opinion in Pharmacology* 34 (2017): 83–90.

Montier, T., P. Delepine, C. Pichon, C. Ferec, D. J. Porteous, and P. Midoux. "Non-viral vectors in cystic fibrosis gene therapy: progress and challenges." Trends in Biotechnology 22, no. 11 (2004): 586–592.

Nonnenmacher, Mathieu, and Thomas Weber. "Intracellular transport of recombinant adeno-associated virus vectors." *Gene Therapy* 19, no. 6 (2012): 649–658.

O'Riordan, Stephen MP, Paul D. Robinson, Kim C. Donaghue, and Antoinette Moran. "Management of cystic fibrosis-related diabetes in children and adolescents." *Pediatric Diabetes* 10 (2009): 43–50.

Pezzulo, Alejandro A., Xiao Xiao Tang, Mark J. Hoegger, Mahmoud H. Abou Alaiwa, Shyam Ramachandran, Thomas O. Moninger, Phillip H. Karp et al. "Reduced airway surface pH impairs bacterial killing in the porcine cystic fibrosis lung." *Nature* 487, no. 7405 (2012): 109–113.

Prickett, Michelle, and Manu Jain. "Gene therapy in cystic fibrosis." *Translating Gene Therapy to the Clinic* 161, (2015): 247–260.

Quinton, Paul M. "Chloride impermeability in cystic fibrosis." *Nature* 301, no. 5899 (1983): 421–422.

Quon, Bradley S., Nicole Mayer-Hamblett, Moira L. Aitken, Alan R. Smyth, and Christopher H. Goss. "Risk factors for chronic kidney disease in adults with cystic fibrosis." *American Journal of Respiratory and Critical Care Medicine* 184, no. 10 (2011): 1147–1152.

Ratjen, Felix, Scott C. Bell, Steven M. Rowe, Christopher H. Goss, Alexandra L. Quittner, and Andrew Bush. "Cystic fibrosis." *Nature Reviews. Disease Primers* 1 (2015 May): 15010. https://doi.org/10.1038/nrdp.2015.10

Rosenstein, Beryl J., and Garry R. Cutting. "The diagnosis of cystic fibrosis: a consensus statement." *The Journal of Pediatrics* 132, no. 4 (1998): 589–595.

Saban, Susan D., Mariena Silvestry, Glen R. Nemerow, and Phoebe L. Stewart. "Visualization of α-helices in a 6-Ångstrom resolution cryoelectron microscopy structure of adenovirus allows refinement of capsid protein assignments." *Journal of Virology* 80, no. 24 (2006): 12049–12059.

Safaeian M, Robbins HA, Berndt SI, Lynch CF, Fraumeni JF Jr, Engels EA. Risk of colorectal cancer after solid organ transplantation in the United States. *American Journal of Transplantation* 2016; 16: 960–67.

Segura, Maria Mercedes, Mathias Mangion, Bruno Gaillet, and Alain Garnier. "New developments in lentiviral vector design, production and purification." Expert Opinion on Biological Therapy 13, no. 7 (2013): 987–1011.

Shanthikumar, Shivanthan, and John Massie. "A review of treatments that improve cystic fibrosis transmembrane conductance regulator function." *Clinical Medicine Insights: Therapeutics* 9 (2017): 1–7.

Shei, Ren-Jay, Jacelyn E. Peabody, Niroop Kaza, and Steven M. Rowe. "The epithelial sodium channel (ENaC) as a therapeutic target for cystic fibrosis." *Current Opinion in Pharmacology* 43 (2018): 152–165.

Stern, Robert C. "The diagnosis of cystic fibrosis." *New England Journal of Medicine* 336, no. 7 (1997): 487–491.

Suk, Jung Soo, Samuel K. Lai, Nicholas J. Boylan, Michelle R. Dawson, Michael P. Boyle, and Justin Hanes. "Rapid transport of muco-inert nanoparticles in cystic fibrosis sputum treated with N-acetyl cysteine." *Nanomedicine* 6, no. 2 (2011): 365–375.

Suzuki, Shingo, R. Geoffrey Sargent, Beate Illek, Horst Fischer, Alaleh Esmaeili-Shandiz, Michael J. Yezzi, Albert Lee et al. "TALENs facilitate single-step seamless SDF correction of F508del CFTR in airway epithelial submucosal gland cell-derived CF-iPSCs." *Molecular Therapy--Nucleic Acids* 5 (2016): e273.

Tabori, Harold, Christin Arnold, Anke Jaudszus, Hans-Joachim Mentzel, Diane M. Renz, Steffen Reinsch, Michael Lorenz et al. "Abdominal symptoms in cystic fibrosis and their relation to genotype, history, clinical and laboratory findings." *PLoS One* 12, no. 5 (2017): e0174463.

Thursfield, Rebecca M., and Jane C. Davies. "Cystic fibrosis: therapies targeting specific gene defects." *Paediatric Respiratory Reviews* 13, no. 4 (2012): 215–219.

Thursfield, Rebecca M., and Jane C. Davies. "Genotype-specific small-molecule therapy for cystic fibrosis." Breathe 9, no. 3 (2013): 176–186.

Tsui, Lap-Chee, Manuel Buchwald, David Barker, Jeffrey C. Braman, Robert Knowlton, James W. Schumm, Hans Eiberg et al. "Cystic fibrosis locus defined by a genetically linked polymorphic DNA marker." *Science* 230, no. 4729 (1985): 1054–1057.

Turnbull, Emma L., Meredith FN Rosser, and Douglas M. Cyr. "The role of the UPS in cystic fibrosis." *BMC Biochemistry* 8, no. 1 (2007): 1–10.

U.S. Food & Drug Administration. 2019. "FDA Approves New Breakthrough Therapy for Cystic Fibrosis." 2019. https://www.fda.gov/news-events/press-announcements/fda-approves-new-breakthrough-therapy-cystic-fibrosis. (accessed on March 24, 2021)

Volpers, Christoph, and Stefan Kochanek. "Adenoviral vectors for gene transfer and therapy." *The Journal of Gene Medicine: A Cross-Disciplinary Journal for Research on the Science of Gene Transfer and its Clinical Applications* 6, no. S1 (2004): S164–S171.

Waddington, Simon N., M. Gabriela Kramer, Ruben Hernandez-Alcoceba, Suzanne MK Buckley, Michael Themis, Charles Coutelle, and Jesus Prieto. "In utero gene therapy: current challenges and perspectives." *Molecular Therapy* 11, no. 5 (2005): 661–676.

Weiss, Daniel J. "Delivery of gene transfer vectors to lung: obstacles and the role of adjunct techniques for airway administration." *Molecular Therapy* 6, no. 2 (2002): 148–152.

Wikipedia Contributors. "Respiratory epithelium," *Wikipedia, The Free Encyclopedia*, (accessed April 18, 2021).

Wilschanski, Michael. "Small molecules to treat cystic fibrosis." *Proceedings of the American Thoracic Society* 7, no. 6 (2010): 399–403.

Wine, Jeffrey J., Gunnar C. Hansson, Peter König, Nam Soo Joo, Anna Ermund, and Mario Pieper. "Progress in understanding mucus abnormalities in cystic fibrosis airways." *Journal of Cystic Fibrosis* 17, no. 2 (2018): S35–S39.

Yan, Ziying, Nicholas W. Keiser, Yi Song, Xuefeng Deng, Fang Cheng, Jianming Qiu, and John F. Engelhardt. "A novel chimeric adenoassociated virus 2/human bocavirus 1 parvovirus vector efficiently transduces human airway epithelia." *Molecular Therapy* 21, no. 12 (2013): 2181–2194.

Yan, Ziying, Paul B. McCray Jr, and John F. Engelhardt. "Advances in gene therapy for cystic fibrosis lung disease." *Human Molecular Genetics* 28, no. R1 (2019): R88–R94.

Yang, Q., A. R. Soltis, G. Sukumar, X. Zhang, H. Caohuy, J. Freedy, C. L. Dalgard et al. "Gene therapy-emulating small molecule treatments in cystic fibrosis airway epithelial cells and patients." *Respiratory Research* 20, no. 1 (2019): 1–15.

Yeung, William H., Judy Palmer, Daniel Schidlow, Michael R. Bye, and Nancy N. Huang. "Evaluation of a paper-patch test for sweat chloride determination." *Clinical Pediatrics* 23, no. 11 (1984): 603–607.

Zabner, Joseph, Seng H. Cheng, David Meeker, Jan Launspach, Rosemary Balfour, Michael A. Perricone, James E. Morris et al. "Comparison of DNA-lipid complexes and DNA alone for gene transfer to cystic fibrosis airway epithelia in vivo." *The Journal of Clinical Investigation* 100, no. 6 (1997): 1529–1537.

Zemanick, Edith T., Brandie D. Wagner, Charles E. Robertson, Richard C. Ahrens, James F. Chmiel, John P. Clancy, Ronald L. Gibson et al. "Airway microbiota across age and disease spectrum in cystic fibrosis." *European Respiratory Journal* 50, no. 5 (2017):1-13.

Zhdanov, R. I., O. V. Podobed, and V. V. Vlassov. "Cationic lipid–DNA complexes—lipoplexes—for gene transfer and therapy." Bioelectrochemistry 58, no. 1 (2002): 53–64.

Ziady, Assem G., Pamela B. Davis, and Michael W. Konstan. "Non-viral gene transfer therapy for cystic fibrosis." *Expert Opinion on Biological Therapy* 3, no. 3 (2003): 449–458.

12 Non-Viral Delivery of Genome-Editing Nucleases for Gene Therapy

Sonia Barua
Chung-Ang University, Seoul, South Korea

Yashwant Pathak
University of South Florida, Tampa, FL, USA
Airlangga University, Surabaya, Indonesia

CONTENTS

12.1 INTRODUCTION

In the current era, gene-editing technologies produce a tremendous effect to cure, treat, or prevent various inherited or acquired diseases. These technologies provide flexibility in localization of correct genes for knockdown or recovery of gene expression, insertion of a therapeutic transgene, or correction of mutations associated with genetic diseases [1]. Gene editing technologies can be divided into four types, including meganucleases, zinc finger nucleases, transcription activator-like effector nucleases, and clustered, regularly interspaced, short palindromic repeat-associated nucleases, such as Cas9 [1, 2]. ZFNs, TALENs, and CRISPR/Cas Zinc-finger nucleases and transcription activator-like effector nucleases (TALENs) are hybrid restriction enzymes composed of a DNA-binding domain and a DNA-cleavage domain based on FokI endonuclease [2] These nucleases introduce precise and specific changes of the genome sequence at any genome locus of interest [2–4] [Figure 12.1]. However, the efficient delivery of these therapeutic genes is challenging. Moreover, the potential therapeutic effect mainly relies on safe and efficient delivery of nucleases into the nucleus [Table 12.1]. Among the various types of nucleases, the CRISPR-associated protein 9 (CRISPR/Cas9) system is considered as a one of the most exciting tools for gene editing. The therapeutic efficacy of CRISPR/Cas 9 system depends on base-pairing between the single-guide RNA (sgRNA) and the target DNA [2, 5]. The nuclease can be delivered mainly by two methods, physical and non-viral based [Table 12.1]. Although physical methods, including electroporation and microinjection, produce high transfection efficiency, low cell viability and cell specific delivery

DOI: 10.1201/9781003186083-12

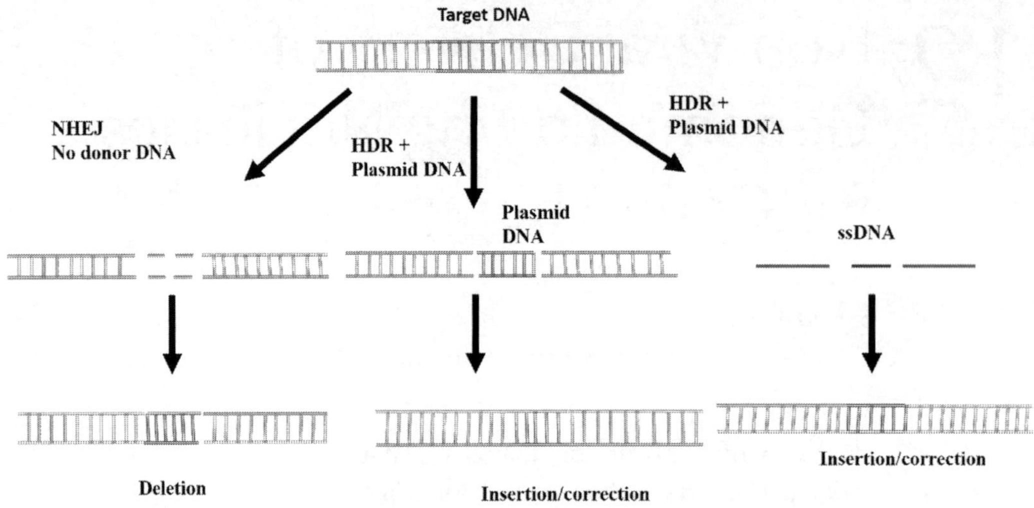

FIGURE 12.1 Mechanism of therapeutic gene-editing in DNA repair pathway.

(Adapted from Ates et al. 2020. NHEJ-Nonhomologous end joining; HDR-Homology-directed repair pathway).

TABLE 12.1
Approaches to Delivery of Nonviral Vehicles-Gene Editing Nucleases

Delivery Methods	Exogenous Materials	Approaches
Electroporation	Electroporation DNA plasmid, mRNA, RNP	Ex vivo
Hydrodynamic delivery	DNA plasmid , Ribonucleoprotein (RNP)	In vivo
Lipid nanoparticles	DNA plasmid, mRNA, RNP	In vivo
Cell penetrating peptides	Protein, RNP	In vivo

may commonly occur with these delivery systems, they also are also difficult for in vivo application [1–5]. Similarly, viral vectors achieve good performance on delivery of CRISPR/Cas9, but are limited by restricted packaging capacity or unwanted genetic mutations and immunogenicity [5, 6], posing concerns about safety in clinical translation [1–3]. In comparison, nonviral delivery methods via nanoparticles have the potential to overcome many of these limitations, particularly with respect to safety, large loading capacity, and in vivo application [5–8]. The first use of gene editing nucleases in humans was discovered in 2009. Zinc-finger nucleases were delivered to the human CD4 T cells isolated from patients with chronic aviremic HIV infection (NCT00842634) to block the functions of CCR5, a major coreceptor of HIV-1. The resistance of CD4 cells to HIV viruses was achieved by modifying the 11% to 28% of alleles with ZFNs, which showed long persistence after administration to the patients [1, 8].

Studies reported that α-helical polypeptide PPABLG based PEGylated nanoparticles achieved efficient cellular internalization and endosomal escape of CRISPR/Cas9. The CRISPR/Cas9 delivery system could reach 47.3% gene editing in cells, 35% gene deletion in vivo, and HeLa tumor growth suppression >71%, demonstrating an advantage over the existing conventional polycationic transfection reagents [1, 2, 5]. They are also capable in knock-in and gene activation, and this CRISPR/Cas9 delivery system has been shown to be a most acceptable delivery vehicle for gene editing in vitro and in vivo [5]. Recently, other approaches, such as cell penetrating peptides, have been incorporated for efficient delivery of gene-editing nucleases to improve membrane destabilization, promote endosomal escape, and enhance overall transfection efficiency [2, 9]. The CPPs, based on materials for membrane destabilization, such as HIV-TAT, GALA, and oligoarginine, have been commonly incorporated into delivery vectors for membrane destabilization to increase uptake, promote endosomal escape, and improve overall transfection efficiency. Studies reported that using CPP is more advantageous for eventual translation. However, a single CPP is lack of adequate cationic charge and too small in size. In addition, conjugation with nucleic acids neutralizes the CPPs and causes blockage of the activity of short oligo peptides, decreasing their membrane activities as well as transfection efficiencies. Therefore, CPPs sometimes only function in a supportive role to enhance the delivery efficiencies of existing systems. However, the transfection efficiencies of cell penetrating peptides may be reduced by lack of cationic charges or neutralization upon conjugation with nucleic acid [9].

The functional activity of gene-editing nuclease is achieved also by delivering into the cell by using viral vectors and nonviral vectors. In this case, enzyme complexes encoded by nucleic acid introduces into the target cells by viruses or nonviral vectors. Although viral vectors are highly effective, their therapeutic capacity is limited by complexity and challenge associated with scale-up of virus production, potential for insertional mutagenesis, and possible immune responses against the viral vector [1, 10]. Non-viral delivery methods have been taken as a promising alternative of viral vectors because they can be engineered to largely avoid these problems. Protein or nucleic acid degradation, opsonization, and immune avoidance, as well as cells specific delivery are the major challenges to nonviral delivery.

12.2 METHODS OF DELIVERY AND NONVIRAL APPROACHES FOR GENE EDITING NUCLEASES

12.2.1 Electroporation

Electroporation is known as a physical transfection method in which high voltage current is applied to the cells to enhance the membrane permeability to nucleic acids, chemicals, or proteins. Due to application of the electrical impulse, small gaps between the cell membranes appear that are resealable allowing the entry of substances [3]. Initial in vitro studies proved that the electroporation method delivered the plasmid DNA containing Cas9 efficiently to the cells which are considered difficult to transfect, such as primary fibroblasts [3, 11], human embryonic stem cells [3, 11], pluripotent stem cells , and neurons. However, therapeutic application of electroporation mediated-delivery is only limited to ex vivo. Currently, clinical trials by this method are taking place [Table 12.2]. Many groups have utilized electroporation to directly deliver nucleic acids or proteins for ZFNs, TALENs, or CRISPR ribonucleoproteins, and many ongoing clinical trials are utilizing electroporation for delivery to T cells. [3, 11–13]

12.2.1.1 Selections of Cargoes for Gene Therapy

12.2.1.1.1 Plasmid DNA

Plasmid DNA vectors are widely chosen for delivery of genome-editing complexes. This vector offers flexibility in design, allowing easy incorporation of DNA into plasmid by simple molecular cloning techniques [2]. However, gene-editing efficiency is often limited by the efficiency of nuclear delivery and gene expression that is required to generate the final gene-editing protein complexes. Conjugation of plasmid –DNA with nanoparticles such as highly positive surface charge reduces the delivery of therapeutic genes, owing to after administration adsorption by the serum proteins, aggregates, and release premature cargos. It is necessity that after injection, DNA nanoparticles need to overcome the systemic barriers before delivering to the targeted tissue [14, 15] [Figure 12.2]. Using the polymeric material such as polyethylene glycol, which is also known as antifouling agent, to

TABLE 12.2
Nonviral Delivery Systems for Gene Editing Nucleases Under Clinical Investigation

Delivery Systems	Drugs	Diseases	Target Genes	Clinical Trial Phase	Status in Clinical Trials	Company
DOTAP-cholesterol	DOTAP-Chol-fus1	Non-small cell lung carcinoma	—	I	Completed	MD Anderson cancer center
GAP-DMORIE–DPyPE	Tetravalent dengue vaccine	Dengue disease vaccine	—	I	Active	US Army Medical Research and Materiel Command
GL67A–DOPE–DMPE–PEG	pGM169/GL67A	Custic fibrosis	—	II	Active	Imperial College London
PEI	BC-819/PEI	Bladder cancer	—	II	Active	BioCancell Ltd.
	BC-819	Ovarian cancer	—	I/II	Completed	BioCancell Ltd.
	DTA-H19	Pancreatic cancer		I/II	completed	BioCancell Ltd
	SNS01-T	Multiple myeloma and B cell lymphoma	—	I/II	Ongoing	Senesco Technologies, Inc
PEG–PEI–cholesterol	EGEN-001	Ovarian, tubal and peritoneal cancers	—	I	Ongoing	Gynecologic Oncology Group
Poloxamer CRL1005–benzalkonium chloride	VCL-CB01	CMV vaccine	—	II	Completed	Astellas Pharma Inc
Lipid-based nanoparticles	ALN-VSP02	Solid tumours	KIF11 and VEGF	I	Completed	Alnylam Pharmaceuticals
	siRNA–EphA2–DOPC	Advanced cancers	EPHA2	I	Active	MD Anderson Cancer Center
	Atu027	Advanced solid cancers	PKN3	I/II	On going	Silence Therapeutics
	PRO-040201	Hypercholesterolemia	APOB	I	Terminated	Tekmira Pharmaceuticals Corporation
	ALN-TTR02	TTR-mediated amyloidosis	TTR	III	Ongoing	Alnylam Pharmaceuticals
CDP-based nanoparticle	CALAA-01	Solid tumours	RRM2	I	Terminated	Calando Pharmaceuticals
Dynamic Poly- Conjugate	ARC-520	Two conserved regions of HBV transcripts	Hepatitis B	I/II	Ongoing	Arrowhead Research Corporation

Adapted From Yin et al. (2014).

coat the surface of the nanoparticles reducing the immune stimulation and increase circulation time after delivering to the body. However, this approach is limited, as various clinical trials [Table 12.2] have proven that repeated administration of PEGylated nanoparticles accelerated the clearance of nanoparticles. Nano-particle size is another important feature because molecules smaller than 5.5

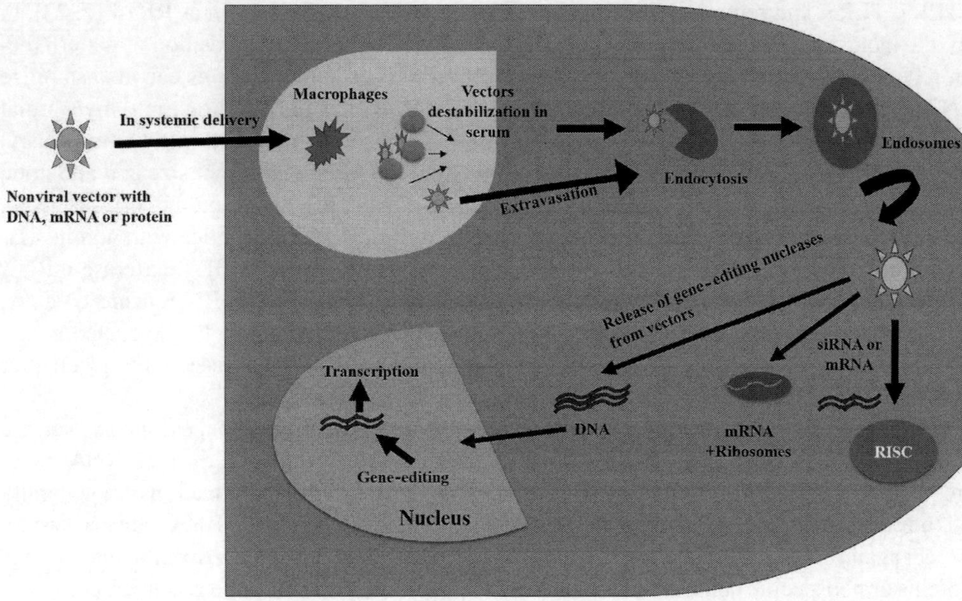

FIGURE 12.2 Schematic presentation of delivery of nonviral vectors for gene editing nucleases in the systemic circulation.

(Adapted from Yin et al 2014).

nm in diameter are subject to rapid clearance from the kidneys [16, 17]. The effective nanoparticle sizes fall between 100-250 nm, which limits the renal filtration of DNA. Other than this, DNA sequences [Figure 12.1] also carry a substantial risk of unintended genomic integration, which can induce insertional mutagenesis [Figure 12.1] due to incorporation of highly active promoter elements into chromosomal DNA or disruption of tumor-suppressor genes. Although the risk of insertional mutagenesis with non-viral delivery of plasmid DNA is generally much lower than with DNA viral vectors, this risk must be taken into account for translationally relevant therapies [3, 18]. However, plasmid DNA delivery is problematic to immune cells because T cells can recognize the intracellular presence of foreign nucleic acids, initiating the innate immune response [3, 17, 18]

12.2.1.1.2 mRNA

mRNA is another tool which acts as a cargo to deliver nucleic acid. Although it is less stable than DNA, one of its advantages is reduced immunogenicity. mRNA also does not require nuclear localization for expression. This cargo can be synthesized from a DNA template using in vitro transcription. mRNA effectively delivers the nucleic acid to the targeted cells and controls the protein expression, which is detectable as quickly as 4–6 h post-transfection [19, 20]. As compared to DNA, mRNA is shortening the protein expression, thus reducing the risk of insertional mutagenesis and also decreasing the probability of off-target effects [2]. It has been reported that mRNA blocks the Cas 9 protein expression for 72 hours after in vitro post transfection and 24 hours post injection in vivo [2, 21]. However, studies demonstrated that, due to different lengths and kinetics of expression, the delivery of Cas9 mRNA and sgRNA may need specialized materials such nanocarriers [2, 22]. One study showed that the optimal condition for an RNA-mediated CRISPR/Cas9 editing system required Cas9 mRNA to be delivered 24 hours before sgRNA delivery [22]. Wang and colleagues demonstrated that ZFNs and a viral vector are needed to deliver the donor DNA to the targeted cells [22, 23]

Studies reported that the modification of mRNA with a combination of 2-thiouridine and 5-methylcytidine effectively reduced the immune stimulation through pattern recognition receptors, such as

TLR3, TLR7, TLR8, and retinoic acid receptor responder protein 3 also known as RIG-I [2, 23]. In addition, the inclusion of pseudouridine in the mRNA (Ψ-mRNA) blocked activation of pattern recognition receptors18 and 2′-5′-oligoadenylate synthetase19. These modifications can also stabilize the mRNA against cleavage and ultimately increase expression rates [24]. A direct intramyocardial injection of vascular endothelial growth factor A (VEGF-A)-encoding modified mRNA (modRNA) complexed with RNAiMAX in myocardial infarction mice decreased the infract size and apoptotic cell frequency, as compared to the control group [3, 23, 24]. In another study, pulmonary surfactant associated protein deficient mice showed 0% survival by five days, whereas mice with administration of modified PSPB-encoding mRNA showed >80% survival by day 30 [3]. Luciferase mRNA successfully delivered into mice after intranasal administration in which mRNA made complex transfection reagent Stemfect or to a hydrophobic poly(β-amino ester) that had been coated with a positively charged lipid layer.[25, 26]. This complex also could express luciferase in the spleen after IV administration [3, 26].

Modification of mRNA molecules is required to overcome the major challenges, such as surface charge of the cell membrane and gene or protein expression after systemic delivery. mRNA molecules are too large in size, hydrophilic, and negatively charged, therefore modifications are generally required to bring therapeutic mRNA to its site of action [2, 3, 24]. The ideal mRNA delivery system must protect against serum endonucleases, elude immune detection, prevent nonspecific interactions with proteins or nonspecific cells, avoid renal clearance, permit extravasation to reach target tissues, and promote cell entry (Figure 12.2). Methods developed to overcome obstacles for RNA delivery include chemical modification, direct injection, and the use of nanoparticles as carriers [2, 26].

12.2.1.1.3 Proteins

Delivery of gene-editing protein complexes eliminates the necessity of intracellular transcription and translation. Protein expression or gene editing can occur immediately after intranuclear delivery. Gene editing protein complexes prompt the gene editing efficacy in postmitotic or hard-to-transfect cells, overcoming the challenges associated with lower gene editing efficacy after use of plasmid DNA or mRNA. On the other hand, this method also reduces the ability of the cargo to enable cell type-specificity, owning to unnecessary transcriptional targeting [27]. In the case of CRISPR/Cas9, there is an additional concern that the Cas9 protein variants from the bacterial species, Staphylococcus pyogenes and S. aureus, have both been shown to be recognized by antibodies in >60% of human patients [27], which could result in rapid clearance of these proteins upon systemic delivery. Furthermore, Cas9 and TAENS have less cell permeability effect, although ZFNs has been reported to be of high membrane permeability efficacy [28]. Because of the nucleic acid binding nature of Cas9 protein, it possesses +20 net charge of unmodified SpCas9, which becomes further positively charged after the addition of nuclear localization signal (NLS) peptide. This positive charge of Cas9 can be neutralized via protein with a glutamate tag up to 20 amino acids in length, thus enhancing the cytosolic delivery of Cas9/sgRNA conjugated with arginine–gold nanoparticles [29]. Another alternative approaches to enhance the plasma-membrane permeability of genome-editing enzymes is conjugation of protein surfaces with multiple viral SV40 NLS domains, which was reported to enable editing without additional vector material both in vitro and in vivo [29, 30]. Likewise, conjugation of Cas9 and sgRNA with cell penetrating peptides facilitates the uptake and endosomal escape as self-condensed cationic nanoparticles, although these nanocarriers only show <10% knockout efficiency in vitro in HEK293T cells [2, 31]

12.3 HYDRODYNAMIC DELIVERY

Hydrodynamic delivery is a physical delivery method discovered in 1999, in which naked DNA is directly delivered into the cells [32]. Various gene editing proteins or their DNA or RNA precursor directly introduces into the cells by this technique. In this method, gene editing nucleases physically disrupt the cells, but they are only applicable to ex vivo delivery. In this technique, a large volume of gene solution is injected into the body to create a high pressure in systemic circulation. This leads to a reversible cell membrane permeation effect, which initiates the intracellular gene transfer [3, 33]. Studies reported that after tail vein injection in rodents, the solution enters into circulation through the inferior vena cava and flows through the heart. However, because of the excessive load in the hepatic vein, the solution is forced out to the tissues. Due to this delivery mechanism, the procedure is especially effective for liver-targeted deliveries [34, 35]. One studies reported that for non-liver tissues specificity of gene editing nucleases can be achieved by directly injected into the desired organ, for example, successful HD targeting to the kidneys can be accomplished by injecting in the renal vein, carotid artery for the brain, and injection to the vena cava for myocardium [3, 35, 36].

Taking all findings together, physical delivery methods exert high potential use in ex vivo delivery of exogenous nucleases. Ex vivo editing of isolated circulating lymphocytes, effectively transferred into a patient for gene therapy. Moreover, application of ex vivo hydrodynamic injection induces expression of Cas9 and sgRNA from plasmid DNA in the liver of rodents. However, this method is not applicable to humans. The HD approach is advantageous in terms of simplicity and cost effectiveness, and it allows the in vivo delivery without the use of viral vectors [3, 34–36]

Apart from this method, other physical delivery methods include micro constriction, which mechanically disrupts the cellular membrane [37, 38], and osmotic potential inducting the cellular uptake and endosomal escape of Cas9 RNP complexes [38]. D'Astolfo and colleagues demonstrated that hypertonic solutions stimulate the macropinocytosis in vitro and subsequently, endosomal disruption by zwitterionic propane betaines effectively delivered the Cas9 RNP complexes, taken as a potential alternative to electroporation or physical membrane disruption method. [3, 39]

12.4 LIPID NANOPARTICLES

Lipid based nanovectors were first discovered in 1987. Cationic lipids consist of a cationic head group and hydrophobic tail, with a linker between these two domains [2, 40]. The cationic headgroup binds to the negatively charged nucleic acid and allows passage through the cell membrane. The exogenous nucleases surrounded by lipid layers protect from RNases and cell degradation enzymes [2, 41]. Cationic lipid vectors are largely utilized for the delivery of siRNA and mRNA, and are currently under clinical trials [Table 12.2] [40–47]. However, limitations of this delivery vehicle include poor rates of transfection [48, 49] and targeting specific tissues. Lipid based nonviral therapeutic approaches widely apply for the biomaterials that are packaged into the vectors and deliver to the target cells. Many lipids and lipid like materials have been formulated for intracellular delivery of gene editing nucleases, such as oligonucleotides, mRNA, and plasmid DNA [2, 49, 50]. Invitro and in vivo applications of Plasmid DNA encoding a Cas9–sgRNA complex, effectively targeted the VEGF when packaged in PEG–PEI– cholesterol lipopolymer and achieved 50 % gene knockout. On the other hand, 30% gene knockout in vitro and 20% in vivo was found with CRISPR DNA assembled with a CD68 promoter encapsulated in lipid-containing PEG–poly(lactic-co-glycolic acid) (PLGA) nanoparticles [2, 51–54]. Zuris and colleagues developed Lipofectamine 2000TM Cas9 and TALENs fused to a negative surface charge of GFP protein. They reported that 24% and 13% gene knockout was found in neuron-derived mouse embryonic stem cells in vitro and in mouse cochlea hair cells in vivo, respectively [55, 56]. Similarly, Wang and colleagues investigated the 70% knockout invitro after application of charge-neutralized Cas9 RNP complexes encapsulated in bioreducible lipid nanoparticles. They also exhibit high expression of gene editing after systemic delivery, primarily in the liver [57]. Yin and colleagues modified the lipid cKK-E12 lipid nanoparticle to improve the delivery of Cas9 mRNA and sgRNA to the hepatocytes, demonstrating 80% gene knockout, whereas 6% editing corrections was found while using AAV encoding a sgRNA and HDR DNA template [58]. A cancer derived exosome has been developed by Kim et al. for delivery of a plasmid DNA encoding a CRISPR/Cas system showing efficient and targeted editing in an ovarian cancer model [2, 3]. An exosome–liposome hybrid vector developed by Lin and colleagues showed CRISPR interference in mesenchymal stem cells, which could not be transfected using liposomes alone [59]. These findings reveal the great potential of lipid nanoparticles for the delivery of genome editing nucleases to treat various diseases, rivaling that of virus-mediated delivery for some tissues such as the liver. A clinical trial report exhibited that LNP formulation consists of cationic lipid (AtuFECT01), a helper lipid (DPhyPE), and a PEG–lipid (PEG–DSPE) in a 50:49:1 ratio, with siRNAs120 shown to internalize into mouse vascular endothelium after intravenous injection [60]. The AtuPLEX-based formulation Atu027 features an siRNA that targets the protein kinase N3 (PKN3) transcript and is under evaluation for the treatment of patients with advanced solid cancer [2, 61, 62]. On the other hand, MD Anderson Cancer Center has initiated a clinical trial for targeting EPHA2, which encodes a tyrosine kinase to treat patients with advanced cancer by using their formulations in which siRNA is incorporated in neutral liposomes composed of DOPC [2, 63].

The siRNA based lipid vectors showed targeted delivery of siRNA to silence the target genes with a nanoparticle size of 200nm [64]. It was first discovered in 2005 to treat hepatitis B virus (HBV) in a mouse model HBV replication [65]. In two separate clinical trials, efficacy of LNP-based siRNA delivery has been evaluated in patients with hepatocellular carcinoma. Tekmira pharmaceuticals corporation investigated the antitumor activity of drug TKM-080301 targeted to suppress the polo-like kinase 1 (PLK1). On the other hand, Alnylam pharmaceuticals developed ALN-VSP02 to target KIF11 (which encodes kinesin spindle protein) and VEGF. In addition, ALN-TTR02 (Alnylam pharmaceuticals) based siRNA formulations are shown to have the ability of silencing the transthyretin (TTR) in TTR-mediated amyloidosis. Patisiran is a second-generation SNALP formulation ALN-TTR02, which is a DLinDMA analogue. In preclinical studies, it showed a tenfold increase in therapeutic activity [1, 66].

12.5 POLYMER BASED NANOPARTICLES

Polymeric materials have been employed to deliver the gene editing nucleases for targeted delivery. They act as a multicomponent delivery system to promote endosomal escape of the delivered exogenous materials [2, 67]. Studies reported that upon the coating of Cas9–sgRNA RNPs with endosomal disruptive polymer poly(N-(N-(2-aminoethyl)-2-aminoethyl) aspartamide is easily adsorbed onto gold nanoparticles and corrects the dystrophin gene in vivo, using a mouse model suffering from Duchenne muscular dystrophy. As compared to scramble nanoparticles, these showed increased levels of muscular strength [2, 67]. CRISPR loaded in gold nanoparticles reduce mRNA and protein levels drastically (up to 50%) after local injection in a rodent brain. Sun and colleagues used an alternative approach to achieve increased levels of gene modification (EGFP, CRISP)) in vivo after administration of the nanoparticles. With this method, DNA nanoclews formed by using rolling-circle amplification, which allows encapsulation of RNP complexes in a DNA-based particle core further coated with a cationic polymer to attain efficient gene editing by exogenous nucleases [68]. Recently, Cheng and Leong reported that coating with cationic polymer, such as cationic a-helical polypeptide, efficiently delivered the Cas9 and sgRNA plasmids to Hela cells, leading to 67% knockdown of targeted protein after repeated intratumoral injections. Both the exogenous gene editing nucleases reduced the tumor growth by 71 % in vivo using xenograft mouse model [69]. However, in scale up and sufficient transport to the targeted cells, a large amount of cationic polymer is required, which is cost effective and difficult to formulate. However, a large amount of cationic polymer is required for efficient delivery of drugs to the targeted cell. Formulation of this kind of nanoparticles is cost effective.

It has been reported that PLL poly (I-lysine), which is a homopolypeptide of the basic amino acid, has the ability to condense the DNA into nanoparticles via electrostatic interactions. One study indicated that PLL complexed with asialoorosomucoid glycoprotein could potentially be applied in non-viral liver-targeted gene delivery [70–72]. In contrast, without a lysosomal disruption agent such as chloroquine, PLL shows poor transfection function, due to the fact that at physiological pH, its surface charge becomes positive, thus low endosomal buffering and lysis are exhibited by this polymer [1]. Moreover, unmodified polymer possessed marked invitro cytotoxicity [73–75]. To overcome this barrier, PLL covered by hydrophilic polymer PEG increases circulation time and minimizes the nonspecific interaction with the serum components [76, 77]. PEGylated PLL is being investigated in a phase I clinical trial to demonstrate the safety and tolerability of these DNA nanoparticles and potential therapeutic vectors for gene transfer [1].

The inclusion of Zwittherionic materials is another approach to systemic delivery of therapeutic genes. The addition of siRNA therapeutic in genome-editing nucleic acid or protein is a new advance to nonviral delivery; especially, zwitter-ionic materials, including cationic quaternary ammonium sulfonamide aminolipids, and a zwitterionic phosphorylcholine-based polymer such as diblock polymer, have been shown to improve systemic delivery, providing more efficient alternatives to PEGylation [78, 79].

Chang and colleagues reported a protein delivery system composed of a dendrimer end-capped with guanidyl groups to facilitate protein binding through hydrogen bonding and salt bridges, as well as phenyl groups to promote endocytosis and endosomal escape [79]. They efficiently transfer the functionally intact proteins invitro after encapsulating the protein of different sizes and surface charges [2, 80]. Overall, these strategies help to increase the cell membrane permeability of gene-editing protein complexes, which are mostly impermeable to cell membranes on their own and are not able to load cationic polymers, due to slightly cationic surface charges on their surface This polymer successfully encapsulated proteins of different sizes and surface charges and promoted the efficient intracellular delivery of functionally intact proteins in vitro. Yan and colleagues implemented a different approach in which they covalently bound thin polymer shells with a protein core in situ, showing that within this shell, the protein molecule was protected from protease degradation and exhibited efficient cellular uptake in vitro and in vivo. Formulated vectors do not rely on the electrostatic charge-based interactions, thus overcoming the challenges of delivery of proteins and gene editing nucleases, such as Cas9 RNP complexes, which contain regions of varying charge [81]. Recently, zinc/imidazole based metal-organic frameworks (MOF) are new advances for intracellular delivery of protein. These MOF nanoparticles efficaciously prevent degradation of protein by protease digestion [82, 83]. However, studies reported that CRISPR RNPs showed low genome editing efficiency by up to 30% in CHO cells invitro. Therefore, further optimization of this vector is necessary[84].

Numerous other polymers, such as poly[(2-dimethylamino) ethyl methacrylate] (pDMAEMA), poly(β-amino ester)s, and various carbohydrate-based polymers and dendrimers, are currently under preclinical trials account for DNA delivery. A polymeric vector consists of non-ionic poloxamer CRL1005, and the cationic surfactant benzalkonium chloride has been pushed to clinical development. This formulation is being under investigation in a Phase II/III clinical trial for a genetic vaccine to prevent CMV infection in patients with allogeneic hematopoietic cell transplant [1, 85, 86] [Table 12.2]

12.6 CELL PENETRATING PEPTIDES

These are protein transduction domains, first discovered in 1988, that reveal the transcription trans-activating protein isolated from the HIV virus that could cross cellular membranes. It consists of small, 6–30 amino acid long, synthetic, or naturally occurring peptides that readily convey the cargos across the cell membranes with the intact functional form [86]. Various exogenous materials, such as peptides, proteins, nucleic acid including DNA and RNA, nanocarriers such as liposomes and nanoparticles, imaging agents' radioisotopes, and other fluorescent probes, are effectively delivered to the targeted sites by these peptides. They also successfully introduce ZFNs, TALENs, and Cas9 in cells in vitro [87–89].

Intradermal or intramuscular injection of stearyl-transportan 10 complexed with plasmid-DNA produced non-cytotoxic, non-immunogenic, and induced transfer/expression of Luciferase reporter plasmid [90]. Modification of S4 (13)-PV with a 5 histidine residue attached to its N-terminus successfully encapsulate plasmid DNA and siRNA invitro, leading to successful gene silencing of Survivin compared to unmodified SV(13)-PV [91]. The cytotoxic effect of Poly-lysine homo-polymers was reduced by condensing with an angiotensin type II receptor, using calcium chloride encapsulated in a plasmid. This procedure also has shown to efficient a tumor targeting effect and decrease the tumor volume in vivo [91, 92].

Dual targeting is another approach to enhanced delivery of gene editing nuclease. In this technique, conjugation of tumor targeting peptide(CREKA)and a homo-polymer of arginine with avidin moiety efficiently condense with plasmid DNA, carrying the p53 gene and showing reduced tumor growth in mice [93]. The conjugation of peptides with siRNA successfully silenced CIP2A in vivo in mice with oral cancer [94]. Polyethylene shell is a tumor pH responsive polymer core that effectively delivers the siRNA, silences the bromodomain 4, and inhibits the prostate cancer growth significantly. This core contained the tumor targeting RGD peptide, along with the homo-polymer of arginine. These complexes bound with siRNA by electrostatic interactions, and thus enhanced the siRNA concentration within the polymeric shell [95, 96].

In the tumor microenvironment, lower pH CPP loaded with siRNA effectively silenced the Rac1, reduced the hepatic metastases in colon cancer, and suppressed c-MYC gene expression in breast cancer cell lines in vivo [97]. Antiangiogenic activity of siRNA was enhanced by encapsulating in stearylated poly-arginine through electrostatic interaction modified with histidine. In addition, this peptide loaded with fausildil, a selective Rho-Kianse inhibitor, showed strong cellular uptake and reduced tumor growth [98, 99]. CPP also can complex with nanoparticles to deliver siRNA to the targeted site. Studies reported that modified CPP, with gelatin silica NP comprised with Tat and influenza hemaglutanin A2, successfully conveys the plasmid DNA to the cells in vivo [100, 101]. A tumor activatable CPP, dual-triggered by lowered pH and matrix metalloproteinase 2, was used to label NPs carrying dual anti-tumor therapies, doxorubicin and siRNA, targeting vascular endothelial growth factor [102–105].

PEGylation of glycosaminoglycan-binding peptides coupled with DNA through electrostatic charge interactions, which formed NPs that targeted bronchial epithelial cell lines with precision cut lung slices in vitro, showing that PEGylation rates of >40% were the optimal formulation [106]. Additionally, other supra-magnetic iron oxide NPs or gold nanoparticles have shown to internalization into tumor or stem cells upon modification of CPP [107–110]. Cytotoxicity of liposomal CPP can be reduced by coating the LP surface with neutral PEG, which also stabilized the NPS, and these specialized CPP nanoparticles also enhanced the gene silencing capacity invitro [111]. Another studies showed that modification of TAT peptide labeled with LPS containing doxorubicin enhanced the efficacy of targeted delivery of therapeutic molecules, resulting in an increase in anti-tumor activity and decrease in systemic side effects [111–113].

12.7 DNA ORIGAMI

DNA nanostructures were first discovered by nature at Nadrian Seeman. The structure of DNA origami is formed by the folding of multiple single strand DNA oligomers to create a nanoscale of arbitrary two- and three-dimensional shapes. The specificity of the interactions between complementary base pairs make DNA a useful construction material, through design of its base sequences [114]. Recently, it has been considered as a potential tool in delivery of various therapeutic molecules. Rothemund developed scaffolded DNA origami by using long ssDNA derived from M13 bacteriophage [115]. The addition of hundreds of short ssDNAs help to produce nanostructures with arbitrary shapes. Short ssDNA has the ability to create different two and three dimensional shapes of DNA origami [116, 117]. Together, these methods treat DNA as a biopolymer, and its information-containing structure is used to drive nanostructure assembly. These DNA origami platforms can be decorated with biomolecules, including peptides, proteins, and functional molecules containing fluorophores, aptamers, quantum dots, and gold nanoparticles, allowing them the ability to target cells [118, 119]. Decoration of DNA origami with the iron transport protein transferrin increases the number of transferrin molecules and helps the efficient delivery of chemotherapeutic agents and gene-editing nucleases. Studies reported that doxorubicin and P53 gene efficiently deliver to the targeted sites after addition of aptamers to the triangular structure of DNA nanostructure [120]. However, use of DNA nanotechnology in CPPs are still limited. One study reported that transduction efficiency of gold nanoparticles could be enhanced by decoration of filamentous DNA origami, with CPP providing 3D superstructure. This strategy is considered an efficient way to improve drug delivery and cellular imaging [9]. In addition, various cationic and amphipathic CPPs, such as TAT, penetratin, and MAP are successfully decorated with DNA oligomer for efficient delivery of siRNA and other exogenous genome-editing nucleases to the targeted sites [121].

Recently, many crucial steps have been taken by researchers and industrialists to achieve potential nonviral based genome editing delivery systems. Various new strategies employed in nonviral delivery systems, such as PEG conjugation with nanoparticles, modification of mRNA structure, and surface modification CPP, are leading to advances in gene therapy. These approaches also improve the potency and stability of nucleic acid. Moreover, advances in non-viral genome editing have been broadening our knowledge, particularly in the genetic basis of diseases, and have laid out a range of new targets for genetic medicine. The efficacy of exogenous nucleic acid may provide a platform for implementing the therapeutic nucleases in genetic disorders. It is mandatory to extend beyond this application of non-viral gene-based cell specific delivery, or the pathology behind nanocarriers-gene interaction, to expand their target specificity, such as including more tissues or organs. In addition, these delivery vehicles provide precise and permanent correction of disease genes. The development in therapeutic gene delivery facilitates the ability to employ gene-based therapy in different tissues and to silence, correct, or introduce specific genes with less adverse effects.

REFERENCES

1. Yin H, Rosemary LK, Ahmed A, et al. Non-viral vectors for gene-based therapy. *Nat. Rev. Genet.* 2014;15:541–555.
2. Rui Y, Wilson DR, and Green J. Non-viral delivery to enable genome editing. *Trends Biotechnol.* 2019;37:281–293.
3. Ates I, Rathbone T, Stuart C, et al. Delivery approaches for therapeutic genome editing and challenges. *Genes (Basel)* 2020;11:1–32.
4. Wang M, Glass ZA, and Xu Q. Non-viral delivery of genome-editing nucleases for gene therapy. *Gene Ther.* 2017;24:144–150.
5. Wang HX, Song Z, Lao YH, et al. Nonviral gene editing via CRISPR/Cas9 delivery by membrane-disruptive and endosomolytic helical polypeptide. *Proc. Natl. Acad. Sci. U. S. A.* 2018;115:4903–4908.
6. Wu Z, Yang H, and Colosi P. Effect of genome size on AAV vector packaging. *Mol. Ther.* 2010;18:80–86.
7. Wells DJ. Gene therapy progress and prospects: electroporation and other physical methods. *Gene Ther.* 2004;11:1363–1369.
8. Maurisse R, Semir DD, Emamekhoo H, et al. Comparative transfection of DNA into primary and transformed mammalian cells from different lineages. *BMC Biotechnol.* 2010;10:9.
9. Taylor RE, and Zahid M. Cell penetrating peptides, novel vectors for gene therapy. *Pharmaceutics* 2020;12:1–21.
10. Beals N, Kasibhatla N, and Basu S. Efficient delivery of plasmid DNA using incorporated nucleotides for precise conjugation of targeted nanoparticles. *ACS Appl. Bio Mater.* 2019;2:717–727.
11. Neumann E, Schaefer-Ridder M, Wang Y, et al. Gene transfer into mouse lyoma cells by electroporation in high electric fields. *EMBO J.* 1982;1:841–845.
12. Ye L, Wang J, Beyer AI, et al. Seamless modification of wild-type induced pluripotent stem cells to the natural CCR5Delta32 mutation confers resistance to HIV infection. *Proc. Natl. Acad. Sci. U. S. A.* 2014;111:9591–9596.
13. Kim S, Kim D, Cho SW, et al. Highly efficient RNA-guided genome editing in human cells via delivery of purified Cas9 ribonucleoproteins. *Genome Res.* 2014;24:1012–1019.

14. Liu J, Zeng F, Allen C, et al. Influence of serum protein on polycarbonate-based copolymer micelles as a delivery system for a hydrophobic anti-cancer agent. *J. Control. Release* 2005;103:481–497.
15. Jones SW, Roberts RA, Robbins GR, et al. Nanoparticle clearance is governed by Th1/Th2 immunity and strain background. *J. Clin. Invest.* 2013;123:3061–3073.
16. Suk JS, Xu Q, Kim N, et al. PEGylation as a strategy for improving nanoparticle-based drug and gene delivery. *Adv. Drug Deliv. Rev.*2016;99:28–51.
17. Ando H, Lila ASA, Kawanishi M, et al. Reactivity of IgM antibodies elicited by PEGylated liposomes or PEGylated lipoplexes against auto and foreign antigens. *J. Control. Release* 2018;270:114–119.
18. Choi HS, Liu W, Misra P, et al. Renal clearance of nanoparticles. *Nat. Biotechnol.* 2007;25:1165–1170.
19. Jiang C, Mei M, Li B, et al. A non-viral CRISPR/Cas9 delivery system for therapeutically targeting HBV DNA and pcsk9 in vivo. *Cell Res.* 2017;27:440.
20. Liang X, Potter J, Kumar S, et al. Rapid and highly efficient mammalian cell engineering via Cas9 protein transfection. *J. Biotechnol.* 2015;208:44–53.
21. Miller JB, Zhang S, Kos P, et al. Non-viral CRISPR/Cas gene editing in vitro and in vivo enabled by synthetic nanoparticle co-delivery of Cas9 mRNA and sgRNA. *Angew. Chem. Int. Ed.* 2017;56:1059–1063.
22. Wang J. Clercq D, Hayward SB, et al. Highly efficient homology-driven genome editing in human T cells by combining zinc-finger nuclease mRNA and AAV6 donor delivery. *Nucleic Acids Res.* 2016;44:e30.
23. Zang L, Lui OK, Gise AV, et al. Modified mRNA directs the fate of heart progenitor cells and induces vascular regeneration after myocardial infarction. *Nature Biotech.* 2013;31:898–907.
24. Kormann MS, Hasenpousch G, Aneja MK, et al. Expression of therapeutic proteins after delivery of chemically modified mRNA in mice. *Nature Biotech.* 2011;29:154–157.
25. Su X, Fricke J, Kavanagh DG, et al. In vitro and in vivo mRNA delivery using lipid-enveloped pH-responsive polymer nanoparticles. *Mol. Pharm.* 2011;8:774–787.
26. Phua KK, Leong K, and Nair SK. Transfection efficiency and transgene expression kinetics of mRNA delivered in naked and nanoparticle format. *J. Control. Release* 2013;166:227–233.
27. Charlesworth CT, Deshpande PS, Dever DP, et al. Identification of pre-existing adaptive immunity to Cas9 proteins in humans. *Nat. Med.* 2019;25(2):249–254.
28. Mout R, Ray M, Tonga GY, et al. Direct cytosolic delivery of CRISPR/Cas9- ribonucleoprotein for efficient gene editing. *ACS Nano* 2017;11:2452–2458.
29. Liu J, Gaj T, Wallen MC, et al. Improved cell-penetrating zinc-finger nuclease proteins for precision genome engineering. *Mol. Ther. Nucl. Acids* 2015;4:e232.
30. Staahl BT, Benekareddy M, Bainier MB, et al. Efficient genome editing in the mouse brain by local delivery of engineered Cas9 ribonucleoprotein complexes. *Nat. Biotechnol.* 2017;35:431–434.
31. Ramakrishna S, Dad ABK, Bloor J, et al. Gene disruption by cell-penetrating peptide-mediated delivery of Cas9 protein and guide RNA. *Genome Res.* 2014;24:1020–1027.
32. Liu F, Song YK, and Liu D. Hydrodynamics-based transfection in animals by systemic administration of plasmid DNA. *Gene Ther.* 1999;6(7):1258–1266.
33. Zhang G, Budker V, and Wolff JA. High levels of foreign gene expression in hepatocytes after tail vein injections of naked plasmid DNA. *Hum. Gene Ther.* 1999;10(10):1735–1737.
34. McCaffrey AP, Meuse L, and Pham TTT. RNA interference in adult mice. *Nature* 2002;418:38–39.
35. Suda T, and Liu D. *Hydrodynamic Delivery*; Elsevier Ltd.: Amsterdam, The Netherlands, 2015;89:89–111.
36. Crespo A, Peydró A, and Dasí F. Hydrodynamic liver gene transfer mechanism involves transient sinusoidal blood stasis and massive hepatocyte endocytic vesicles. *Gene Ther.* 2005;12(11):927–935.
37. Yin H, Xue W, Chen S, et al. Genome editing with Cas9 in adult mice corrects a disease mutation and phenotype. *Nat. Biotechnol.* 2014;32:551–553.
38. Han X, Liu Z, Chan M, et al. CRISPR-Cas9 delivery to hard-to-transfect cells via membrane deformation. *Sci. Adv.* 2015;1:e1500454.
39. D'Astolfo DS, Pagliero RJ, Pras A, et al. Efficient intracellular delivery of native proteins. *Cell* 2015;161:674–690.
40. Mintzer MA, and Simanek EE. Nonviral vectors for gene delivery. *Chem. Rev.* 2009;109:259–302.
41. Liu Q, Zhu H, Liu Y, et al. BRAF peptide vaccine facilitates therapy of murine BRAF-mutant melanoma. *Cancer Immunol. Immunother.* 2018;67:299–310.
42. Basha G, Ordobadi M, Scott WR, et al. Lipid nanoparticle delivery of siRNA to osteocytes leads to effective silencing of SOST and inhibition of sclerostin in vivo. *Mol. Ther. Nucleic Acids* 2016;5:e363.
43. Kranz LM, Diken M, Haas H, et al. Systemic RNA delivery to dendritic cells exploits antiviral defence fo rcancer immunotherapy. *Nature* 2016;534:396–401.

44. Magini D, Giovani C, and Mangiavacchi S. Self-amplifying mRNA vaccines expressing multiple conserved influenza antigens confer protection against homologous and heterosubtypic viral challenge. *PLoS One* 2016;11:e0161193.
45. Bahl K, Senn JJ, Yuzhakov O, et al. Preclinical and clinical demonstration of immunogenicity by mRNA vaccines against H10N8 and H7N9 influenza viruses. *Mol. Ther.* 2017;25:1316–1327.
46. Coelho T, Adams D, Silva A, et al. Safety and efficacy of RNAi therapy for transthyretin amyloidosis. *N. Engl. J. Med.* 2013;369:819–829.
47. Richner JM, Himansu S, Dowd KA, et al. Modified mRNA vaccines protect against zika virus infection. *Cell* 2017;168(6):1114–1225.
48. Miller JB, Zhang S, Kos P, et al. Non viral CRISPR/Cas gene editing in vitro and in vivo enabled by synthetic nanoparticle co-delivery of Cas9 mRNA and sgRNA. *Angew. Chem. Int. Ed. Engl.* 2017;56:1059–1063.
49. Yin H, Song CQ, Dorkin JR, et al. Therapeutic genome editing byc ombined viral and non-vivo. *Nat. Biotechnol.* 2016;34:328–333.
50. Wang Y, Miao L, Satterlee A, et al. Delivery of oligonucleotides with lipid nano-particles. *Adv. Drug Deliv. Rev.* 2015;87:68–80.
51. Hajj KA, and Whitehead KA. Tools for translation: non-viral materials for therapeutic mRNA delivery. *Nat. Rev. Mater* 2017;2:17056.
52. de Lima MCP, Simoes S, Faneca H, et al. Cationic lipid–DNA complexes in gene delivery: from biophysics to biological applications. *Adv. Drug Deliv. Rev.* 2001;47:277–294.
53. Liang C, Li F, Wang L, et al Tumor cell-targeted delivery of CRISPR/ Cas9 by aptamer-functionalized lipopolymer for therapeutic genome editing of VEGFA in osteosarcoma. *Biomaterials* 2017;147:68–85.
54. Luo YL , Xu CF, Li HJ, et al. Macrophage-specific in vivo gene editing using cationic lipid-assisted polymeric nanoparticles. *ACS Nano* 2018;12:994–1005.
55. Zuris JA, Thompson DB, Shu Y, et al Cationic lipid-mediated delivery of proteins enables efficient protein-based genome editing in vitro and in vivo. *Nat. Biotechnol.* 2015;33:73–80.
56. Gao X, Tao Y, Lamas V, et al. Treatment of autosomal dominant hearing loss by in vivo delivery of genome editing agents. *Nature* 2018;553:217–221.
57. Wang M, Zuris JA, Meng F, et al. Efficient delivery of genome-editing pro-teins using bioreducible lipid nanoparticles. *Proc. Natl. Acad. Sci.* 2016;113:2868.
58. Yin H. Song CQ, Suresh S, et al. Structure-guided chemical modification of guide RNA enables potent non-viral in vivo genome editing. *Nat. Biotechnol.* 2017;35:1179.
59. Lin Y, Wu J, Gu W, et al. Exosome–liposome hybrid nanoparticles deliver CRISPR/Cas9 system in MSCs. *Adv. Sci.* 2018;5:1700611.
60. Santel A, Aleku M, Keil O, et al. A novel siRNA–lipoplex technology for RNA interference in the mouse vascular endothelium. *Gene Ther.* 2006;13:1222–1234.
61. Santel A, Aleku M, Keil O, et al. RNA interference in the mouse vascular endothelium by systemic administration of siRNA–lipoplexes for cancer therapy. *Gene Ther.* 2006;13:1360–1370.
62. Aleku M, Schulz P, Keil O, et al. Atu027, a liposomal small interfering RNA formulation targeting protein kinase N3, inhibits cancer progression. *Cancer Res.* 2008;68:9788–9798.
63. Landen CN Jr, Reyes AC, Bucana C, et al. Therapeutic EphA2 gene targeting in vivo using neutral liposomal small interfering RNA delivery. *Cancer Res.* 2005;65:6910–6918.
64. Wheeler JJ, Palmer L, Ossanlou M, et al. Stabilized plasmid–lipid particles: construction and characterization. *Gene Ther.* 1999;6:271–281.
65. Morrissey DV, Lockridge JA, Shaw L, et al. Potent and persistent in vivo anti-HBV activity of chemically modified siRNAs. *Nature Biotech* 2005;23:1002–1007.
66. Jayaraman M, Mui BL, Ansel SM, et al. Maximizing the potency of siRNA lipid nanoparticles for hepatic gene silencing in vivo. *Angew. Chem. Int. Ed. Engl.* 2012;51:8529–8533.
67. Lee K, Conboy M, Park HM, et al. Nanoparticle delivery of Cas9 ribonucleopro-tein and donor DNA in vivo induces homology-directed DNA repair. *Nat. Biomed. Eng.* 2017;1:889–901.
68. Lee B, Lee K, Panda S, et al. Nanoparticle delivery of CRISPR into the brain rescues a mouse model of fragile X syndrome from exaggerated repetitive behaviours. *Nat. Biomed. Eng.* 2018;2:497.
69. Sun W, Ji W, Hall JM, et al. Self-assembled DNA nanoclews for the efficient delivery of CRISPR-Cas9 for genome editing. *Angew. Chem. Int. Ed.* 2015;54:12029–12033.
70. Wang HX, Song Z, Lao YH, et al. Nonviral gene editing via CRISPR/Cas9 delivery by membrane-disruptive and endosomolytic helical poly-peptide. *Proc. Natl. Acad. Sci.* 2018;115(19):4903–4908.

71. Olins DE, Olins AL, and Von Hippel PH. Model nucleoprotein complexes: studies on the interaction of cationic homopolypeptides with DNA. *J. Mol. Biol.* 1967;24:157–176.
72. Laemmli UK. Characterization of DNA condensates induced by poly(ethylene oxide) and polylysine. *Proc. Natl. Acad. Sci. U. S. A.* 1975;72:4288–4292.
73. Wu GY, and Wu CH. Receptor-mediated in vitro gene transformation by a soluble DNA carrier system. *J. Biol. Chem.* 1987;262:4429–4432.
74. Wu GY, and Wu CH. Receptor-mediated gene delivery and expression in vivo. *J. Biol. Chem.* 1988;263:14621–14624.
75. Choi YH, Liu F, Kim JS, et al. Polyethylene glycol-grafted poly-L-lysine as polymeric gene carrier. *J. Control. Release* 1998;54:39–48.
76. Alexis F, Pridgen E, and Molnar LK. Factors affecting the clearance and biodistribution of polymeric nanoparticles. *Mol. Pharm.* 2008;5:505–515.
77. Bazile D, Prudhomme C, Bassoullet C, et al. Stealth Me. PEG–PLA nanoparticles avoid uptake by the mononuclear phagocytes system. *J. Pharm. Sci.* 1995;84:493–498.
78. Miller JB, Cos P, Tieu V, et al. Development of cationic quaternary ammonium sulfonamide amino lipids for nucleic acid delivery. *ACS Appl. Mater. Interfaces* 2018;10:2302–2311.
79. Jackson MA, Thomas A, Elizabeth J, et al. Zwitterionic nanocarrier surface chemistry improves siRNA tumor delivery and silencing activity relative to polyethylene glycol. *ACS Nano* 2017;11:5680–5696.
80. Chang H, Lv J, Gao X, et al. Rational design of a polymer with robust efficacy for intracellular protein and peptide delivery. *Nano Lett.* 2017;17:1678–1684.
81. Yan M, Du J, Gu Z, et al. A novel intracellular protein delivery platform based on single-protein nanocapsules. *Nat. Nanotechnol.* 2009;5:48.
82. Chen T, Yi JT, Zhao YY, et al. Biomineralized metal–organic framework nanoparticles enable intracellular delivery and endo-lysosomal release of native active proteins. *J. Am. Chem. Soc.* 2018;140:9912–9920.
83. Cheng G, Li W, Ha L, et al. Self-assembly of extracellular vesicle-like metal–organic framework nanoparticles for protection and intra-cellular delivery of biofunctional proteins. *J. Am. Chem. Soc.* 2018;140:7282–7291.
84. Alsaiari SK, Patil S, Alaymi M, et al. Endosomal escape and delivery of CRISPR/Cas9 genome editing machinery enabled by nanoscale zeolitic imidazolate framework. *J. Am. Chem. Soc.* 2017;140:143–146.
85. Pack DW, Hoffman A, Pun S, et al. Design and development of polymers for gene delivery. *Nat. Rev. Drug Discov.* 2005;4:581–593.
86. Mintzer MA, and Simanek EE. Nonviral vectors for gene delivery. *Chem. Rev.* 2009;109:259–302.
87. Frankel AD, and Pabo CO. Cellular uptake of the tat protein from human immunodeficiency virus. *Cell* 1988;55:1189–1193.
88. Gaj T, Guo J, Kato Y, et al. Targeted gene knockout by direct delivery of zinc-finger nuclease proteins. *Nat. Methods* 2012;9:805–807.
89. Ramakrishna S, Kwaku DAB, Beloor J, et al. Gene disruption by cell-penetratingpeptide-mediated delivery of Cas9 protein and guide RNA. *Genome Res.* 2014;24:1020–1027.
90. Lehto T, Simonson OE, Mäger I, et al. A peptide-based vector for efficient gene transfer in vitro and in vivo. *Mol. Ther.* 2011;19:1457–1467.
91. Zhang W, Song J, Liang R, et al. Stearylated antimicrobial peptide [D]-K6L9 with cell penetrating property for efficient gene transfer. *Peptides* 2013;46:33–39.
92. Cardoso AM, Trabulo S, and Cardoso AL. Comparison of the efficiency of complexes based on S413-PV cell-penetrating peptides in plasmid DNA and siRNA delivery. *Mol. Pharm.* 2013;10:2653–2666.
93. Alhakamy NA, Ishiguro S, Uppalapati D, et al. AT2R gene delivered by condensed polylysine complexes attenuates lewis lung carcinoma after intravenous injection or intratracheal spray. *Mol. Cancer Ther.* 2015;15, 209–218.
94. Qu W, Chen WH, Kuang Y, et al. Avidin–biotin interaction mediated peptideassemblies as efficient gene deliveryvectors for cancer therapy. *Mol. Pharm.* 2012;10:261–269.
95. Gong C, Pan D, Qiu F, et al. Selective DNA delivery to tumor cells using an oligoarginine-LTVSPWY peptide. *PLoS One* 2014;9:e110632.
96. Wang H, Wang H, Liang J, et al. Cell-penetrating apoptotic peptide/p53 DNA nanocomplex as adjuvant therapy for drug-resistant breast cancer. *Mol. Pharm.* 2014;11:3352–3360.
97. Bao Y, Guo H, Lu Y, et al. Blocking hepatic metastases of colon cancer cells using an shRNA against Rac1 delivered by activatable cell-penetrating peptide. *Oncotarget* 2016;7:77183–77195.

98. Yang Y, Xia X, Dong W, et al. . Acid sensitive polymeric micelles combining folate and bioreducible conjugate for specific intracellular siRNA delivery. *Macromol. Biosci.* 2016;16:759–773.
99. Liu Y, Wu X, Gao Y, et al. Aptamer-functionalized peptide H3CR5C as a novel nanovehicle for codelivery of fasudil and miRNA-195 targeting hepatocellular carcinoma. *Int. J. Nanomedicine* 2016;11:3891–3905.
100. Ye S, Tian MM, Wang TX, et al. Synergistic effects of cell-penetrating peptide Tat and fusogenic peptide HA2-enhanced cellular internalization and gene transduction of organosilica nanoparticles. *Nanomed. Nanotechnol. Biol. Med.* 2012;8:833–841.
101. Huang S, Shao K, Liu Y, et al. Tumor-targeting and microenvironment-responsive smart nanoparticles for combination therapy of antiangiogenesis and apoptosis. *ACS Nano* 2013;7:2860–2871.
102. Yang Y, Xie X, Yang Y, et al. Polymer nanoparticles modified with photo- and pH-dual-responsive polypeptides for enhanced and targeted cancer therapy. *Mol. Pharm.* 2016;13:1508–1519.
103. Osman G, Rodriguez J, Chan SY, et al. PEGylated enhanced cell penetrating peptide nanoparticles for lung gene therapy. *J. Control. Release* 2018;285:35–45.
104. Dowaidar M, Abdelhamid HN, Hällbrink M, et al. Magnetic nanoparticle assisted self-assembly of cell penetrating peptides-oligonucleotides complexes for gene delivery. *Sci. Rep.* 2017;7:9159.
105. Ben Djemaa S, David S, Aubert KH, et al. Formulation and in vitro evaluation of a siRNA delivery nanosystem decorated with gH625 peptide for triple negative breast cancer theranosis. *Eur. J. Pharm. Biopharm.* 2018;131:99–108.
106. Osman G, Rodriguez J, Chan SY, et al. PEGylated enhanced cell penetrating peptide nanoparticles for lung gene therapy. *J. Control. Release* 2018;285:35–45.
107. Dowaidar M, Abdelhamid HN, Hällbrink M, et al. Magnetic nanoparticle assisted self-assembly of cell penetrating peptides-complexes for gene delivery. *Sci. Rep.* 2017;7:9159.
108. Ben Djemaa S, David S, and Aubert KH. Formulation and in vitro evaluation of a siRNA delivery nanosystem decorated with gH625 peptide for triple negative breast cancer theranosis. *Eur. J. Pharm. Biopharm.* 2018;131:99–108.
109. Peng LH, Niu J, Zhang CZ, et al. TAT conjugated cationic noble metal nanoparticles for gene delivery to epidermal stem cells. *Biomaterials* 2014;35:5605–5618.
110. Park H, Tsutsumi H, and Mihara H. Cell penetration and cell-selective drug delivery using α-helix peptides conjugated with gold nanoparticles. *Biomaterials* 2013;34:4872–4879.
111. Fisher RK, Mattern-Schain SI, Best MD, et al. Improving the efficacy of liposome-mediated vascular gene therapy via lipid surface modifications. *J. Surg. Res.* 2017;219:136–144.
112. Fu H, Shi K, Hu G, Yang Y, et al. Tumor-targeted paclitaxel delivery and enhanced penetration using TAT-decorated liposomescomprising redox-responsive poly(ethylene glycol). *J. Pharm. Sci.* 2015;104:1160–1173.
113. Liu Z, Xiong M, Gong J, et al. Legumain protease-activated TAT-liposome cargo for targeting tumours and their microenvironment. *Nat. Commun.* 2014;5:4280.
114. Seeman NC Nucleic acid junctions and lattices. *J. Theor. Biol.* 1982;99:237–247.
115. Rothemund PWK. Folding DNA to create nanoscale shapes and patterns. *Nature* 2006;440:297–302.
116. Ke Y, Ong LL, Sun W, et al. DNA brick crystals with prescribed depths. *Nat. Chem.* 2014;6:994–1002.
117. Ke Y, Ong LL, William M, et al. Three-dimensional structures self-assembled from DNA bricks. *Science* 2012;338:1177–1183.
118. Sacca B, and Niemeyer CM. Functionalization of DNA nanostructures with proteins. *Chem. Soc. Rev.* 2011;40:5910.
119. Sacca B, Meyer R, Erkelenz M, et al. Orthogonal protein decoration of DNA origami. *Angew. Chem. Int. Ed. Engl.* 2010;49:9378–9383.
120. Schaffert DH, Okholm AH, and Sørensen RS. Intracellular delivery of a planar DNA origami structure by the transferrin-receptor internalization pathway. *Small* 2016;12:2634–2640.
121. Liu J, Song L, Liu S, et al. ADNA-based nanocarrier for efficient gene delivery and combined cancer therapy. *Nano Lett.* 2018;18:3328–3334.

13 CRISPER Gene Therapy Recent Trends and Clinical Applications

Prachi Pandey
Babaria Institute of Pharmacy, Vadodara, India

Jayvadan Patel
Sankalchand Patel University, Visnagar, India

Samarth Kumar
Sun Pharmaceutical Industries Ltd, Vadodara, India

CONTENTS

13.1 INTRODUCTION

Clustered regularly-interspaced short palindromic repeats (CRISPR) is a recent and powerful genetic editing tool. In the past decade, CRISPR–Cas9 technology is widely acknowledged for playing a substantial role in the field of biotechnology for editing the genome of any organism in the treatment of a variety of complex systemic diseases and for many other purposes. 'CRISPR–Cas' acronym stands for clustered regularly interspaced short palindromic repeats–CRISPR-associated genes. It is interesting to know that CRISPR sequences are an important component of the immune

DOI: 10.1201/9781003186083-13

systems of simple life forms like bacteria and other microbes. The immune system is in charge of protecting the health and wellbeing of an organism. This genetic organization prevails in prokaryotic organisms and assists in the development of adaptive immunity because of a protein called Cas9 nuclease, which splits a specific target nucleic acid sequence of unknown invaders and destroys them. This mode of action has acquired the attention of researchers to understand the prospects of CRISPR–Cas9 technology. There are many future aspects and potential applications of CRISPR–Cas9 technology required mainly for the treatment of dreadful diseases, crop improvement, and genetic improvement in human beings. However, safety measures are implemented on this technology to avoid misuse or ethical issues [1].

CRISPR is such a powerful genetic editing tool mainly because of its distinct structure and design. But, surprisingly, the creation of CRISPR relies on an immune system of bacteria. When scientists began their research, they found that some genes in the bacteria did not perform any function for constructing the cells and helping the survival of bacteria [2, 3]. Eventually, as scientists realized that this "useless" gene is commonly present in bacteria and other microbes, they started to study more about the gene; this led to a discovery that this exceptional gene in bacteria is coding for a vital function of protection of the bacteria from the invasion by viruses [4]. It is also a very primitive form of CRISPR. Due to its effectiveness and affinity, CRISPR has been contemplated to be a remarkable invention in the area of genetic-editing and molecular biology. When compared with many other genetic-editing techniques, CRISPR has an incomparable advantage, which is its accessibility. Besides, the application of CRISPR does not need training for many years, as well as a high laboratory cost [5–12]. It only requires some basic training and the expense only for buying the CRISPR molecule, which allows researchers to study genetic editing and execute an experiment. The development of CRISPR will follow the invention of numerous applications in the fields of agriculture, environmental science, biology, and public health care. There will be an immense commercial interest in this field, which will enhance the development of those applications. Even though CRISPR promises a bright future, it also has its challenges, and the biggest one is the ethics problem.[13–24]

13.2 CRISPR–CAS9 SYSTEM

Today, we know that clustered regularly-interspaced short palindromic repeats (CRISPR) are an advanced and potent genetic editing tool [25]. The mechanism starts when a cell is infected with a virus, which consists of a guide molecule, healthy DNA copy, and DNA cleaving enzyme. The action begins with RNA, which is a molecule that can read the genetic information in the DNA.

The first step consists of the identification of specific DNA strands of the virus by the guide RNA that is composed of crRNA and tracrRNA. Following which RNAse III is needed for producing crRNA from pre-crRNA.

In the second step, this guide RNA steers Cas9 to the exact spot-on DNA where we want to split it.

Cas9 then locks onto the double-stranded DNA and unwinds it. This favors the guide RNA to pair up with some part of the DNA it has targeted. Cas9 splits the DNA at this spot. This produces a break in both strands of the DNA molecule. Thus, in the third step, RNase III is guided by crRNA: tracrRNA: Cas9 complex to cleave target DNA.

The cell immediately senses a problem and begins to repair the break. After splitting of DNA by CRISPR-Cas9, DNA can be repaired by Non-Homologous End Joining (NHEJ) and Homology Directed Repair (HDR) in the fourth step (Figure 13.1). NHEJ is simple and random, in contrast to HDR, which is much more complex and accurate

13.2.1 Structure of CRISPR Loci

CRISPR loci were reported to be almost palindromic and were first detected in the *iap* gene (gene instigating the alkaline phosphatase isozyme conversion) of Escherichia coli inside an intergenic region upstream to the gene. These DNA repeats were also noticed in many bacterial species, like

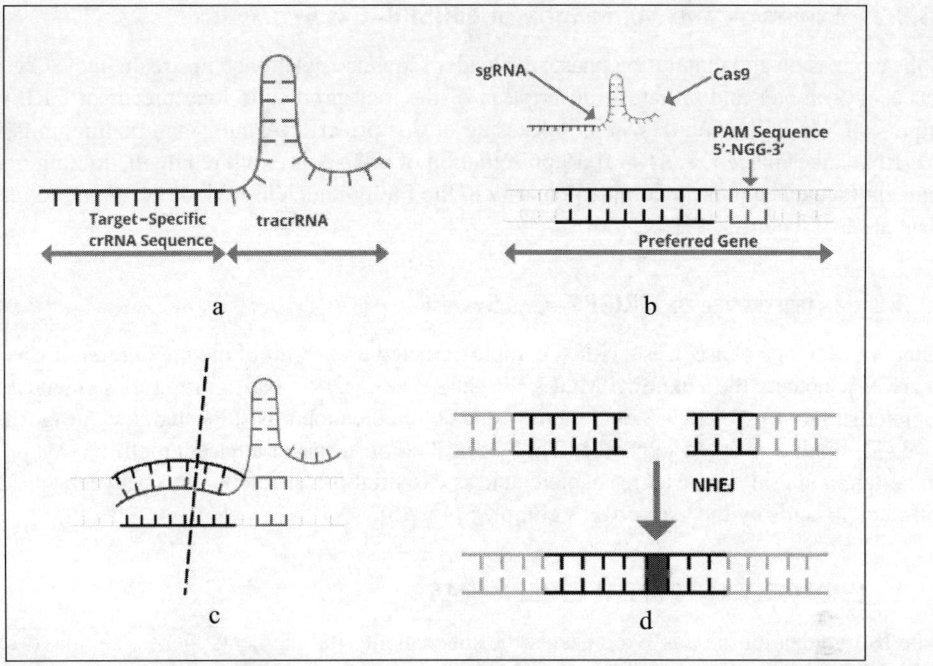

FIGURE 13.1 (a) sgRNA(single guide RNA) (b) 2sgRNA + Cas9 protein (c) Target specific cleavage (d) Cellular error repair "knocks-out "gene.

Mycobacterium tuberculosis and *Streptococcus pyogenes*, as well as in a few archaeal species; namely *Haloferax mediterranei*, *Haloferax volcanii, Thermotoga Maritima*, along with filamentous cyanobacterium Anabaena species.

A series of repeats exist in CRISPR loci, which flank the 'spacer' sequence, and this spacer sequence exactly matches the sequences present in virus, plasmid, or any other pathogen genomic elements [26–32]. Usually, an AT-rich leader sequence is situated in the upstream position of the CRISPR array. On one end of the array, a set of conserved genes coding for various types of Cas proteins, called CRISPR-associated (Cas) genes, are present [33, 34].

13.3 THE MECHANISM INVOLVED IN THE CRISPR–CAS SYSTEM

The mechanism involved in CRISPR–Cas organization works in three conspicuous phases:

1. Adaptation,
2. Expression and Maturation, and
3. Interference [35, 36].

13.3.1 Adaptation of CRISPR–Cas Spacer Sequences

The adaptation phase takes place in two steps; first, Cas proteins of the bacterium recognize the invader and obtain specific sequences from foreign nucleic acids(these sequences are labeled as 'protospacer'). Second, the protospacer is merged in the extremity of the leader sequence in the CRISPR array as 'spacer,' which results in the first repeat of the CRISPR array to be extended [37, 38]. These spacers are accountable for creating immunological memory to archaea and bacteria for defense in case they encounter the mobile genetic elements or MGEs for the second time. Cas1 and Cas2 are crucially involved in this phase [37, 39, 40].

13.3.2 Expression and Maturation of CRISPR–Cas System

In the expression and maturation phase, the leader sequence positioned upstream to the CRISPR loci acts as a promoter and initiates transcription of the loci, producing long precursor CRISPR RNA or pre-crRNA. Following this step, processing of this pre-crRNA into small mature units, referred as crRNA takes place [38, 41, 42]. Representation of crRNA is exhibited by the joining of a spacer segment (sequence present complementarily to the foreign nucleic acid) at the 5′ end to repeat segment at the 3′ end [43, 44].

13.3.3 Interference of CRISPR–Cas System

In the interference phase, Cas–crRNA complex created as a result of the integration of Cas proteins to crRNA, detects the unknown MGEs via the Watson–Crick base pairing of sequences that are complementary to the crRNA, and thereby, the targeted element is subjected to cleavage. The presence of a small conserved sequence (2 to 5 bp) called protospacer adjacent motif (PAM) juxtaposed to the target site in the invading nucleic acid is essential for differentiation between self and non-self-nucleic acids by the Cas–crRNA complex [45, 46].

13.4 BIOLOGY OF TYPE II CRISPR–CAS9

Type II organization of Cas 9 represents the subsequent vital elements: genes, i.e., cas1, cas2, and cas9, CRISPR array, along with a tracrRNA that shows correspondence to the sequence of the CRISPR repeat. All the Cas signature proteins are associated during the acquisition of spacers, while in the interference stage, only the Cas9 plays a key role. Class 2 type II organization codes for endonucleases like Cas9 signature protein, as well as a non-coding RNA called tracrRNA, in addition to crRNA. Base pairing of crRNA and tracrRNA results in the formation of a crRNA: tracrRNA hybrid, after which RNase III slices the hybrid, thereby forming a developed dual-RNA hybrid. Afterward, recruitment of Cas9 proteins takes place.

Chimeric single guide RNA (sgRNA) is built by hybridization of the 5′ end of the tracrRNA with that of the 3′ end of the crRNA, subsequently forming a single guide RNA, which has a great potential in genome engineering as it can decompose any target DNA sequence. Type II is made up of two parts, Cas9 and sgRNA. The sgRNA guides the Cas9 endonuclease and identifies G-rich PAM (i.e., 5′-NGG) and then recognizes the target DNA sequence that are located in the upstream position of the PAM sequence and causes dissolving of the target DNA. As a result of this, the strands undergo directional separation, upstream to the PAM forming an R-loop. Later, the siRNA strand is incorporated and, thereby, the formation of the RNA–DNA heteroduplex takes place. The duplex is created by base pairing of the ~ 20nt spacer sequence of the sgRNA with that of the proto spacer of the target DNA because they are complementary to each other. One domain of the Cas9 enzyme, HNH, slices the DNA sequence that is complementary to the sequence (target sequence) present in guide RNA, while the other domain of the enzyme, RuvC, slices the sequence that is non-complementarily to the sequence (non-target sequence) present in guide RNA. This way, the two domains cleave the RNA–DNA hybrid at a spot three bp upstream to the PAM, and the consequence of cleavage is the formation of a double-strand break (DSB) with blunt ends [42–45, 47–51].

13.5 RECENT TRENDS IN CRISPER GENE THERAPY

13.5.1 CRISPR System Accurately Cut Target DNA

This defense system of bacteria can destroy the invading virus in two steps. The invader with a specific strand of DNA of the virus is attacked in the first step. Then the next step is to cut the target DNA strand of the virus by using a nuclease. As described in the figure, in step one, to identify the location of the viruses and target them, the defense system needs help from the guide RNA. The guide RNA, which is made of crRNA and tracrRNA, is in charge of detecting and targeting the unknown invaders of the bacteria by identifying specific nucleotides of the invading viruses [52–67].

The Guide RNA has to go through an elaborate process to find a piece of DNA sequence inside the viral genome. When viruses start to occupy the living cells of bacteria, they will let out their invading DNA strands into the bacterial cells for transforming the invaded cells into factories that manufacture more viruses.[68–70] At this instant, the immune system of bacteria will produce a copy of one DNA strand of the invading virus, and then drive it into a group of the genome of bacteria, which is referred to as CRISPR array [71–73]. The DNA of the virus will be secured in the spacers among the two palindromic repeats in the CRISPR array. Then a selected portion of the CRISPR array will be transcribed by the bacteria to make a pre-crRNA strand that also incorporates multiple complementary spacers and palindromic repeats. For the process of transforming pre-crRNA to crRNA, a tracrRNA, as well as a double-stranded RNA-specific ribonuclease, is essential. TracrRNA will integrate into one of the palindromic replicates of the pre-crRNA, and the ribonuclease III will split the tracrRNA. At the end, such separated pre-crRNA is matured and referred to as Cr-RNA. Later on, the crRNA and tracrRNA will adhere to the cas9 nuclease and lead to the formation of a crRNA: tracrRNA: Cas9 complex. Cr-RNA in the complex is entirely corresponding to a DNA sequence of the invading virus so that crRNA guides the cas9 nuclease to the viral DNA sequence that is entirely corresponding to the crRNA [74–77].

Significantly, a proto spacer adjacent motif (PAM) is necessary during this process. PAM is a sequence of nucleotides that are situated near to the sequences complementary to the crRNA and plays a role in the initiation of attaching crRNA to its corresponding complementary strand in the virus. Once crRNA: tracrRNA: Cas9 complex is actively initiated to the invading virus's DNA strand, the location of the virus is recognized. The tool that destroys the virus is the Cas9 nuclease. After crRNA: tracrRNA: Cas9 complex and the targeted strand form the bond, the Cas9 nuclease will cut the DNA strands of the virus close to the PAM so that the virus will eventually have incomplete DNA strands, which will conquer the viruses [78–80].

13.5.2 Non-Homologous End Joining (NHEJ) and (Homology Directed Repair) HDR

CRISPR's ability to cut genomes of each living life along with viruses leads to the further evolution of genetic editing in other organisms. Inside a bacteria, it is because of the presence of Cas9 nuclease and crRNA:tracrRNA: Cas9 complex that the CRISPR can kill invading viruses by tracing and cutting their DNA strands. This exact tracing and cutting procedure can also be availed in cutting the DNA strand of any other organism. Once we insert CRISPR into a specific site of a DNA strand, it is capable of cutting the DNA strand by launching a DNA Strands Breaks (DSB) to the site, which is exactly similar to how the CRISPR can kill the invading viruses of the bacteria.

The purpose of carrying out the CRISPR research is not to cut the DNA strand, but CRISPR is used basically to initiate the DSB to those DNA strands to change the genetic sequence of the nucleotides on that DNA strand. CRISPR will not be useful for such a purpose of changing the specific nucleotide because the only function it can perform is cutting the DNA strand. Thus, along with this, a procedure of fixing and editing the fragmented DNA strands is essential for the DNA to perform its function again and initiate changes to the targeted DNA sequences. There are two distinct mechanisms to edit and fix the broken DNA sequence, Non-Homologous End Joining (NHEJ) and Homology Directed Repair (HDR) [81, 82].

NHEJ is a natural way of healing the impairment of DNA strands in most organisms. For example, when UV lights injure the DNA strands of the skin cells, our body can use NHEJ to re-join two broken DNA strands so that they can function again. Commonly, such a method is relatively simple and efficient for fixing damages of genes because this fixing procedure does not require a homologous template to repair the DNA. In NHEJ, the Ku protein will integrate with two ends of the broken DNA strands and form a Ku DNA end complex. This Ku DNA end complex will then associate with a DNA-PKcs complex to slice away overhangs of nucleotides close by the end of the two broken strands. Then, with a support from the XLF: XRCC4 DNA ligase IV, the two fragmented ends of DNA strands will join together and re-join each other via ligation (23). Due to the launching of DSB and some nucleotide deletions, during repair, the original DNA sequence is permanently altered after NHEJ.

Ultimately, repairing DNA strands via using NHEJ is a simple method to initiate some unpredicted mutations to the targeted DNA strand because we cannot control which nucleotide is affected and knocked out and which is not. Primarily, as this mechanism is simple and efficient, it may indeed have a positive future in the field of non-human genetic editing, like using CRISPR in the field of agriculture or fixing the ecosystem [83, 84].

13.5.3 On-Target Activity and Off-Target Activity

The wide range of Cas9 specificity is directed to the bases within the ~20 nucleotide sequence of the sgRNA, but during hybridization of sgRNA and target DNA, multiple mismatches may occur, and Cas9 can tolerate about five such mismatches. These mismatches are the consequence of off-target sites (sites different from actual target sites for a few bases) located within the target DNA that remains briefly bounded to the sgRNA sequence. In studies, it is reported that Cas9 also enables the binding of sgRNA with off-target sites, and as an outcome, Cas9 splits these sites to form DSBs. To conduct genome editing efficiently, a Cas9 should cleave the target DNA sequence accurately, and the off-target effect has to be minimized mandatorily [85].

13.5.4 CRISPR Tools

Genome scan is carried out utilizing a variety of bioinformatics tools as these tools function to detect accurate target sites as well as off-target sites and design sgRNA for attaining high specificity for cleaving the target site. Bioinformatic tools are listed below in the Table 13.1.

13.5.5 CRISPR in Gene Editing

Researchers have achieved the ability to change the genome sequences using CRISPR. This possibility sounds very promising to the fields that are associated with therapy because there are ample opportunities that may be explored by CRISPR. In agriculture, CRISPR will be utilized to alter the genome of crops and plants to supply them phenotypes that provide resistance to bacteria, in order that less pesticide are going to be required to treat it, which permits economizing for farmers and lowers the issues from overuse of pesticides. CRISPR may be employed in the treatment of genetic diseases that are triggered by some mutations in DNA strands. For instance, the mutation on HTT genes is liable for Huntington's disease, which is usually represented by having incorrect nucleotide pairs on the genome. If CRISPR will be utilized to substitute the correct nucleotides on the HTT gene for the wrong ones, Huntington's disease is effortlessly treated. In the field of microbiology, the CRISPR–Cas9 system has also been observed to be very effective in mutating or altering genes in yeasts, molds, and filamentous fungi, which have pathogenicity in addition as industrial importance. Studies show that CRISP-Cas9 system technology has prospects within the treatment of neurodegenerative disease like dystrophy. Research studies show that with the applying of technology, viral diseases like hepatitis B might even be treated. Studies also show that antimicrobial treatment-mediated CRISPR–Cas9 is a great option, when compared with traditional antibiotics. However, the biggest challenge is within the advancement of delivery strategies of the CRISPR–Cas9 system of CRISPR-based antimicrobial. Somatic genome altering mediated by the CRISPR–Cas9 system has surfaced as a replacement pathway in cancer modeling that still has to be used in cancer detection

TABLE 13.1
CRISPER Tools

Name of the Tool	Type of the Tool	Purpose of the Tool
CRISPResso2	Computational tool	Analyses the results attained from the experiments of genome editing; evaluate as well as determines the differences amid the experiments performed, requires less time for programming [86].
CRISPR-ERA (editing, repression, and activation)	Computational tool	Detects sgRNA binding sites by scanning the genome; determines sgRNA binding specificity as well as efficiency [87].
WU-CRISPR	Computational tool	Picks genomic gRNA for Cas9 and develops the competence of the design of CRISPR assay [88].
CRISPR-P	Webtool	Selects appropriate target sites within desired DNA which has high specificity for Cas9; helps in predicting the off-target loci; determines restriction sites along with off-target sites [89].
CRISPR-P 2.0	Webtool	Can predict the on-target and off-target effect of sgRNA on the target DNA; evaluates microhomology score and secondary structure of sgRNA; aids in the visualization of GC content with the sgRNA [90].
CRISPR seek	Webtool	It is a Bioconductor package; can design gRNA specificity for target sites in a genome and studies the off-target sites [91]
COSMID (CRISPR off-target sites with mismatches, insertions, and deletions)	Webtool	Distinguishes potential off-target sites throughout the genome; evaluates the mismatch bases along with the excluded or incorporated bases [92]
CHOPCHOP v2	Webtool	Designs sgRNA; predicts possible off-target sites; assists in targeting a wide range of sequences of the specified genome by sgRNA [93]
Cas-Designer	Webtool	Recognizes target sites located in a genome of interest, which is later cleaved by Cas9; gene knockout is accomplished selectively of appropriate target site; from the supplied sequences of DNA, the record for all favorable gRNA sequences likewise as off-target sites can be determined [94]
E-CRISP	Webtool	Designs sequences of gRNA; establishes target sites that show complementarity to the gRNA; followed by which cleaving of the dsDNA by Cas9 endonuclease takes place [95]
CRISPR MultiTargeter	Webtool	Classifies potential target sites for sgRNA; distinguishes target sites which are complementary and non-complementary to the supplied DNA sequences [96]
CRISPy	Webtool	Selects sgRNA target sequences of chosen DNA; assists in viewing the image of the target DNA; stores data about off-target sites [97]
CRISPy-web	Webtool	Can predict target sites; designs sgRNA that is capable of targeting the DNA of interest is supplied a sequence of the genome. [98]
EuPaGDT (eukaryotic pathogen gRNA design tool)	Webtool	Designing gRNA CRISPR–Cas system to counter pathogenic eukaryotic organism; aids in gRNA library construction [99]
stop (CRISPR/Cas9 target online prediction)	Webtool	Identifies target sites for sgRNA; useful for gene inactivation; widely used for HDR as well as NHEJ for repairing DSBs [100]

during the initial stages of the disease in a person, thanks to the ability of Cas13a to differentiate mutations which might result in cancer.[65, 101]

13.6 CLINICAL APPLICATION OF CRISPER GENE THERAPY

The first CRISPR Phase 1 trial within the US opened in 2018 with the intent to use CRISPR/Cas9 to edit autologous T cells for cancer immunotherapy against several cancers with relapsed tumors and no further curative treatment options. These include myeloma, melanoma, synovial sarcoma, and myxoid/round cell liposarcoma. This trial was approved by the US Food and Drug Administration (FDA) after careful consideration of the risk to benefit ratios of this first application of CRISPR gene therapy into the clinic. During this trial, T lymphocytes were collected from the patients' blood and ex vivo engineered with CRISPR/Cas9 to knockout the α and β chains of the endogenous T cell receptor (TCR), which recognizes a particular antigen to mediate an immune reaction, and therefore the programmed cell death-1 (PD-1) protein, which attenuates reaction. The cells were then transduced with lentivirus to deliver a gene encoding a TCR specific for a NY-ESO-1 antigen, which has been shown to be highly upregulated within the relapsed tumors, and thus can function as a therapeutic target. Since then, many trials have opened for CRISPR-mediated cancer immunotherapy, which is currently the foremost employed strategy for CRISPR gene therapy (Table 13.2). An attempt implementing this strategy using other tools had already been conducted in both preclinical and clinical settings, but this was the primary time CRISPR/Cas9 was utilized to generate the genetically modified T cells. The moderate transition of switching only the tool used for an already approved therapeutic strategy may be key to paving the road for using CRISPR's novel abilities for gene manipulation, like targeted gene disruption.[102]

13.6.1 Gene Disruption

The first clinical trials in the US using CRISPR to catalyze gene disruption for therapeutic benefit were for patients with sickle-cell anemia (SCD) and later β-thalassemia, by Vertex Pharmaceuticals and CRISPR Therapeutics. This therapy, named CTX001, increases fetal hemoglobin (HbF) levels, which can occupy one or two of four hemoglobin binding pockets on erythrocytes, and thereby provides clinical benefit for major β-hemoglobin diseases such as SCD and β-thalassemia. The trial involved collecting autologous hematopoietic stem and progenitor cells from peripheral blood and using CRISPR/Cas9 to disrupt the intronic erythroid-specific enhancer for the *BCL11A* gene (NCT03745287) as disruption of this gene increases HbF expression. Genetically modified hematopoietic stem cells with *BCL11A* disruption are delivered by IV infusion after myeloablative conditioning with busulfan to destroy unedited hematopoietic stem cells in the bone marrow. Preliminary findings from two patients receiving this treatment seem promising. One SCD patient was reported to have 46.6% HbF and 94.7% erythrocytes expressing HbF after 4 months of CTX001 transfusions and one β-thalassemia patient is expressing 10.1 g/dL HbF out of 11.9 g/dL total hemoglobin, and 99.8% erythrocytes expressing HbF after 9 months of the therapy. Results from the clinical trial that has opened for this therapy (NCT04208529) to assess the long-term risks and benefits of CTX001 will dictate whether this approach can provide a novel therapeutic opportunity for a disease that otherwise has limited treatment options.

13.6.2 *In vivo* CRISPR Gene Therapy

While the aforementioned trials rely on *ex vivo* editing and subsequent therapy with modified cells, *in vivo* approaches have been less extensively employed. An exciting step forward with CRISPR gene therapy has been recently launched with a clinical trial using *in vivo* delivery of CRISPR/Cas9 for the first time in patients. While *in vivo* editing has been largely limited by inadequate accessibility to the target tissue, a few organs, such as the eye, are accessible. Leber congenital amaurosis (LCA) is a debilitating monogenic disease that results in childhood blindness caused by a bi-allelic loss-of-function mutation in the Centrosomal Protein 290 *(CEP290* gene), with no treatment options. This therapy, named EDIT-101, delivers CRISPR/Cas9 directly into the retina of LCA patients specifically with the intronic IVS26 mutation, which drives aberrant splicing resulting in a non-functional protein. The therapy uses an AAV5 vector to deliver nucleic acid instructions for *Staphylococcus aureus* Cas9 and two guides targeting the ends of the *CEP290* locus containing the IVS26 mutation. The DSB induced by Cas9 and both guides result in either a deletion or inversion of the IVS26 intronic region, thus preventing the aberrant splicing caused by the genetic mutation and enabling subsequent translation of the functional protein. Potential immunotoxicity or Off target editing (OTEs) arising from nucleic acid viral delivery will have to be closely monitored. Nonetheless, a possibly curative medicine for genetic blindness using an *in vivo* approach marks an important advancement for CRISPR gene therapy.

13.6.3 CRISPR Editing in Human Embryos and Ethical Considerations

While somatic editing for CRISPR therapy has been permitted after careful consideration, human germline editing for therapeutic intent still remains controversial. With somatic edition, the potential risks would be probable in a private patient who participates in the study after being informed. Embryonic editing not only removes autonomy within the decision-making process of the born individuals, but also allows unpredictable and permanent adverse effects which will be transmitted through generations. A pilot study involved gene-splicing of the C-C chemokine receptor type 5 (CCR5) gene in human embryos, with the intention of conferring HIV-resistance, as seen by a present CCR5Δ32 mutation in a very few individuals [103]. On the idea of limited evidence, CRISPR/Cas9 is reported to be utilized to target this gene, but rather than replicating the naturally observed and beneficial 32-base deletion, this gene editing induces DSBs at one end of the deletion, thereby allowing NHEJ to repair the damaged DNA and also introducing random mutations. it's not certain that the resultant protein can function similarly to the present CCR5Δ32 protein and confer HIV resistance. Additionally, just one of the two embryos, termed with the pseudonym Nana, had undergone successful gene editing in both copies of the CCR5 gene, whereas the opposite embryo, with pseudonym Lulu, had successful editing in precisely one copy.

Moreover, in recent studies, it's observed that the mechanism for infection of some variants of the highly mutable HIV virus may heavily depend upon the C-X-C chemokine receptor type 4 (CXCR4) co-receptor. With no attempts at editing CXCR4, this adds yet one more layer of skepticism toward achieving HIV resistance by this strategy. Additionally, OTEs, particularly over the lifetime of an individual, remain a significant concern for applying this technology in humans. The recent advances within the editing tool to limit OTEs, like using high fidelity Cas9 variants, has not been exploited. Furthermore, the rationale for choosing HIV prevention for the primary use of CRISPR in implanted human embryos contributes to the poor risk to learn ratio of this study, considering HIV patients can live long, healthy lives on a drug regimen. A more appropriate first attempt is to use this technology for a more severe disease. as an example, correction of the myosin-binding protein C (MYBPC3) gene is arguably an improved target for embryonic gene editing, as mutations in MYBPC3 can cause cardiomyopathy (HCM), a cardiopathy liable for most sudden cardiac deaths in people under the age of 30. Gene correction for this pathological mutation was achieved recently for the first time within the US in viable human embryos using the HDR-mediated CRISPR/Cas9 system.[104, 105]

However, these embryos were edited for basic research purposes and not intended for implantation. During this study, sperm carrying the pathogenic MYBPC3 mutation and therefore the CRISPR/Cas9 machinery as an RNP complex were microinjected into healthy donor oocytes arrested at MII, achieving 72.4% homozygous wild type embryos, as against 47.4% in untreated embryos. The HDR-mediated gene correction was observed at considerably high frequencies with no detectable OTEs in selected blastomeres, likely because of the direct microinjection delivery of the RNP complex within the early zygote. Interestingly, the maternal wildtype DNA was used preferentially

TABLE 13.2
Clinical Intervention of Crisper Gene Therapy

Affiliation	Type of Disease	Gene Target	Clinial Trial ID	CRISPR-Cas9 Mediated Intervention
University of Pennsylvania/Parker Institute for Cancer Immunotherapy/ Immunity	Multiple Myeloma, Melanoma, Synovial Sarcoma, Myxoid/ Round Cell Liposarcoma	TCRα, TCRβ, PDCD1	NCT03399448	NY-ESO-1 redirected autologous T cells with CRISPR edited endogenous TCR and PD-1
Affiliated Hospital to Academy of Military Medical Sciences/Peking University/Capital Medical University	HIV-1	CCR5	NCT03164135	CD34+ hematopoietic stem/ progenitor cells from donor are treated with CRISPR/Cas9 targeting CCR5 gene
CRISPR Therapeutics AG	Multiple Myeloma	TCRα, TCRβ, B2M	NCT04244656	CTX120 B-cell maturation antigen (BCMA)-directed T-cell immunotherapy comprised of allogeneic T cells genetically modified *ex vivo* using CRISPR-Cas9 gene editing components
Crispr Therapeutics/Vertex	Beta-Thalassemia, Thalassemia, Genetic Diseases Inborn, Hematologic Diseases, Hemoglobinopathies	BCL11A	NCT03655678	CTX001 (autologous CD34+ hHSPCs modified with CRISPR-Cas9 at the erythroid lineage-specific enhancer of the BCL11A gene)
Crispr Therapeutics	B-cell Malignancy Non-Hodgkin Lymphoma B-cell Lymphoma	TCRα, TCRβ	NCT04035434	CTX110 (CD19-directed T-cell immunotherapy comprised of allogeneic T cells genetically modified *ex vivo* using CRISPR-Cas9 gene editing components)
Editas Medicine, Inc./ Allergan	Leber Congenital Amaurosis 10	CEP290	NCT03872479	Single escalating doses of AGN-151587 (EDIT-101) administered via subretinal injection
Vertex Pharmaceuticals Incorporated/CRISPR Therapeutics	Sickle Cell Disease, Hematological Diseases, Hemoglobinopathies	BCL11A	NCT03745287	CTX001 (autologous CD34+ hHSPCs modified with CRISPR-Cas9 at the erythroid lineage-specific enhancer of the BCL11A gene)
Allife Medical Science and Technology Co., Ltd.	Thalassemia	HBB	NCT03728322	Investigate the safety and efficacy of the gene correction of HBB in patient-specific iHSCs using CRISPR/Cas9
Yang Yang, The Affiliated Nanjing Drum Tower Hospital of Nanjing University Medical School	Stage IV Gastric Carcinoma, Stage IV Nasopharyngeal Carcinoma, T-Cell Lymphoma Stage IV, Stage IV Adult Hodgkin Lymphoma, Stage IV Diffuse Large B-Cell Lymphoma	PDCD1	NCT03044743	CRISPR-Cas9 mediated PD-1 knockout-T cells from autologous origin

(*continued*)

TABLE 13.2 (Continued)
Clinical Intervention of Crisper Gene Therapy

Affiliation	Type of Disease	Gene Target	Clinial Trial ID	CRISPR-Cas9 Mediated Intervention
First Affiliated Hospital, Sun Yat-Sen University/ Jingchu University of Technology	Human Papillomavirus-Related Malignant Neoplasm	HPV16 and HPV18 E6/E7 DNA	NCT03057912	Evaluate the safety and efficacy of TALEN-HPV E6/E7 and CRISPR/Cas9-HPV E6/E7 in treating HPV Persistency and HPV-related Cervical Intraepithelial Neoplasia I
Sichuan University/ Chengdu MedGenCell, Co., Ltd.	Metastatic Non-small Cell Lung Cancer	PDCD1	NCT02793856	CRISPR-Cas9 mediated PD-1 knockout-T cells from autologous origin
Peking University	Metastatic Renal Cell Carcinoma	PDCD1	NCT02867332	CRISPR-Cas9 mediated PD-1 knockout-T cells from autologous origin
Peking University	Hormone Refractory Prostate Cancer	PDCD1	NCT02867345	CRISPR-Cas9 mediated PD-1 knockout-T cells from autologous origin
Peking University	Invasive Bladder Cancer Stage IV	PDCD1	NCT02863913	CRISPR-Cas9 mediated PD-1 knockout-T cells from autologous origin
Hangzhou Cancer Hospital/Anhui Kedgene Biotechnology Co., Ltd	Esophageal Cancer	PDCD1	NCT03081715	CRISPR-Cas9 mediated PD-1 knockout-T cells from autologous origin
Chinese PLA General Hospital	Solid Tumor, Adult	TCRα, TCRβ, PDCD1	NCT03545815	Evaluate the feasibility and safety of CRISPR-Cas9 mediated PD-1 and TCR gene-knocked out chimeric antigen receptor (CAR) T cells in patients with mesothelin positive multiple solid tumors
Baylor College of Medicine/The Methodist Hospital System	T-cell Acute Lymphoblastic Leukemia, T-cell Acute Lymphoblastic Lymphoma, T-non-Hodgkin Lymphoma	CD7	NCT03690011	CRISPR-Cas9 mediated CD7 knockout-T cells from autologous origin
Chinese PLA General Hospital	B Cell Leukemia, B Cell Lymphoma	PDCD1	NCT03398967	Determine the safety of the allogenic CRISPR-Cas9 gene-edited dual specificity CD19 and CD20 or CD22 CAR-T cells

for templated repair over the provided exogenous ssODN template [106–108]. While evidence for gene correction was promising, non-homologous end joining (NHEJ) mediated DNA repair was still observed in many embryos, highlighting the necessity to enhance HDR efficiency before clinical application is considered. Although strategies are developed to boost HDR, like chemical inhibitors of NHEJ, such techniques may have varying outcomes in embryonic cells and side effects that will arise from treatment that has to be investigated. Germline gene editing will remain ethically unfavorable in its current state, and its discussions might not be considered until sufficient long-term studies of the continued somatic CRISPR therapy clinical trials are evaluated.

13.7 CONCLUSION

The knowledge of the biology of the CRISPR–Cas9 system is required for genome editing. The intensification in cleavage specificity of Cas9, and also the lessening of off-target activity of this enzyme, precisely enables to acknowledge particular target DNA sequences so as to edit or modify the genome accurately. CRISPR databases and tools provide detailed information and proper facility for altering, modifying, or visualizing genomes to conduct correct genome editing experiments. It is seen that CRISPR–Cas9 technology has been utilized in various fields, including disease treatment associated with genetic disorders or pathogens, recombinant DNA technology, agriculture, and clinical applications. However, many challenges have to be overcome.

Additional studies on CRISPR–Cas9 organization and possible applications of this technology will help us overcome these challenges, thereby resulting in better quality within the lives of humans within society by conquering complex diseases and improving crop field. Improvement of CRISPR tools is indispensable so off-target cleavage activity by Cas9 is decreased effectively. Advancement in exact delivery strategies of CRISPR–Cas9 system into cells of higher-class organisms like plants and mammals are crucially important. Identification of drug targets accurately by the utilization of CRISPR–Cas9 technology will play a very important role within the development of current drugs for the impending dreadful diseases. Insertion of altered transposable elements or TEs by this technology can result in crop improvement, and therefore the ability to form new ornamental plants.

As CRISPR–Cas system is found in bacteria, it undertakes evolution rapidly and eventually may produce new Cas genes which will encode new proteins. Thus, these proteins may have the budding potential for genome editing or other similar applications within the near future. Of late, Cas12a has been known to have great purpose in genome editing. CRISPR technology has also been of great use to treat cancer and other diseases. CRISPR–Cas9 technology experiments are normally performed in vitro in model organisms and stem cells like human pluripotent stem cells (hPSCs). Gene editing by this technology is also conducted in human embryonic stem cells (ESCs) in vitro to correct mutations, but research in ESCs gives rise to many ethical issues. However, it is undeniable that the genome editing of ESCs has the potential to give rise to organisms possessing outstanding desirable qualities. Safety measures are essential while using this technology to prevent its misuse or minimize the risk of the negative impact of genome editing.

We can conclude that CRISPR is still a young and very influential genetic editing tool. Its mechanism lets scientists edit any DNA sequences within any organism, which is remarkable for genetic editing. The invention of CRISPR may well lead to a golden age of genetic editing. In the future, CRISPR and genetic editing will be pragmatic in almost every aspect of our life.

REFERENCES

1. Moon SB, Ko JH, and Kim YS. Recent advances in the CRISPR genome editing toolset. *Experimental & Molecular Medicine* 2019 Nov 5;51(11):1–11.
2. Ishino Y, Shinagawa H, Makino K, Amemura M, and Nakata A. Nucleotide sequence of the iap gene, responsible for alkaline phosphatase isozyme conversion in Escherichia coli, and identification of the gene product. *Journal of Bacteriology* 1987;169(12):5429–33.
3. Lim KI. Recent advances in developing molecular tools for targeted genome engineering of mammalian cells. *BMB Reports* 2015 Jan;48(1):6.
4. Barrangou R, Fremaux C, Deveau H, Richards M, Boyaval P, Moineau S, Romero DA et al. CRISPR provides acquired resistance against viruses in prokaryotes. *Science* 2007 Mar 23;315(5819):1709–12.
5. Hwang WY, Fu Y, Reyon D, Maeder ML, Tsai SQ, Sander JD, Peterson RT et al. Efficient genome editing in zebrafish using a CRISPR-Cas system. *Nature Biotechnology* 2013 Mar;31(3):227–9.
6. Deveau H, Garneau JE, and Moineau S. CRISPR/Cas system, and its role in phage-bacteria interactions. *Annual Review of Microbiology* 2010 Oct 13;64:475–93.
7. Deveau H, Barrangou R, Garneau JE, Labonté J, Fremaux C, Boyaval P, Romero DA et al. Phage response to CRISPR-encoded resistance in Streptococcus thermophilus. *Journal of Bacteriology* 2008 Feb 15;190(4):1390–400.

8. Bhaya D, Davison M, and Barrangou R. CRISPR-Cas systems in bacteria and archaea: versatile small RNAs for adaptive defense and regulation. *Annual Review of Genetics* 2011 Dec 15;45:273–97.
9. Brouns SJ, Jore MM, Lundgren M, Westra ER, Slijkhuis RJ, Snijders AP, Dickman MJ et al. Small CRISPR RNAs guide antiviral defense in prokaryotes. *Science* 2008 Aug 15;321(5891):960–4.
10. Karvelis T, Gasiunas G, Miksys A, Barrangou R, Horvath P, and Siksnys V. crRNA and tracrRNA guide Cas9-mediated DNA interference in Streptococcus thermophilus. *RNA Biology* 2013 May 1;10(5):841–51.
11. Carte J, Christopher RT, Smith JT, Olson S, Barrangou R, Moineau S, Glover III CV et al. The three major types of CRISPR Cas systems function independently in CRISPR RNA biogenesis in Streptococcus thermophilus. *Molecular Microbiology* 2014 Jul;93(1):98–112.
12. Deltcheva E, Chylinski K, Sharma CM, Gonzales K, Chao Y, Pirzada ZA, Eckert MR et al. CRISPR RNA maturation by trans-encoded small RNA and host factor RNase III. *Nature* 2011 Mar;471(7340):602–7.
13. Swarts DC, Mosterd C, Van Passel MW, and Brouns SJ. CRISPR interference directs strand specific spacer acquisition. *PLoS One* 2012 Apr 27;7(4):e35888.
14. Cencic R, Miura H, Malina A, Robert F, Ethier S, Schmeing TM, Dostie J et al. Protospacer adjacent motif (PAM)-distal sequences engage CRISPR Cas9 DNA target cleavage. *PLoS One* 2014 Oct 2;9(10):e109213.
15. Lee SH, and Bae S. Structural and dynamic views of the CRISPR-Cas system at the single-molecule level. *BMB Reports* 2016 Apr 30;49(4):201.
16. Jiang W, Bikard D, Cox D, Zhang F, and Marraffini LA. RNA-guided editing of bacterial genomes using CRISPR-Cas systems. *Nature Biotechnology* 2013 Mar;31(3):233–9.
17. Iliakis G, Wang H, Perrault AR, Boecker W, Rosidi B, Windhofer F, Wu W et al. Mechanisms of DNA double-strand break repair and chromosome aberration formation. *Cytogenetic and Genome Research* 2004;104(1–4):14–20.
18. Gu J, and Lieber MR. Mechanistic flexibility as a conserved theme across 3 billion years of nonhomologous DNA end-joining. *Genes & Development* 2008 Feb 15;22(4):411–5.
19. Aravind L, and Koonin EV. Prokaryotic homologs of the eukaryotic DNA-end-binding protein Ku, novel domains in the Ku protein, and prediction of a prokaryotic double-strand break repair system. *Genome Research* 2001 Aug 1;11(8):1365–74.
20. Zhu S, and Peng A. Non-homologous end-joining repair in Xenopus egg extract. *Scientific Reports* 2016 Jun 21;6:27797.
21. Riballo E, Woodbine L, Stiff T, Walker SA, Goodarzi AA, and Jeggo PA. XLF-Cernunnos promotes DNA ligase IV–XRCC4 re-adenylation following ligation. *Nucleic Acids Research* 2009 Feb 1;37(2):482–92.
22. Makarova KS, Wolf YI, Alkhnbashi OS, Costa F, Shah SA, Saunders SJ, Barrangou Ret al. An updated evolutionary classification of CRISPR–Cas systems. *Nature Reviews Microbiology* 2015 Nov;13(11):722–36.
23. Shmakov S, Smargon A, Scott D, Cox D, Pyzocha N, Yan W, Abudayyeh OO et al. Diversity and evolution of class 2 CRISPR–Cas systems. *Nature Reviews Microbiology* 2017 Mar;15(3):169–82.
24. Koonin EV, Makarova KS, and Zhang F. Diversity, classification, and evolution of CRISPR-Cas systems. *Current Opinion in Microbiology* 2017 Jun 1;37:67–78.
25. Shen S, Loh TJ, Shen H, Zheng X, and Shen H. CRISPR as a strong gene-editing tool. *BMB Reports* 2017;50(1):20.
26. Makarova KS, Haft DH, Barrangou R, Brouns SJ, Charpentier E, Horvath P, Moineau S et al. Evolution and classification of the CRISPR–Cas systems. *Nature Reviews Microbiology* 2011 Jun;9(6):467–77.
27. Makarova KS, Wolf YI, and Koonin EV. The basic building blocks and evolution of CRISPR–Cas systems. *Biochemical Society Transactions* 2013 Dec 1;41(6):1392–400.
28. Bolotin A, Quinquis B, Sorokin A, and Ehrlich SD. Clustered regularly interspaced short palindrome repeats (CRISPRs) have spacers of extrachromosomal origin. *Microbiology* 2005 Aug 1;151(8):2551–61.
29. Van der Oost J, Jore MM, Westra ER, Lundgren M, and Brouns SJ. CRISPR-based adaptive and heritable immunity in prokaryotes. *Trends in Biochemical Sciences* 2009 Aug 1;34(8):401–7.
30. Horvath P, and Barrangou R. CRISPR/Cas, the immune system of bacteria and archaea. *Science* 2010 Jan 8;327(5962):167–70.
31. Terns MP, and Terns RM. CRISPR-based adaptive immune systems. *Current Opinion in Microbiology* 2011 Jun 1;14(3):321–7.
32. Deveau H, Garneau JE, and Moineau S. CRISPR/Cas system, and its role in phage-bacteria interactions. *Annual Review of Microbiology* 2010 Oct 13;64:475–93.

33. Jansen R, Embden JD, Gaastra W, and Schouls LM. Identification of genes that are associated with DNA repeats in prokaryotes. *Molecular Microbiology* 2002 Mar;43(6):1565–75.
34. Marraffini LA, and Sontheimer EJ. CRISPR interference: RNA-directed adaptive immunity in bacteria and archaea. *Nature Reviews Genetics* 2010 Mar;11(3):181–90.
35. Amitai G, and Sorek R. CRISPR–Cas adaptation: insights into the mechanism of action. *Nature Reviews Microbiology* 2016 Feb;14(2):67.
36. Puschnik AS, Majzoub K, Ooi YS, and Carette JE. A CRISPR toolbox to study virus-host interactions. *Nature Reviews Microbiology* 2017 Jun;15(6):351–64.
37. Pourcel C, Salvignol G, and Vergnaud G. CRISPR elements in Yersinia pestis acquire new repeats by preferential uptake of bacteriophage DNA and provide additional tools for evolutionary studies. *Microbiology* 2005 Mar 1;151(3):653–63.
38. Yosef I, Goren MG, and Qimron U. Proteins, and DNA elements essential for the CRISPR adaptation process in Escherichia coli. *Nucleic Acids Research* 2012 Jul 1;40(12):5569–76.
39. Mojica FJ, Díez-Villaseñor C, García-Martínez J, and Almendros C. Short motif sequences determine the targets of the prokaryotic CRISPR defence system. *Microbiology* 2009 Mar 1;155(3):733–40.
40. Mojica FJ, García-Martínez J, and Soria E. Intervening sequences of regularly spaced prokaryotic repeats derive from foreign genetic elements. *Journal of Molecular Evolution* 2005 Feb 1;60(2):174–82.
41. Pougach K, Semenova E, Bogdanova E, Datsenko KA, Djordjevic M, Wanner BL, and Severinov K. Transcription, processing and function of CRISPR cassettes in Escherichia coli. *Molecular Microbiology* 2010 Sep;77(6):1367–79.
42. Wei Y, Chesne MT, Terns RM, and Terns MP. Sequences spanning the leader-repeat junction mediate CRISPR adaptation to phage in Streptococcus thermophilus. *Nucleic Acids Research* 2015 Feb 18;43(3):1749–58.
43. Garneau JE, Dupuis MÈ, Villion M, Romero DA, Barrangou R, Boyaval P, Fremaux C et al. The CRISPR/Cas bacterial immune system cleaves bacteriophage and plasmid DNA. *Nature* 2010 Nov;468(7320):67–71.
44. Barrangou R. Diversity of CRISPR-Cas immune systems and molecular machines. *Genome Biology* 2015 Dec 1;16(1):247.
45. Deveau H, Barrangou R, Garneau JE, Labonté J, Fremaux C, Boyaval P, Romero DA et al. Phage response to CRISPR-encoded resistance in Streptococcus thermophilus. *Journal of Bacteriology* 2008 Feb 15;190(4):1390–400.
46. Westra ER, Semenova E, Datsenko KA, Jackson RN, Wiedenheft B, Severinov K, and Brouns SJ. Type IE CRISPR-Cas systems discriminate the target from non-target DNA through base pairing-independent PAM recognition. *PLoS Genetics* 2013 Sep 5;9(9):e1003742.
47. Gasiunas G, Barrangou R, Horvath P, and Siksnys V. Cas9–crRNA ribonucleoprotein complex mediates specific DNA cleavage for adaptive immunity in bacteria. *Proceedings of the National Academy of Sciences* 2012 Sep 25;109(39):E2579–86.
48. Chylinski K, Makarova KS, Charpentier E, and Koonin EV. Classification and evolution of type II CRISPR-Cas systems. *Nucleic Acids Research* 2014 Jun 2;42(10):6091–105.
49. Clement K, Rees H, Canver MC, Gehrke JM, Farouni R, Hsu JY, Cole MA et al. CRISPResso2 provides accurate and rapid genome editing sequence analysis. *Nature Biotechnology* 2019 Mar;37(3):224–6.
50. Doudna JA, and Charpentier E. The new frontier of genome engineering with CRISPR-Cas9. *Science* 2014 Nov 28;346(6213):1258.
51. Wong N, Liu W, and Wang X. WU-CRISPR: characteristics of functional guide RNAs for the CRISPR/Cas9 system. *Genome Biology* 2015 Dec;16(1):1–8.
52. Lei Y, Lu L, Liu HY, Li S, Xing F, and Chen LL. CRISPR-P: a web tool for synthetic single-guide RNA design of CRISPR-system in plants. *Molecular Plant* 2014 Sep 1;7(9):1494–6.
53. Bae S, Kweon J, Kim HS, and Kim JS. Microhomology-based choice of Cas9 nuclease target sites. *Nature Methods* 2014 Jul;11(7):705–6.
54. Zhu LJ, Holmes BR, Aronin N, and Brodsky MH. CRISPRseek: a Bioconductor package to identify target-specific guide RNAs for CRISPR-Cas9 genome-editing systems. *PLoS One* 2014 Sep 23;9(9):e108424.
55. Cradick TJ, Qiu P, Lee CM, Fine EJ, and Bao G. COSMID: a web-based tool for identifying and validating CRISPR/Cas off-target sites. *Molecular Therapy--Nucleic Acids* 2014 Jan 1;3:e214.
56. Labun K, Montague TG, Gagnon JA, Thyme SB, and Valen E. CHOPCHOP v2: a web tool for the next generation of CRISPR genome engineering. *Nucleic Acids Research* 2016 Jul 8;44(W1):W272–6.

57. Park J, Bae S, and Kim JS. Cas-Designer: a web-based tool for the choice of CRISPR-Cas9 target sites. *Bioinformatics* 2015 Dec 15;31(24):4014–6.
58. Heigwer F, Kerr G, and Boutros M. E-CRISP: fast CRISPR target site identification. *Nature Methods* 2014 Feb;11(2):122–3.
59. Prykhozhij SV, Rajan V, Gaston D, and Berman JN. CRISPR multitargeted: a web tool to find common and unique CRISPR single guide RNA targets in a set of similar sequences. *PLoS One* 2015 Mar 5;10(3):e0119372.
60. Ronda C, Pedersen LE, Hansen HG, Kallehauge TB, Betenbaugh MJ, Nielsen AT, and Kildegaard HF. Accelerating genome editing in CHO cells using CRISPR Cas9 and CRISPy, a web-based target finding tool. *Biotechnology and Bioengineering* 2014 Aug;111(8):1604–16.
61. Blin K, Pedersen LE, Weber T, and Lee SY. CRISPy-web: an online resource to design sgRNAs for CRISPR applications. *Synthetic and Systems Biotechnology* 2016 Jun 1;1(2):118–21.
62. Peng D, and Tarleton R. EuPaGDT: a web tool tailored to design CRISPR guide RNAs for eukaryotic pathogens. *Microbial Genomics* 2015 Oct;1(4).
63. Stemmer M, Thumberger T, del Sol Keyer M, Wittbrodt J, and Mateo JL. CCTop: an intuitive, flexible, and reliable CRISPR/Cas9 target prediction tool. *PLoS One* 2015 Apr 24;10(4):e0124633.
64. Wu X, Scott DA, Kriz AJ, Chiu AC, Hsu PD, Dadon DB, Cheng AW et al. Genome-wide binding of the CRISPR endonuclease Cas9 in mammalian cells. *Nature Biotechnology* 2014 Jul;32(7):670–6.
65. Träger U, Andre R, Lahiri N, Magnusson-Lind A, Weiss A, Grueninger S, McKinnon C et al. HTT-lowering reverses Huntington's disease immune dysfunction caused by NFκB pathway dysregulation. *Brain* 2014 Mar 1;137(3):819–33.
66. Kang X, He W, Huang Y, Yu Q, Chen Y, Gao X, Sun X et al. Introducing precise genetic modifications into human 3PN embryos by CRISPR/Cas-mediated genome editing. *Journal of Assisted Reproduction and Genetics* 2016 May 1;33(5):581–8.
67. Bordignon C, Notarangelo LD, Nobili N, Ferrari G, Casorati G, Panina P, et al. Gene therapy in peripheral blood lymphocytes and bone marrow for ADA- immunodeficient patients. *Science* 1995;270:470–5. 10.1126/science.270.5235.470
68. Nishimasu H, Ran FA, Hsu PD, Konermann S, Shehata SI, Dohmae N, Ishitani R et al. Crystal structure of Cas9 in complex with guide RNA and target DNA. *Cell* 2014 Feb 27;156(5):935–49.
69. Wilkinson RA, Martin C, Nemudryi AA, and Wiedenheft B. CRISPR RNA-guided autonomous delivery of Cas9. *Nature Structural & Molecular Biology* 2019 Jan;26(1):14–24.
70. Fathema Uddin,Charles M. Rudin, and Triparna Sen, CRISPR gene therapy: applications, limitations, and implications for the future. *Frontiers in Oncology* 2020;10:1387.
71. Lopalco L. CCR5: from natural resistance to a new anti-HIV strategy. *Viruses* 2010 Feb; 2(2):574–600.
72. Sibbald B. Death but one unintended consequence of gene-therapy trial. *CMAJ* 2001;164:1612.
73. Raper SE, Chirmule N, Lee FS, Wivel NA, Bagg A, Gao GP, et al. Fatal systemic inflammatory response syndrome in a ornithine transcarbamylase deficient patient following adenoviral gene transfer. *Molecular Genetics and Metabolism* 2003;80:148–58.
74. Nunes FA, Furth EE, Wilson JM, and Raper SE. Gene transfer into the liver of nonhuman primates with E1-deleted recombinant adenoviral vectors: safety of readministration. *Human Gene Therapy* 1999;10:2515–26.
75. Cavazzana-Calvo M, Hacein-Bey S, De Saint Basile G, Gross F, Yvon E, Nusbaum P, et al. Gene therapy of human severe combined immunodeficiency (SCID)-X1 disease. *Science* 2000;288:669–72.
76. Herzog RW. Gene therapy for SCID-X1: round 2. *Molecular Therapy* 2010;18:1891.
77. Hacein-Bey-Abina S, Von Kalle C, Schmidt M, McCormack MP, Wulffraat N, Leboulch P, et al. LMO2-associated clonal T cell proliferation in two patients after gene therapy for SCID-X1. *Science* 2003;302:415–9.
78. Lee CS, Bishop ES, Zhang R, Yu X, Farina EM, Yan S, et al. Adenovirus-mediated gene delivery: potential applications for gene and cell-based therapies in the new era of personalized medicine. *Genes & Diseases* 2017;4:43–63.
79. Goswami R, Subramanian G, Silayeva L, Newkirk I, Doctor D, Chawla K, et al. Gene therapy leaves a vicious cycle. *Frontiers in Oncology* 2019;9:297.
80. Sadelain M. Insertional oncogenesis in gene therapy: how much of a risk? *Gene Therapy* 2004;11:569–73.
81. Nayak S, and Herzog RW. Progress and prospects: immune responses to viral vectors. *Gene Therapy* 2010;17:295–304.

82. Ramamoorth M, and Narvekar A. Non viral vectors in gene therapy – an overview. *Journal of Clinical and Diagnostic Research* 2015;9:Ge01–06.
83. Bibikova M, Beumer K, Trautman JK, and Carroll D. Enhancing gene targeting with designed zinc finger nucleases. *Science* 2003;300:764.
84. Chandrasegaran S, and Carroll D. Origins of programmable nucleases for genome engineering. *Journal of Molecular Biology* 2016;428:963–89.
85. Tan Y, Chu A, Bao S, Hoang DA, Kebede FT, Xiong W, et al. Rationally engineered Staphylococcus aureus Cas9 nucleases with high genome-wide specificity. *Proceedings of the National Academy of Sciences of the United States of America* 2019;116:20969–76.
86. Luan B, Xu G, Feng M, Cong L, and Zhou R. Combined computational–experimental approach to explore the molecular mechanism of SaCas9 with a broadened DNA targeting range. *Journal of the American Chemical Society* 2019;141:6545–52.
87. Chatterjee P, Jakimo N, and Jacobson JM. Minimal PAM specificity of a highly similar SpCas9 ortholog. *Science Advances* 2018;4:eaau0766.
88. Semenova E, Jore MM, Datsenko KA, Semenova A, Westra ER, Wanner B, et al. Interference by clustered regularly interspaced short palindromic repeat (CRISPR) RNA is governed by a seed sequence. *Proceedings of the National Academy of Sciences of the United States of America* 2011;108:10098–103.
89. Liu X, Homma A, Sayadi J, Yang S, Ohashi J, and Takumi T. Sequence features associated with the cleavage efficiency of CRISPR/Cas9 system. *Scientific Reports* 2016;6:19675.
90. Bruegmann T, Deecke K, and Fladung M. Evaluating the efficiency of gRNAs in CRISPR/Cas9 mediated genome editing in poplars. *International Journal of Molecular Sciences* 2019;20:3623.
91. Fu Y, Sander JD, Reyon D, Cascio VM, and Joung JK. Improving CRISPR-Cas nuclease specificity using truncated guide RNAs. *Nature Biotechnology* 2014;32:279–84.
92. Heigwer F, Kerr G, and Boutros M. E-CRISP: fast CRISPR target site identification. *Nature Methods* 2014;11:122–3.
93. Yan J, Chuai G, Zhou C, Zhu C, Yang J, Zhang C, et al. Benchmarking CRISPR on-target sgRNA and design. *Briefings in Bioinformatics* 2017;15:1–4.
94. Cui Y, Xu J, Cheng M, Liao X, and Peng S. Review of CRISPR/Cas9 sgRNA design tools. *Interdisciplinary Sciences: Computational Life Sciences* 2018;10:455–65.
95. Hiranniramol K, Chen Y, Liu W, and Wang X. Generalizable sgRNA design for improved CRISPR/Cas9 editing efficiency. *Bioinformatics* 2020;36:2684–89.
96. Lino CA, Harper JC, Carney JP, and Timlin JA. Delivering CRISPR: a review of the challenges and approaches. *Drug Delivery* 2018;25:1234–57.
97. Lau CH, Suh Y. *In vivo* genome editing in animals using AAV-CRISPR system: applications to translational research of human disease. *F1000Res* 2017;6:2153.
98. Kim HK, Lee S, Kim Y, Park J, Min S, Choi JW, et al. High-throughput analysis of the activities of xCas9, SpCas9-NG and SpCas9 at matched and mismatched target sequences in human cells. *Nature Biomedical Engineering* 2020;4:111–24.
99. O'Connell MR, Oakes BL, Sternberg SH, East-Seletsky A, Kaplan M, and Doudna JA. Programmable RNA recognition and cleavage by CRISPR/Cas9. *Nature* 2014;516:263–6.
100. Bratovič M, Fonfara I, Chylinski K, Gálvez EJC, Sullivan TJ, Boerno S, et al.. Bridge helix arginines play a critical role in Cas9 sensitivity to mismatches. *Nature Chemical Biology* 2020;16:587–95.
101. Walker FO. Huntington's disease. *The Lancet* 2007 Jan 20;369(9557):218–28.
102. Baylis F, and McLeod M. First-in-human phase 1 CRISPR gene editing cancer trials: are we ready? *Current Gene Therapy* 2017;17:309–19.
103. Ye L, Wang J, Beyer AI, Teque F, Cradick TJ, Qi Z, et al. Seamless modification of wild-type induced pluripotent stem cells to the natural CCR5Delta32 mutation confers resistance to HIV infection. *Proceedings of the National Academy of Sciences of the United States of America* 2014;111:9591–6.
104. Strutt SC, Torrez RM, Kaya E, Negrete OA, Doudna JA. RNA-dependent RNA targeting by CRISPR-Cas9. *eLife* 2018;7:e32724.
105. Konermann S, Lotfy P, Brideau NJ, Oki J, Shokhirev MN, and Hsu PD. Transcriptome engineering with RNA-targeting type VI-D CRISPR effectors. *Cell* 2018;173:665–76.
106. Greely HT. CRISPR'd babies: human germline genome editing in the 'He Jiankui affair'. *Journal of Law and the Biosciences* 2019;6:111–83.
107. Alkhatib G. The biology of CCR5 and CXCR4. *Current Opinion in HIV and AIDS* 2009;4:96–103.
108. Ma H, Marti-Gutierrez N, Park SW, Wu J, Lee Y, Suzuki K, et al. Correction of a pathogenic gene mutation in human embryos. *Nature* 2014;548:413–9.

14 Clinical Applications of Gene Therapy for Immuno-Deficiencies

Khushboo Faldu, Sakshi Gurbani and Jigna Shah
Nirma University, Ahmedabad, Gujarat, India

CONTENTS

14.1 PRIMARY IMMUNODEFICIENCIES (PIDS)

Primary immunodeficiencies (PIDS) are the inherited birth defects of the immune system (innate or adaptive) that affects 1:10,000 born individuals [1, 2]. There exists >400 PIDS and have been categorized as immune regulatory disorders, combined immunodeficiencies, defects in natural immunity, lack of antibodies, deficiency of the complement system, defective phagocytosis, and autoinflammatory diseases [3]. The immune system deficiencies are mostly diagnosed in juvenile and adolescent patients, but the mild phenotype of PID may not present itself until adulthood. Thus, increased numbers of individuals are being diagnosed in adulthood [4].

Clinical presentation of PIDs is diverse, while frequent infections that are severe and fatal are the most commonly encountered. Cytopenias, serious autoinflammatory complications (hemophagocytic lymphohistiocytosis (HLH)), autoimmune phenomena (idiopathic thrombocytopenic purpura), or hematological malignancies (lymphoid malignancies) can all point toward an underlying PID [5]. PID clinical severity is also disparate, with a child suffering from severe combined immunodeficiency (SCID) requiring compelling allogeneic hematopoietic stem cell transplantation (alloHSCT),

DOI: 10.1201/9781003186083-14

while an adult with common variable immunodeficiency (CVID) thrives on basic treatment involving immunoglobulin (Ig)-replacement.

Mild phenotypic PID individuals may only require supportive care for maintenance of the quality of life and life expectancy, while PIDs resulting in fatal complications (HLH, malignancy, or infections) require a reliable treatment for prevention of morbidity and mortality. AlloHSCT is imperative for the survival of infants suffering from SCID. AlloHSCT is a part of the primary treatment for aggressive PIDs like Wiskott–Aldrich syndrome (WAS) and chronic granulomatous disease (CGD). The course of action for rare PIDs involves transplantation only in the case of the development of fatal complications. The alloHSCT field has progressed for over 50 years and has achieved a greater than 95% survival rate in SCID patients [6].

The major limitation of transplantation is graft-versus-host disease (GvHD). Graft-versus-malignancy effect provides an advantage to malignant PIDs cases, as it eliminates residual malignant cells from the system. AlloHSCT should provide lasting engraftment and resolve clinical presentation of the PID, while also minimizing the chances of GvHD. Human leukocyte antigen (HLA)-compatibility served as a limitation for alloHSCT, as only 25% of patients had an HLA-identical sibling donor, and fewer than 70% had an HLA-compatible unrelated donor, which was overcome by the development of haploidentical alloHSCT procedures that utilize T-cell depletion to prevent GvHD, even if the donor and recipient have a difference of the entire haplotype [7, 8]. The improved protocols utilizing transplant manipulation, together with depletion of B- and T-cells or post-transplant cyclophosphamide (PTCy), have increased overall survival (OS) and decreased GvHD in non-malignant PID patients [9–11]. To achieve better outcomes in PID patients, prompt transplantation is preferred, but the development of reduced-intensity conditioning (RIC) regimens for adult PID cases has improved survival rates and transplant outcomes for alloHSCT [12, 13]. While the alloHSCT has come a long way, it still can lead to complications like GvHD, graft failure and rejection, and transplantation-related mortality (TRM) [5, 14, 15]. AlloHSCT with prolonged immunosuppression is not a viable option for adult PID patients with end-stage organ damage [16].

Autologous hematopoietic stem cell gene therapy (HSC-GT) does not require immunosuppression as prophylaxis for GvHD and the requirement of an HLA-compatible donor [17]. HSC-GT utilizes retroviruses as vectors for targeting PID due to the monogenetic nature of the defect and clinical presentation of the immune dysfunction. It can readily repair the immune cells defect by manipulating HSC and confers a survival advantage to genetically modified cells. Hematopoietic and progenitor cells demonstrating CD34, a cell surface marker, are harvested by bone marrow purification or by extracting mobilized stem cells by aphaeresis of the peripheral blood with the use of a plerixafor (selective chemokine receptor (CXCR4) antagonist) and granulocyte colony-stimulating factor (G-CSF) combination. For sickle cell disease patients, plerixafor alone is used, instead of the combination to overcome disease-specific limitations in progenitor and hematopoietic cells' mobilization [18–20]. Leukapheresis harvests HSCs from the blood circulation. The extracted CD34+ cells are carefully chosen and cultivated under artificial media that support repopulation, genetic editing, and transmission of the correction of the genetic defect [21–23]. Cell division ensures that the genetic modification is inherited by the subsequent progeny of hematopoietic lineages [24]. The patient is pre-conditioned, and the gene edited HSCs are transfused into the patients. These gene edited HSCs gather into the bone marrow, where they preserve their self-renewal aptitude, and the genetic correction is passed on to the future immune-hematopoietic progenies via cell differentiation [25].

HSC-GT is making progress for the treatment of PID, with Strimvelis being the first HSC-GT, a SIN-γRV (self-inactivating (SIN) LTR modification of γ-retrovirus vector(γRV)) approved in 2016 in Europe for the treatment of adenosine deaminase-deficient SCID (ADA-SCID) in patients who do not have an HLA compatible supporter for alloHSCT [26]. HSC-GT carries the risk of γRV induced insertional mutagenesis caused by the gamma-retroviral vector and toxicities related to erroneous genetic modification of non-targeted sites by the techniques used for gene editing [27].

14.1.1 Severe Combined Immunodeficiencies

14.1.1.1 Adenosine Deaminase-Deficient Severe Combined Immunodeficiency (ADA-SCID)

ADA-SCID, a PID that lacks enzyme adenosine deaminase, causing immunodeficiency due to absenteeism of natural killer (NK), B, and T-cells, with fatal outcomes in infancy without alloHSCT therapy. HSC-GT clinical study established that transduction of T-lymphocytes was possible with the use of γRV vectors containing ADA enzyme. The treated patients produced ADA enzyme endogenously that resulted in improvement of the clinical phenotype [28, 29]. This method overcame the limitation of multiple transfusions of modified T-lymphocytes for the maintenance of therapeutic outcomes [29–31]. Busulfan conditioning improved ADA multilineage engraftment [32, 33] and reduced the number of patients requiring enzyme replacement therapy [34]. Strimvelis is the first HSC-GT, a SIN-γRV (self-inactivating (SIN) LTR modification of γ-retrovirus vector (γRV)) approved in 2016 in Europe for the treatment of adenosine deaminase-deficient SCID (ADA-SCID) who lack a suitable donor for alloHSCT [26], which is exclusively available in Milan, Italy, and requires patients globally to undertake a medical holiday to receive Strimvelis, as it is fresh cell product. When γRV vectors were utilized for X-linked SCID (X-SCID), they resulted in insertional mutagenesis [35–37] that led to the development of SIN-LV vectors that have an enhanced safety and efficiency profile when compared to γRV vectors [38]. HSC-GT is now considered one of the first-line therapies for ADA-SCID in Europe [39].

14.1.1.2 X-linked Severe Combined Immunodeficiency (X-SCID)

X-SCID is an X-linked disorder that mutates the common cytokine receptor gamma chain [interleukin 2 receptor subunit gamma (IL2RG) gene], prevents normal development and functioning of the lymphocytes, and produces deficits in cellular and humoral immunity. Infants suffering from severe infections are the first signs of clinical presentation. Treatment involves alloHSCT or HSC-GT for the prevention of mortality [40]. γRV vectors were initially utilized, which provided successful immune reconstitution and persistence of gene-marked cells, but led to the development of T-cell lymphoblastic leukemia [41, 42]. To overcome insertional mutagenesis, SIN- γRV and SIN-LV vectors were developed. HSC-GT is being evaluated in numerous clinical studies for X-SCID.

14.1.1.3 Artemis SCID and Recombinase-activating Gene 1 (RAG1) Deficiency

RAG1 gene defects impair the functioning of T-cells due to variable (V), diversity (D), and joining (J) genes-VDJ recombination faults and, consequently, lead to the development of abnormal T-cell receptor (TCR) and Ig receptor reorganization. Additionally, these patients are predisposed to develop malignancies [43]. Treatment with alloHSCT is challenging, as patients suffering from Artemis SCID bear a mutation that confers on them sensitivity to radiotherapy or alkylating chemotherapy, which is utilized in preconditioning to achieve optimal immune reconstitution [44]. SIN-LV vector carries Artemis cDNA and endogenous promoter. It potentially restores B and T-cells after transducing the CD34+ cells of Artemis SCID patients, and corrects radiosensitivity in patients' fibroblasts [44, 45].

14.1.2 Combined Immunodeficiencies

14.1.2.1 Wiskott–Aldrich Syndrome (WAS)

WAS, X-linked inheritable PID presents itself as autoimmunity, recurrent infections, tendency to develop lymphoid cancer, macrothrombocytopenia, and severe eczema. WAS gene mutation hinders WAS protein (WASp) which is essential for actin polymerization. Actin is crucial for entire hematopoietic lineages, as it plays an important role in formation of immunological synapse, cell migration, and cytotoxicity [46]. AlloHSCT has provided conservative benefits in WAS, with benefits deteriorating for patients aged more than five years [27]. γRV vectors led to the development of leukemogenesis [47]. A SIN-LV vector utilizing a fragment of endogenous WAS gene promoter, together with fludarabine and busulfan conditioning, is showing promise in clinical trials with patients not requiring Ig-replacement therapy [48]. Yet the platelet counts were below normal, maybe attributed to WASp expression in platelet precursors [49, 50].

14.1.3 Disorders of Phagocyte Number and Function

14.1.3.1 Chronic Granulomatous Disease (CGD)

CGD causes mutations in nicotinamide adenine dinucleotide phosphate (NADPH) oxidase complex that compromises the production of reactive oxygen species (ROS), causing impaired functioning of neutrophils, macrophages, phagocytes, and monocytes. The patients usually present themselves with fatal fungal and bacterial infections and hyper inflammation, mostly due to treatment-resistant colitis [51]. There are two variants of CGD. X-linked variation is the most encountered mutation of cytochrome B-245 beta chain (CYBB) gene that codes gp91phox subunit of NADPH-oxidase, while the autosomal recessive form of CGD is rare and affects one-third of all the CGD patient population. The symptoms don't necessarily surface in childhood or adolescence, but can also present themselves in adulthood [52]. Registry trials data suggests that 50–55% of CGD patients survive till the age of 40 years with supportive care, but the condition severely affects the quality of life [53]. AlloHSCT has been therapeutically efficacious, and with RIC regimens, exceptional survival outcomes have been achieved [13]. But early diagnosis and definitive treatment are the keys to preventing mortality and morbidity [12–14, 54].

Similar to other PIDs, chances of malignancy increased with the use of γRV vectors, and those vectors even lacked adequate gene marking in myeloid lineage without pre-conditioning [55]. SIN-LV vectors were safer and more efficacious alternatives to the γRV vectors, as these improved myeloid gene marking with the use of a chimeric promoter that drives the codon-optimized CYBB cDNA [56, 57]. HSC-GT utilizing SIN-LV is in the clinical trial phase and has demonstrated immune reconstitution in older (>18 years) patients [58]. HSC-GT utilizing SIN-LV vectors targeting p47 deficiency of autosomal recessive CGD is being evaluated in clinical trials [59].

14.1.3.2 Leucocyte Adhesion Defect Type 1 (LAD-1)

LAD-1 is a phagocyte functional disease caused due to defects in integrin subunit beta 2 (ITGB2) gene that causes faulty or missing b-integrin molecules (CD11a and CD18) to impair leucocytes' adhesion and migration. Symptoms surface during infancy and are presented as delayed wound healing and deep tissue infections [60]. The preferred treatment is alloHSCT, with overall survival rates of approximately 75% [60]. HSC-GT utilizing SIN-LV vectors, together with pre-conditioning with busulfan, is in the clinical phase of development [61, 62].

14.1.4 Diseases of Immune Dysregulation

14.1.4.1 Familial Hemophagocytic Lymphohistiocytosis (FLH)

Hemophagocytic lymphohistiocytosis (HLH) can be fatal. The symptoms include cytopenias, hyper inflammation, splenomegaly, and uncontrolled immune activation [63]. HLH can be caused due to various underlying conditions, like familial hemophagocytic lymphohistiocytosis (FHL), autoimmunity, infection, or malignancy. Defects in the functioning of cytotoxic T- and NK-cells, together with autosomal recessive alterations in syntaxin 11 (STX11), MUNC 13-4 [Protein unc-13 homolog D (UNC13D)], and perforin (PRF1) result in the development of FHL. The most common cause is PRF1 mutations. AlloHSCT is the preferred treatment modality for genetically mutated or relapsed refractory HLH patients [64, 65]. Etoposide, with or without cyclosporine and glucocorticoids, are a part of the standard of care, but lack adequate disease control in 40% of patients [66]. Emapalumab, an anti-interferon gamma (IFNγ) monoclonal antibody, has been approved as second-line therapy for primary HLH in the USA, as IFNγ plays a central role in pathogenesis in HLH, with levels correlating with active disease [67, 68]. Alemtuzumab and Janus kinase (JAK) inhibitors are being tested for improving remission and allowing progress to HSCT [69–74]. Alemtuzumab with corticosteroids and cyclosporine has provided favorable safety and efficacy in children suffering from primary HLH (91.6% survived to alloHSCT) [74]. HSC-GT and T-cell strategies are being developed for the treatment of FHL [75–79].

14.1.4.2 X-Linked Lymphoproliferative Disease 1 (XLP1)

XLP1 develops due to mutations in the SH2 domain-containing 1A (SH2D1A) gene that leads to the development of defects in an intracellular adaptor protein, SLAM-associated protein (SAP), essential for T-cell and NK-mediated cytotoxicity and T follicular helper cell function [80]. The clinical manifestation of XLP1 is dysregulation of the immune system, with HLH and lymphoma development [81]. The only definitive therapy available is alloHSCT but active disease at the time of transplant worsens survival outcomes [81]. T cell transduction with SIN-LV vectors incorporating SAP cDNA is being developed for XLP1 [82].

14.1.4.3 Immune Dysregulation, Polyendocrinopathy, Enteropathy, X-Linked (IPEX) Syndrome

Regulatory T cells (Tregs) prevent autoimmunity as they control immune response to self-antigens. IPEX syndrome develops due to the defects in transcription factor fork head box P3 (FoxP3) that are essential for the normal development and functioning of Tregs. IPEX syndrome clinically presents itself as eczema, enteropathy, and type I diabetes mellitus. Treatment involves prolonged immunosuppression and alloHSCT, but immune-mediated complications can limit its efficacy [83]. The T-cell approach lacks capability for the generation of an adequate amount of Tregs to improve clinical phenotype [84]. HSC-GT is challenging as FoxP3 expression in HSCs prevents proliferation and differentiation of the stem cells. The SIN-LV vector incorporating the endogenous FoxP3 promoter and regulatory elements has shown favourable results that induced lineage-specific expression in progeny of transduced HSCs [84–86].

14.2 ACQUIRED IMMUNODEFICIENCIES (AIDS)

Human Immunodeficiency Virus (HIV) causes the development of Acquired Immune Deficiency Syndrome (AIDS). HIV is infectious and spreads through sexual contact [87]. It is categorized into HIV type 1 (HIV-1) and HIV type 2 (HIV-2) [88]. Later stages of viral infection, wherein the body's immune system is severely compromised, is known as AIDS. HIV patients are at an increased risk of developing AIDS-related syndromes and hematological malignancies, like multiple myeloma, Hodgkin lymphoma, non-Hodgkin lymphoma (NHL), and leukemia [89, 90].

Strategies employed to cure HIV are subdivided into two groups: "sterilizing cure" eradicates HIV-1 from the patients while "functional cure" controls HIV prognosis in terms of virus replication and CD4 T-lymphocytes in the absence of ART therapy. AlloHSCT in the absence of ART therapy has cured three patients globally that received stem cells from donors that carried the homozygous CCR5Δ32/Δ32 mutation [91, 92]. The patients suffered homological cancer that required alloHSCT [93]. Gene therapy with various strategies, like adoptive transfer of chimeric antigen receptor T cells or T cells with affinity-enhanced T cell receptors, fusion inhibitors, trans-dominant proteins, ribozymes, endoribonucleases, antisense strategies, combinatorial strategies, genome editing, epigenome editing, are being developed [93].

14.3 CONCLUSION

HSC-GT can become a part of the standard of care for certain PIDs. PIDs like ADA-SCID, which have an excellent prognosis with alloHSCT, may not witness preference to HSC-GT in the primary treatment protocol. HSC-GT may be preferred in conditions that carry high risk with alloHSCT, like the X-linked inhibitor of apoptosis (XIAP) and Artemis SCID. HSC-GT is promising, as it provides a treatment option to patients who may have limited treatment options. The widespread adoption of HSC-GT is limited due to high cost and lack of availability, as compared to alloHSCT. Non-genotoxic conditioning agents and gene-editing technologies may reduce risks and improve the efficacy of HSC-GT.

Development of cell and gene therapies for HIV have not yet achieved success due to various factors, For instance, ell-based therapies must consider the reservoirs of latent viruses, virus heterogeneity in the patients, escape mutations, technical challenges related to successful gene transfer or genome editing, as well as the sustenance of the genetically altered cells *in vivo*. The latter includes the potential immunogenicity of expressed transgenes and the cytotoxicity associated with the genetic modification. The choice of the genetic target is demanding, too. Even aiming for highly conserved regions of HIV is not always crowned by success.

The development of HSC-GT may better equip us to provide a better treatment alternative that is efficacious and devoid of side effects, and also circumvents the difficulties of alloHSCT for PIDs and AIDs.

REFERENCES

1. Picard C, Bobby Gaspar H, Al-Herz W, Bousfiha A, Casanova J-L, Chatila T, et al. International union of immunological societies: 2017 primary immunodeficiency diseases committee report on inborn errors of immunity. *J Clin Immunol* 2017 381 [Internet]. 2017 Dec 11 [cited 2021 Aug 2];38(1): 96–128. Available from: https://link.springer.com/article/10.1007/s10875-017-0464-9
2. Notarangelo LD. Primary immunodeficiencies. *J Allergy Clin Immunol*. 2010 Feb 1;125(2):S182–94.
3. Bousfiha A, Jeddane L, Picard C, Al-Herz W, Ailal F, Chatila T, et al. Human inborn errors of immunity: 2019 update of the IUIS phenotypical classification. *J Clin Immunol* 2020 401 [Internet]. 2020 Feb 11 [cited 2021 Aug 2];40(1):66–81. Available from: https://link.springer.com/article/10.1007/s10875-020-00758-x
4. Morris EC, and Albert MH. Allogeneic HSCT in adolescents and young adults with primary immunodeficiencies. *Front Pediatr* [Internet]. 2019 Oct 24 [cited 2021 Aug 2];7:437. Available from: /pmc/articles/PMC6821713/
5. Pai S-Y. Treatment of primary immunodeficiency with allogeneic transplant and gene therapy. *Hematology* [Internet]. 2019 Dec 6 [cited 2021 Aug 2];2019(1):457–65. Available from: http://ashpublications.org/hematology/article-pdf/2019/1/457/1546234/hem2019000052c.pdf
6. Gennery AR, and Lankester A. (EBMT) IEWP (IEWP) of the ES for B and MT. Long term outcome and immune function after hematopoietic stem cell transplantation for primary immunodeficiency. *Front Pediatr*. 2019 Sep;7:381.
7. Rocha V, and Locatelli F. Searching for alternative hematopoietic stem cell donors for pediatric patients. *Bone Marrow Transplant* 2008 412 [Internet]. 2007 Dec 17 [cited 2021 Aug 2];41(2):207–14. Available from: https://www.nature.com/articles/1705963
8. Aversa F, Tabilio A, Velardi A, Cunningham I, Terenzi A, Falzetti F, et al. Treatment of high-risk acute leukemia with T-cell–depleted stem cells from related donors with One fully mismatched HLA haplotype. *N Engl J Med* http://dx.doi.org/101056/NEJM199810223391702 [Internet]. 2009 Jul 15 [cited 2021 Aug 2];339(17):1186–93. Available from: https://www.nejm.org/doi/10.1056/NEJM199810223391702
9. Kurzay M, Hauck F, Schmid I, Wiebking V, Eichinger A, Jung E, et al. T-cell replete haploidentical bone marrow transplantation and post-transplant cyclophosphamide for patients with inborn errors. *Haematologica* [Internet]. 2019 Oct 1 [cited 2021 Aug 2];104(10):e478–82. Available from: https://haematologica.org/article/view/9104
10. Bertaina A, Merli P, Rutella S, Pagliara D, Bernardo ME, Masetti R, et al. HLA-haploidentical stem cell transplantation after removal of αβ+ T and B cells in children with nonmalignant disorders. *Blood* [Internet]. 2014 Jul 31 [cited 2021 Aug 2];124(5):822–6. Available from: http://ashpublications.org/blood/article-pdf/124/5/822/541634/822.pdf

11. Shah RM, Elfeky R, Nademi Z, Qasim W, Amrolia P, Chiesa R, et al. T-cell receptor αβ+ and CD19+ cell–depleted haploidentical and mismatched hematopoietic stem cell transplantation in primary immune deficiency. *J Allergy Clin Immunol.* 2018 Apr 1;141(4):1417–1426.e1.
12. Fox TA, Chakraverty R, Burns S, Carpenter B, Thomson K, Lowe D, et al. Successful outcome following allogeneic hematopoietic stem cell transplantation in adults with primary immunodeficiency. *Blood* [Internet]. 2018 Feb 22 [cited 2021 Aug 2];131(8):917–31. Available from: http://ashpublications.org/blood/article-pdf/131/8/917/1466640/blood807487.pdf
13. Güngör T, Teira P, Slatter M, Stussi G, Stepensky P, Moshous D, et al. Reduced-intensity conditioning and HLA-matched haemopoietic stem-cell transplantation in patients with chronic granulomatous disease: A prospective multicentre study. *Lancet.* 2014 Feb 1;383(9915):436–48.
14. Chiesa R, Party for the EIEW, Wang J, Party for the EIEW, Blok H-J, Party for the EIEW, et al. Hematopoietic cell transplantation in chronic granulomatous disease: A study of 712 children and adults. *Blood.* 2020 Sep 3;136(10):1201–11.
15. Pai SY, and Cowan MJ. Stem cell transplantation for primary immunodeficiency diseases: The North american experience. *Curr Opin Allergy Clin Immunol* [Internet]. 2014 [cited 2021 Aug 2];14(6):521–6. Available from: https://journals.lww.com/co-allergy/Fulltext/2014/12000/Stem_cell_transplantation_for_primary.6.aspx
16. Morris EC, Fox T, Chakraverty R, Tendeiro R, Snell K, Rivat C, et al. Gene therapy for Wiskott-Aldrich syndrome in a severely affected adult. *Blood* [Internet]. 2017 Sep 14 [cited 2021 Aug 2];130(11):1327–35. Available from: http://ashpublications.org/blood/article-pdf/130/11/1327/1402533/blood777136.pdf
17. Kohn DB, and Kuo CY. new frontiers in the therapy of primary immunodeficiency: From gene addition to gene editing. *J Allergy Clin Immunol.* 2017 Mar 1;139(3):726–32.
18. Kuan JW, Su AT, and Leong CF. Pegylated granulocyte-colony stimulating factor versus non-pegylated granulocyte-colony stimulating factor for peripheral blood stem cell mobilization: A systematic review and meta-analysis. *J Clin Apher* [Internet]. 2017 Dec 1 [cited 2021 Aug 2];32(6):517–42. Available from: https://onlinelibrary.wiley.com/doi/full/10.1002/jca.21550
19. DiPersio JF, Micallef IN, Stiff PJ, Bolwell BJ, Maziarz RT, Jacobsen E, et al. Phase III prospective randomized double-blind placebo-controlled trial of plerixafor plus granulocyte colony-stimulating factor compared with placebo plus granulocyte colony-stimulating factor for autologous stem-cell mobilization and transplantation for patients with non-hodgkin's lymphoma. *J Clin Oncol.* https://doi.org/101200/JCO2008207209. 2009 Aug 31;27(28):4767–73.
20. Lagresle-Peyrou C, Lefrère F, Magrin E, Ribeil J-A, Romano O, Weber L, et al. Plerixafor enables safe, rapid, efficient mobilization of hematopoietic stem cells in sickle cell disease patients after exchange transfusion. *Haematologica* [Internet]. 2018 May 1 [cited 2021 Aug 2];103(5):778–86. Available from: https://haematologica.org/article/view/8448
21. Naldini L. Genetic engineering of hematopoiesis: Current stage of clinical translation and future perspectives. *EMBO Mol Med* [Internet]. 2019 Mar 1 [cited 2021 Aug 2];11(3):e9958. Available from: https://www.embopress.org/doi/full/10.15252/emmm.201809958
22. Fares I, Chagraoui J, Lehnertz B, MacRae T, Mayotte N, Tomellini E, et al. EPCR expression marks UM171-expanded CD34+ cord blood stem cells. *Blood* [Internet]. 2017 Jun 22 [cited 2021 Aug 2];129(25):3344–51. Available from: www.ncbi.nlm.nih.gov/geo/
23. Radtke S, Adair JE, Giese MA, Chan Y-Y, Norgaard ZK, Enstrom M, et al. A distinct hematopoietic stem cell population for rapid multilineage engraftment in nonhuman primates. *Sci Transl Med* [Internet]. 2017 Nov 1 [cited 2021 Aug 2];9(414). Available from: https://stm.sciencemag.org/content/9/414/eaan1145
24. Porteus MH. A new class of medicines through DNA editing. *N Engl J Med* https://doi.org/101056/NEJMra1800729 [Internet]. 2019 Mar 6 [cited 2021 Aug 2];380(10):947–59. Available from: https://www.nejm.org/doi/10.1056/NEJMra1800729
25. Morgan RA, Gray D, Lomova A, and Kohn DB. Hematopoietic stem cell gene therapy: Progress and lessons learned. *Cell Stem Cell.* 2017 Nov 2;21(5):574–90.
26. Aiuti A, Roncarolo MG, and Naldini L. Gene therapy for ADA-SCID, the first marketing approval of an ex vivo gene therapy in Europe: Paving the road for the next generation of advanced therapy medicinal products. *EMBO Mol Med* [Internet]. 2017 Jun 1 [cited 2021 Aug 2];9(6):737–40. Available from: https://www.embopress.org/doi/full/10.15252/emmm.201707573
27. Fox TA, and Booth C. Gene therapy for primary immunodeficiencies. *Br J Haematol.* 2021;193(6):1044–59.

28. Kohn DB, Weinberg KI, Nolta JA, Heiss LN, Lenarsky C, Crooks GM, et al. Engraftment of gene–modified umbilical cord blood cells in neonates with adenosine deaminase deficiency. *Nat Med* 1995 110 [Internet]. 1995 [cited 2021 Aug 2];1(10):1017–23. Available from: https://www.nature.com/articles/nm1095-1017
29. Bordignon C, Notarangelo LD, Nobili N, Ferrari G, Casorati G, Panina P, et al. Gene therapy in peripheral blood lymphocytes and bone marrow for ADA– Immunodeficient patients. *Science* (80-) [Internet]. 1995 Oct 20 [cited 2021 Aug 2];270(5235):470–5. Available from: https://science.sciencemag.org/content/270/5235/470
30. Aiuti A, Vai S, Mortellaro A, Casorati G, Ficara F, Andolfi G, et al. Immune reconstitution in ADA-SCID after PBL gene therapy and discontinuation of enzyme replacement. *Nat Med* 2002 85 [Internet]. 2002 [cited 2021 Aug 2];8(5):423–5. Available from: https://www.nature.com/articles/nm0502-423
31. Gaspar HB, Bjorkegren E, Parsley K, Gilmour KC, King D, Sinclair J, et al. Successful reconstitution of immunity in ADA-SCID by stem cell gene therapy following cessation of PEG-ADA and use of mild preconditioning. *Mol Ther*. 2006 Oct 1;14(4):505–13.
32. Kohn DB, Hershfield MS, Carbonaro D, Shigeoka A, Brooks J, Smogorzewska EM, et al. T lymphocytes with a normal ADA gene accumulate after transplantation of transduced autologous umbilical cord blood CD34+ cells in ADA-deficient SCID neonates. *Nat Med* 1998 47 [Internet]. 1998 Jul [cited 2021 Aug 2];4(7):775–80. Available from: https://www.nature.com/articles/nm0798-775
33. Aiuti A, Brigida I, Ferrua F, Cappelli B, Chiesa R, Marktel S, et al. Hematopoietic stem cell gene therapy for adenosine deaminase deficient-SCID. *Immunol Res* 2009 441 [Internet]. 2009 Feb 18 [cited 2021 Aug 2];44(1):150–9. Available from: https://link.springer.com/article/10.1007/s12026-009-8107-8
34. Cicalese MP, Ferrua F, Castagnaro L, Pajno R, Barzaghi F, Giannelli S, et al. Update on the safety and efficacy of retroviral gene therapy for immunodeficiency due to adenosine deaminase deficiency. *Blood* [Internet]. 2016 Jul 7 [cited 2021 Aug 2];128(1):45–54. Available from: http://ashpublications.org/blood/article-pdf/128/1/45/1395155/45.pdf
35. Hacein-Bey-Abina S, Pai S-Y, Gaspar HB, Armant M, Berry CC, Blanche S, et al. A modified γ-retrovirus vector for X-linked severe combined immunodeficiency. *N Engl J Med*. http://dx.doi.org/101056/NEJMoa1404588 [Internet]. 2014 Oct 8 [cited 2021 Aug 2];371(15):1407–17. Available from: https://www.nejm.org/doi/10.1056/NEJMoa1404588
36. Gaspar HB, Cooray S, Gilmour KC, Parsley KL, Adams S, Howe SJ, et al. Long-term persistence of a polyclonal T cell repertoire after gene therapy for X-linked severe combined immunodeficiency. *Sci Transl Med* [Internet]. 2011 Aug 24 [cited 2021 Aug 2];3(97):97ra79–97ra79. Available from: https://stm.sciencemag.org/content/3/97/97ra79
37. Montini E, Cesana D, Schmidt M, Sanvito F, Bartholomae CC, Ranzani M, et al. The genotoxic potential of retroviral vectors is strongly modulated by vector design and integration site selection in a mouse model of HSC gene therapy. *J Clin Invest* [Internet]. 2009 Apr 1 [cited 2021 Aug 2];119(4):964–75. Available from: http://www.jci.org
38. Modlich U, Navarro S, Zychlinski D, Maetzig T, Knoess S, Brugman MH, et al. Insertional transformation of hematopoietic cells by self-inactivating lentiviral and gammaretroviral vectors. *Mol Ther*. 2009 Nov 1;17(11):1919–28.
39. Kohn DB, Hershfield MS, Puck JM, Aiuti A, Blincoe A, Gaspar HB, et al. Consensus approach for the management of severe combined immune deficiency caused by adenosine deaminase deficiency. *J Allergy Clin Immunol*. 2019 Mar 1;143(3):852–63.
40. Mamcarz E, Zhou S, Lockey T, Abdelsamed H, Cross SJ, Kang G, et al. Lentiviral gene therapy combined with low-dose busulfan in infants with SCID-X1. *N Engl J Med*. https://doi.org/101056/NEJMoa1815408 [Internet]. 2019 Apr 17 [cited 2021 Aug 2];380(16):1525–34. Available from: https://www.nejm.org/doi/10.1056/NEJMoa1815408
41. Hacein-Bey-Abina S, Garrigue A, Wang GP, Soulier J, Lim A, Morillon E, et al. Insertional oncogenesis in 4 patients after retrovirus-mediated gene therapy of SCID-X1. *J Clin Invest* [Internet]. 2008 Sep 2 [cited 2021 Aug 2];118(9):3132–42. Available from: http://www.jci.org
42. Howe SJ, Mansour MR, Schwarzwaelder K, Bartholomae C, Hubank M, Kempski H, et al. Insertional mutagenesis combined with acquired somatic mutations causes leukemogenesis following gene therapy of SCID-X1 patients. *J Clin Invest* [Internet]. 2008 Sep 2 [cited 2021 Aug 2];118(9):3143–50. Available from: http://www.jci.org

43. Lagresle-Peyrou C, Benjelloun F, Hue C, Andre-Schmutz I, Bonhomme D, Forveille M, et al. Restoration of human B-cell differentiation into NOD-SCID mice engrafted with gene-corrected CD34+ cells isolated from artemis or RAG1-deficient patients. *Mol Ther*. 2008 Feb 1;16(2):396–403.
44. Punwani Divya Kawahara Misako, Yu Jason Sanford Ukina, Roy Sushmita Patel Kiran, et al. Lentivirus mediated correction of artemis-deficient severe combined immunodeficiency. *Hum Gene Ther*. https://home.liebertpub.com/hum [Internet]. 2017 Jan 1 [cited 2021 Aug 2];28(1):112–24. Available from: https://www.liebertpub.com/doi/abs/10.1089/hum.2016.064
45. Multhaup Megan M, Podetz-Pedersen Kelly M, Karlen Andrea D, Olson Erik R, Gunther Roland, Somia Nikunj V, et al. Role of transgene regulation in ex vivo lentiviral correction of artemis deficiency. *Hum Gene Ther*. https://home.liebertpub.com/hum [Internet]. 2015 Mar 2 [cited 2021 Aug 2];26(4):232–43. Available from: https://www.liebertpub.com/doi/abs/10.1089/hum.2014.062
46. Ochs HD, and Thrasher AJ. The Wiskott-Aldrich syndrome. *J Allergy Clin Immunol*. 2006 Apr 1;117(4):725–38.
47. Braun CJ, Witzel M, Paruzynski A, Boztug K, Von Kalle C, Schmidt M, et al. Gene therapy for Wiskott-Aldrich Syndrome—Long-term reconstitution and clinical benefits, but increased risk for leukemogenesis. *Rare Dis* [Internet]. 2014 Jan [cited 2021 Aug 2];2(1):e947749. Available from: /pmc/articles/PMC4755244/
48. Abina SH-B, Gaspar HB, Blondeau J, Caccavelli L, Charrier S, Buckland K, et al. Outcomes following gene therapy in patients with severe Wiskott-Aldrich syndrome. *JAMA* [Internet]. 2015 Apr 21 [cited 2021 Aug 2];313(15):1550–63. Available from: https://jamanetwork.com/journals/jama/fullarticle/2275447
49. Ferrua F, Cicalese MP, Galimberti S, Giannelli S, Dionisio F, Barzaghi F, et al. Lentiviral haemopoietic stem/progenitor cell gene therapy for treatment of Wiskott-Aldrich syndrome: interim results of a non-randomised, open-label, phase 1/2 clinical study. *Lancet Haematol* 2019 May 1;6(5):e239–53.
50. Candotti F. Gene therapy for Wiskott-Aldrich syndrome: Here to stay. *Lancet Haematol* [Internet]. 2019 May 1 [cited 2021 Aug 2];6(5):e230–1. Available from: http://www.thelancet.com/article/S2352302619300663/fulltext
51. Holland SM. Chronic granulomatous disease. *Clin Rev Allergy Immunol* 2009 381 [Internet]. 2009 Jun 9 [cited 2021 Aug 2];38(1):3–10. Available from: https://link.springer.com/article/10.1007/s12016-009-8136-z
52. Liese JG, Jendrossek V, Jansson A, Petropoulou T, Kloos S, Gahr M, et al. Chronic granulomatous disease in adults. *Lancet*. 1996 Jan 27;347(8996):220–3.
53. Jones LBKR, McGrogan P, Flood TJ, Gennery AR, Morton L, Thrasher A, et al. Special article: chronic granulomatous disease in the United Kingdom and Ireland: a comprehensive national patient-based registry. *Clin Exp Immunol* [Internet]. 2008 [cited 2021 Aug 2];152(2):211. Available from: /pmc/articles/PMC2384093/
54. Chiesa R, Wang J, Blok H-J, Neven B, Moshous D, Schulz A, et al. Allogeneic hematopoietic stem cell transplantation in children and adults with chronic granulomatous disease (CGD): a study of the inborn errors working party (IEWP) of the EBMT. *Blood*. 2018 Nov 29;132(Supplement 1):970–970.
55. Malech HL, Maples PB, Whiting-Theobald N, Linton GF, Sekhsaria S, Vowells SJ, et al. Prolonged production of NADPH oxidase-corrected granulocytes after gene therapy of chronic granulomatous disease. *Proc Natl Acad Sci* [Internet]. 1997 Oct 28 [cited 2021 Aug 2];94(22):12133–8. Available from: https://www.pnas.org/content/94/22/12133
56. Santilli G, Almarza E, Brendel C, Choi U, Beilin C, Blundell MP, et al. Biochemical correction of X-CGD by a novel chimeric promoter regulating high levels of transgene expression in myeloid cells. *Mol Ther*. 2011 Jan 1;19(1):122–32.
57. Brendel C, Rothe M, Santilli G, Charrier S, Stein S, Kunkel H et al. Non-clinical efficacy and safety studies on G1XCGD, a lentiviral vector for ex vivo gene therapy of X-linked chronic granulomatous disease. *Hum Gene Ther Clin Dev*. https://home.liebertpub.com/humc [Internet]. 2018 Jun 1 [cited 2021 Aug 2];29(2):69–79. Available from: https://www.liebertpub.com/doi/abs/10.1089/humc.2017.245
58. Kohn DB, Booth C, Kang EM, Pai S-Y, Shaw KL, Santilli G, et al. Lentiviral gene therapy for X-linked chronic granulomatous disease. *Nat Med* 2020 262 [Internet]. 2020 Jan 27 [cited 2021 Aug 2];26(2):200–6. Available from: https://www.nature.com/articles/s41591-019-0735-5
59. Thrasher AJ, and Williams DA. Evolving gene therapy in primary immunodeficiency. *Mol Ther* [Internet]. 2017 May 3 [cited 2021 Aug 2];25(5):1132. Available from: /pmc/articles/PMC5417846/

60. Horikoshi Y, Umeda K, Imai K, Yabe H, Sasahara Y, Watanabe K, et al. Allogeneic hematopoietic stem cell transplantation for leukocyte adhesion deficiency. *J Pediatr Hematol Oncol* [Internet]. 2018 [cited 2021 Aug 2];40(2):137–40. Available from: https://journals.lww.com/jpho-online/Fulltext/2018/03000/Allogeneic_Hematopoietic_Stem_Cell_Transplantation.10.aspx
61. A clinical trial to evaluate the safety and efficacy of RP-L201 in subjects with leukocyte adhesion deficiency-I – Full Text View – *ClinicalTrials.gov* [Internet]. [cited 2021 Aug 4]. Available from: https://clinicaltrials.gov/ct2/show/NCT03812263
62. A gene therapy trial to evaluate the safety and efficacy of RP-L201 in subjects with leukocyte adhesion deficiency-I – Full Text View – *ClinicalTrials.gov* [Internet]. [cited 2021 Aug 4]. Available from: https://clinicaltrials.gov/ct2/show/NCT03825783
63. Parikh SA, Kapoor P, Letendre L, Kumar S, and Wolanskyj AP. Prognostic factors and outcomes of adults with hemophagocytic lymphohistiocytosis. *Mayo Clin Proc*. 2014 Apr 1;89(4):484–92.
64. Schram AM, and Berliner N. How I treat hemophagocytic lymphohistiocytosis in the adult patient. *Blood* [Internet]. 2015 May 7 [cited 2021 Aug 2];125(19):2908–14. Available from: http://ashpublications.org/blood/article-pdf/125/19/2908/1385109/2908.pdf
65. Henter J-I, Horne A, Aricó M, Egeler RM, Filipovich AH, Imashuku S, et al. HLH-2004: Diagnostic and therapeutic guidelines for hemophagocytic lymphohistiocytosis. *Pediatr Blood Cancer* [Internet]. 2007 Feb 1 [cited 2021 Aug 2];48(2):124–31. Available from: https://onlinelibrary.wiley.com/doi/full/10.1002/pbc.21039
66. Bergsten E, Horne A, Aricó M, Astigarraga I, Egeler RM, Filipovich AH, et al. Confirmed efficacy of etoposide and dexamethasone in HLH treatment: long-term results of the cooperative HLH-2004 study. *Blood* [Internet]. 2017 Dec 21 [cited 2021 Aug 2];130(25):2728–38. Available from: http://ashpublications.org/blood/article-pdf/130/25/2728/1404373/blood788349.pdf
67. Locatelli F, Jordan MB, Allen C, Cesaro S, Rizzari C, Rao A, et al. Emapalumab in Children with Primary Hemophagocytic Lymphohistiocytosis. https://doi.org/101056/NEJMoa1911326 [Internet]. 2020 May 6 [cited 2021 Aug 2];382(19):1811–22. Available from: https://www.nejm.org/doi/full/10.1056/NEJMoa1911326
68. Xu XJ, Tang YM, Song H, Yang SL, Xu WQ, Zhao N, et al. Diagnostic accuracy of a specific cytokine pattern in hemophagocytic lymphohistiocytosis in children. *J Pediatr*. 2012 Jun 1;160(6):984–990.e1.
69. Marsh RA, Allen CE, McClain KL, Weinstein JL, Kanter J, Skiles J, et al. Salvage therapy of refractory hemophagocytic lymphohistiocytosis with alemtuzumab. *Pediatr Blood Cancer* [Internet]. 2013 Jan 1 [cited 2021 Aug 2];60(1):101–9. Available from: https://onlinelibrary.wiley.com/doi/full/10.1002/pbc.24188
70. Marsh RA, and Haddad E. How i treat primary haemophagocytic lymphohistiocytosis. *Br J Haematol* [Internet]. 2018 Jul 1 [cited 2021 Aug 2];182(2):185–99. Available from: https://onlinelibrary.wiley.com/doi/full/10.1111/bjh.15274
71. Albeituni S, Verbist KC, Tedrick PE, Tillman H, Picarsic J, Bassett R, et al. Mechanisms of action of ruxolitinib in murine models of hemophagocytic lymphohistiocytosis. Blood [Internet]. 2019 Jul 11 [cited 2021 Aug 2];134(2):147–59. Available from: http://ashpublications.org/blood/article-pdf/134/2/147/1553926/bloodbld2019000761.pdf
72. Das R, Guan P, Sprague L, Verbist K, Tedrick P, An QA, et al. Janus kinase inhibition lessens inflammation and ameliorates disease in murine models of hemophagocytic lymphohistiocytosis. *Blood* [Internet]. 2016 Mar 31 [cited 2021 Aug 2];127(13):1666–75. Available from: http://ashpublications.org/blood/article-pdf/127/13/1666/1392477/1666.pdf
73. Meyer LK, Verbist KC, Albeituni S, Scull BP, Bassett RC, Stroh AN, et al. JAK/STAT pathway inhibition sensitizes CD8 T cells to dexamethasone-induced apoptosis in hyperinflammation. *Blood* [Internet]. 2020 Aug 6 [cited 2021 Aug 2];136(6):657–68. Available from: http://ashpublications.org/blood/article-pdf/136/6/657/1750748/bloodbld2020006075.pdf
74. Moshous D, Briand C, Castelle M, Dupic L, Morelle G, Abou Chahla W, et al. Alemtuzumab as first line treatment in children with familial lymphohistiocytosis. *Blood*. 2019 Nov 13;134(Supplement 1):80.
75. Carmo M, Risma KA, Arumugam P, Tiwari S, Hontz AE, Montiel-Equihua CA, et al. Perforin gene transfer into hematopoietic stem cells improves immune dysregulation in murine models of perforin deficiency. *Mol Ther*. 2015 Apr 1;23(4):737–45.

76. Dettmer V, Bloom K, Gross M, Weissert K, Aichele P, Ehl S, et al. Retroviral UNC13D gene transfer restores cytotoxic activity of T cells derived from familial hemophagocytic lymphohistiocytosis type 3 patients in vitro. *Hum Gene Ther*. https://home.liebertpub.com/hum [Internet]. 2019 Aug 16 [cited 2021 Aug 2];30(8):975–84. Available from: https://www.liebertpub.com/doi/abs/10.1089/hum.2019.025
77. Soheili T, Durand A, Sepulveda FE, Rivière J, Lagresle-Peyrou C, Sadek H, et al. Gene transfer into hematopoietic stem cells reduces HLH manifestations in a murine model of Munc13-4 deficiency. *Blood Adv* [Internet]. 2017 Dec 26 [cited 2021 Aug 2];1(27):2781. Available from: /pmc/articles/PMC5745141/
78. Soheili T, Rivière J, Ricciardelli I, Durand A, Verhoeyen E, Derrien A-C, et al. Gene-corrected human Munc13-4–deficient CD8+ T cells can efficiently restrict EBV-driven lymphoproliferation in immunodeficient mice. *Blood* [Internet]. 2016 Dec 15 [cited 2021 Aug 2];128(24):2859–62. Available from: http://ashpublications.org/blood/article-pdf/128/24/2859/1358019/blood729871.pdf
79. Takushi SE, Paik NY, Fedanov A, Prince C, Doering CB, Spencer HT, et al. Lentiviral gene therapy for familial hemophagocytic lymphohistiocytosis type 3, caused by UNC13D genetic defects. *Hum Gene Ther*. https://home.liebertpub.com/hum [Internet]. 2020 Jun 12 [cited 2021 Aug 2];31(11–12):626–38. Available from: https://www.liebertpub.com/doi/abs/10.1089/hum.2019.329
80. Parolini S, Bottino C, Falco M, Augugliaro R, Giliani S, Franceschini R, et al. X-linked lymphoproliferative disease. 2b4 molecules displaying inhibitory rather than activating function are responsible for the inability of natural killer cells to kill epstein-barr virus–infected cells. *J Exp Med* [Internet]. 2000 Aug 7 [cited 2021 Aug 2];192(3):337–46. Available from: http://www.jem.org/cgi/content/full/192/3/337
81. Booth C, Gilmour KC, Veys P, Gennery AR, Slatter MA, Chapel H, et al. X-linked lymphoproliferative disease due to SAP/SH2D1A deficiency: a multicenter study on the manifestations, management and outcome of the disease. *Blood* [Internet]. 2011 Jan 6 [cited 2021 Aug 2];117(1):53–62. Available from: http://ashpublications.org/blood/article-pdf/117/1/53/1334971/zh800111000053.pdf
82. Panchal N, Houghton B, Diez B, Ghosh S, Ricciardelli I, Thrasher AJ, et al. Transfer of gene-corrected T cells corrects humoral and cytotoxic defects in patients with X-linked lymphoproliferative disease. *J Allergy Clin Immunol*. 2018 Jul 1;142(1):235–245.e6.
83. Barzaghi F, Amaya Hernandez LC, Neven B, Ricci S, Kucuk ZY, Bleesing JJ, et al. Long-term follow-up of IPEX syndrome patients after different therapeutic strategies: An international multicenter retrospective study. *J Allergy Clin Immunol*. 2018 Mar 1;141(3):1036–1049.e5.
84. Passerini L, Mel ER, Sartirana C, Fousteri G, Bondanza A, Naldini L, et al. CD4+ T cells from IPEX patients convert into functional and stable regulatory T cells by FOXP3 gene transfer. *Sci Transl Med* [Internet]. 2013 Dec 11 [cited 2021 Aug 2];5(215):215ra174–215ra174. Available from: https://stm.sciencemag.org/content/5/215/215ra174
85. Masiuk KE, Laborada J, Roncarolo MG, Hollis RP, and Kohn DB. Lentiviral gene therapy in HSCs restores lineage-specific FOXP3 expression and suppresses autoimmunity in a mouse model of IPEX syndrome. *Cell Stem Cell*. 2019 Feb 7;24(2):309–317.e7.
86. Passerini L, andBacchetta R. Forkhead-box-P3 gene transfer in human CD4+ T conventional cells for the generation of stable and efficient regulatory T cells, suitable for immune modulatory therapy. *Front Immunol* [Internet]. 2017 Oct 12 [cited 2021 Aug 2];8(OCT). Available from: /pmc/articles/PMC5643480/
87. Taylor G. Sexually transmitted infections: Rolling out HIV antiretroviral therapy in sub-Saharan Africa: 2003–2017. *Canada Commun Dis Rep* [Internet]. 2018 Feb 1 [cited 2021 Aug 4];44(2):68. Available from: /pmc/articles/PMC5864409/
88. Sheykhhasan M, Foroutan A, Manoochehri H, Khoei SG, Poondla N, and Saidijam M. Could gene therapy cure HIV? *Life Sci* [Internet]. 2021;277(September 2020):119451. Available from: https://doi.org/10.1016/j.lfs.2021.119451
89. Sutton L, Guénel P, Tanguy M-L, Rio B, Dhedin N, Casassus P, et al. Acute myeloid leukaemia in human immunodeficiency virus-infected adults: epidemiology, treatment feasibility and outcome. *Br J Haematol* [Internet]. 2001 Mar 1 [cited 2021 Aug 4];112(4):900–8. Available from: https://onlinelibrary.wiley.com/doi/full/10.1046/j.1365-2141.2001.02661.x
90. Robbins HA, Shiels MS, Pfeiffer RM, and Engels EA. Epidemiologic contributions to recent cancer trends among HIV-infected people in the United States. *AIDS* [Internet]. 2014 Mar 27 [cited 2021 Aug 4];28(6):881–90. Available from: https://journals.lww.com/aidsonline/Fulltext/2014/03270/Epidemiologic_contributions_to_recent_cancer.10.aspx

91. Hütter G, Nowak D, Mossner M, Ganepola S, Müßig A, Allers K, et al. Long-term control of HIV by CCR5 Delta32/Delta32 stem-cell transplantation. *N Engl J Med.* http://dx.doi.org/101056/NEJMoa0802905 [Internet]. 2009 Dec 10 [cited 2021 Aug 4];360(7):692–8. Available from: https://www.nejm.org/doi/10.1056/NEJMoa0802905
92. Gupta RK, Abdul-Jawad S, McCoy LE, Mok HP, Peppa D, Salgado M, et al. HIV-1 remission following CCR5Δ32/Δ32 haematopoietic stem-cell transplantation. *Nat* 2019 5687751 [Internet]. 2019 Mar 5 [cited 2021 Aug 4];568(7751):244–8. Available from: https://www.nature.com/articles/s41586-019-1027-4
93. Cornu TI, Mussolino C, Müller MC, Wehr C, Kern WV, and Cathomen T. HIV gene therapy: An update. *Hum Gene Ther*. 2021;32(1–2):52–65.

15 Regulatory Challenges for Gene Delivery

Vineet Mahajan
University of Pittsburgh, Pittsburgh, United States

Shruti Saptarshi
University of South Florida, Tampa, United States

Yashwant Pathak
University of South Florida, Tampa, United States
University of Airlangga, Surabaya, Indonesia

CONTENTS

DOI: 10.1201/9781003186083-15

15.1 INTRODUCTION

15.1.1 Gene Delivery

As genome editing quickly advances toward its clinical translation,[1] evaluating the efficiency of current instruments and forms utilized for its benefit-risk evaluation is critical.[2] Despite the fact that current directions may at first give a satisfactory administrative regulatory system, enhancements are suggested and are required to overcome a few existing technology-based security concerns.[2] For the last three decades, gene therapy has been utilized to rectify inborn natural defects, and several gene therapy treatments have received European marketing authorization.[2, 3] Numerous, effective, quality therapy applications have centered on the transfer of restorative qualities, utilizing both integrating and non-integrating viral vector frameworks.[4] The exchanged genes usually complement or supplant the function of the dysfunctional or dysregulated genes, thereby correcting the phenotype of the targeted cells.[1]

15.1.1.1 Somatic Gene Editing

Alternatively, other types of gene therapy focus on targeted genome editing by utilizing designer nucleases.[5] In this setting, a flawed gene is repaired in situ to rectify loss-of-function mutations underlying acquired disorders, or on the other hand, functional genes are edited to bestow resistance to a pathogen.[1, 5] A major disadvantage of genome editing has been relatively lower transduction efficiency in human cells, but advances within the recent years have essentially enhanced gene-editing efficiencies and empowered a quick translation of this innovation to the clinic.[1]

15.1.1.2 Genome Editing Technologies: Viral Vectors and Nonviral Vectors

Viruses and liposomes: Although numerous delivery systems are accessible, retrovirus and lentivirus mediated gene transfer allows integration and long-term gene expression, which may, in turn, advance a solid restorative therapeutic impact.[6, 7, 8, 9, 10] The primary results utilizing retroviral-mediated quality exchange to treat patients with progressed melanoma were reported in 1990.[10] Since then, persistent advancements have been made within the field of quality treatment, extending the strategies and security of gene editing systems (retroviral vectors, adenoviral vectors, and lentiviral vectors), target cells (stem cell, dendritic cell, T cell, etc.).[7] For the eight cell therapy products under investigation, five utilized a third-generation, self-inactivating lentiviral vector framework; one utilized a two-plasmid lentiviral framework (HIV LTR); and two utilized second-generation retroviral vector frameworks for manufacturing.[7] Herpes simplex virus (HSV) based vectors have also been attractive choices, due to their huge capacity to accommodate foreign DNA and their episomal presence within the latent phase in the host organism.[11, 12]

Non-viral methods include engineered strategies that utilize cationic lipids, polymers, or proteins that form complexes with DNA, hence condensing it into particles 100 to 300 nm in size. Cationic lipids can be utilized in clinical trials, but due to their toxicity for intravenous infusions, they have been suggested for intraperitoneal transfer of genes, such as in intraperitoneally metastatic ovarian cancer. Examples of cationic lipids include DDAB, DODAP, DOGS, DOTAP, DC-Chol in 1:1 molar ratios with the zwitterionic and fusogenic DOPE [Abbreviations: DDAB: dimethyldioctadecyl ammonium bromide (same asN,N-distearyl-N,N-dimethylammonium bromide); DODAP:1,2-dioleoyl-3-dimethylammonium propane; DOGS: Dio-ctadecylamidoglycylspermine; DOTAP (same as DOTMA): N-(1-(2,3-dioleoyloxy)propyl)-N,N,N-trimethylammoniumchloride; DC-Chol, 3β-(N-(N′,N′-dimethylaminoethane carbamoyl) cholesterol. DOPE, 1,2-sn-dioleoylphoshatidylethanolamine.] [11, 13, 14] Not only cationic liposomes, but also neutral/zwitterionic liposomes have been investigated for delivery of DNA to cells.[15]

Polymers: Dendrimers are nanometer-sized, highly branched, and monodisperse polymers, with symmetrical architecture that can be used for clinical trial. Nanoparticles have been used in a variety of ways to deliver drugs and genes into cells.[11, 16]

Peptide-DNA Complexes: Basic amphiphilic peptides have been designed as components to bind to nucleic acids, improving plasmid delivery to cells.[11, 17]

Other systems: Endovascular microcoils as interventional methods to treat cerebral aneurysms have been effectively utilized as a gene delivery system.[11, 18]

15.1.2 Clinical Trials in Gene Therapy

15.1.2.1 Clinical Trials in Liver

The liver performs a host of complex functions, such as (a) bile generation and excretion, and (b) excretion of bilirubin, cholesterol, hormones, and drugs. (c) digestion of fats, proteins, and carbohydrates, (d) capacity of glycogen, vitamins, and minerals, (e) blood detoxification and filtration, (f) union of plasma proteins, and clotting components, thus is a high value therapeutic target.[19, 20] Within the liver, pioneering trials in adults with hemophilia B, and more as of late with hemophilia A, have demonstrated a sustained clinical advantage after systemic delivery of adeno-associated infection (AAV) vectors encoding human factor IX or VIII.[21–23] Currently, there are over of 250[20], [24] approved AAV-based trials, with approximately 15% of these trials targeting the liver.[25–27]

15.1.2.2 Clinical Trials in Rare Diseases

One of the most successful stories in rare ocular diseases is the evolution of gene therapy to treat retinal biallelic dystrophy associated with the RPE65 mutation.[28] The RPE65 gene codes for the RPE65 protein, involved in vitamin A metabolism associated with the conversion of retinol in the concurrent vision cycle.[28] Mutations in this gene lead to functional loss or even death of retinal pigmented epithelial cells that causes early vision loss, resulting in blindness.[28] In total, approximately 1,000–2,000 people in the US are affected by hereditary retinal dystrophy.[28] Rare disease trials come with their special set of ethical challenges – restricted information of disease pathogenesis, few or non-validated biomarkers or clinical endpoints, and few patients with tall topographical scattering, with a limited number of experienced clinical examiners.[28] Assessing a quality treatment complicates trial operations, as trial supports must adapt with a long possibility handle to ensure potential trial destinations and expertise to run a quality treatment trial.[28, 29]

15.1.2.3 Clinical Trials in Cancer

Since 2016, 18 clinical trials have been propelled and eight of them are immunotherapies for modified cell death-1 (PD-1) protein. In 2016, Chinese researchers conducted the first human phase I clinical trial based on CRISPR/Cas9, to knock out PD-1 for metastatic Non-small cell Lung Cancer (NCT02793856) treatment. Besides lung cancer, the clinical trials of CRISPR/Cas9 mediated PD-1 knock-out have also been utilized for other types of cancers, including esophageal cancer (NCT0308171), prostate cancer (NCT03525652), EBV (Epstein-Barr virus) positive advanced malignant tumors (NCT03044743), and liver cancer (NCT04417764).

15.1.3 Ethical Considerations for Gene Therapy

After decades of setback, gene therapy has encountered major success.[30] Five gene therapies have received FDA approval since 2017,[30, 31] and over 900 others are in development as of now.[30, 31] Many of these GT targets are extremely life-limiting conditions, such as pediatric diseases, given a lack of effective treatments.[32] As these gene therapies enter early-phase clinical trials, particular ethical challenges stay uncertain in three domains: assessing risks and potential benefits, selecting members reasonably, and locking in with quiet communities.[32] The ethical applications of genome altering hinge upon a risk/benefit proportion that guarantees an advantageous outcome for the patient. Cutting edge genome altering tools might ease a few security concerns due to the targeted nature of the technology, but others continue, for example, the potential off-target effects, where genome editing follows at unintended sites, with possibly destructive results for physiological gene function. These ethical challenges summarized in Table 15.1 are presented from a bioethics point of view and offer points to consider for future gene therapy trials (taken as it is from A.A. Iyer et al.)[30]

15.1.3.1 Regulations for Somatic Gene Editing in United States

There have been four gene therapy products approved by the United States Food and Drug Administration (USFDA) so far, that insert new genetic material into a patient's cells.[33] Approximately 900 investigational new drug (IND) applications for ongoing clinical studies in this field suggest that the agency expects to approve hundreds more in the coming years.[33] According to FDA, this will result in more therapeutic choices for patients and providers. The six final guidance documents were issued by the FDA to provide recommendations on manufacturing issues for product developers and other recommendations for genetic therapies to address specific diseases. By including input from a variety of stakeholders, the six guidance papers have taken a crucial step towards defining the modern structure associated with the production of gene therapies.[33] With these documents, the FDA hopes to advance the field of gene therapy while providing recommendations on how to make sure these innovative products meet FDA safety and effectiveness standards. During scientific reviews of gene therapies, it is essential to evaluate a great deal of information about manufacturing and quality. Clinical trials of these products pose more challenging questions to regulators, such as questions about the durability of therapy response, and they are hard to answer in pre-market tests of reasonable size and duration. Despite some gene therapy products meeting FDA approval standards, some level of uncertainty may exist around the duration of the response when a marketing authorization is approved.[33]

15.1.3.2 Regulations for Somatic Gene Editing in United Kingdom

As genome editing approaches vary considerably depending on tissue origin and how the therapy will be used, regulators are challenged to address these concerns. In the regulatory context, somatic genome editing is defined as an Advanced Therapy Medicinal Product (ATMP), which refers to medicines with genes, tissues, or cells. Since the types and amounts of data required to demonstrate the quality, safety, and efficacy of diverse ATMPs are highly specific, the technical requirements set out in this legislation are high level. In class-specific guidance developed by the European Medicines Agency (EMA), these issues are further explored. It is required that all medicines be approved by the regulatory authorities before they can be distributed to patients. While most medicines are approved at a national level, ATMPs need to go through a centralized process overseen by the EMA in order to ensure that their products are available across all member states through a single license, allowing patients faster access to these treatments. It needs assessment by European Commission authorized EMA's scientific committees.

15.1.3.3 Current Regulatory Framework: Is It Fit for Purpose?

Since the early 1990s, the EU has had a legal and regulatory framework for ATMPs. Fast-evolving innovative therapies such as genome editing, which target highly personalized medical problems, are challenging established guidelines and regulatory framework. ATMPs don't currently face any revisions to their legislation; however, the necessity for adapting the policy has become increasingly important, as recognized by the EMA and EC, who released a joint action plan for fostering future ATMPs in December 2017. In addition to convening an expert group to discuss critical regulatory issues and uncertainties associated with genome editing products, the organization published new guidelines, updated others, and ensured that "the regulatory framework supports – instead of hinders – the development of high-quality genome editing products" development of ATMPs":

Guidelines on quality, nonclinical, and clinical aspects of medicines that consist of genetically modified cells: Genome engineering's unique starting materials and production processes that enable precise gene editing by controlling and identifying modified cells, have been provided.

TABLE 15.1
Determining Whether to Initiate Early-Phase Gene Therapy (GT) Trials

Step	Guiding Question	Assessment
1. Form preliminary expectation for GT's risk–benefit profile	What is the study intervention's expected risk–benefit profile for participants?	Determine the type, magnitude, and likelihood of harms and benefits observed in preclinical studies
2. Gauge uncertainty in the GT's risk–benefit profile by evaluating preclinical evidence quality	Do preclinical studies support a causal relationship between the GT and observed outcomes (internal validity)?	• Evaluate the balance of benefits and harms • Check whether design features support a causal relationship (e.g., controls, randomization, blinding)
	Do preclinical studies accurately represent the clinical conditions in which the GT will be tested (construct validity)?	• Look for similarity between preclinical models and human patients (e.g., clinical presentation, age- and sex-matching for animal models)
	How generalizable are preclinical results (external validity)?	• Check whether preclinical GT was delivered in an analogous way to the intended delivery in humans (e.g., same volume-adjusted dose between animals and humans)
	Overall, what is the uncertainty in the risk–benefit profile?	• Determine whether preclinical safety and efficacy findings have been validated in multiple models and/ or species
3. Refine judgment of the GT's expected risk–benefit profile by considering evidence on similar GTs	Can data on similar interventions refine or complement the above estimates about harms, benefits, and uncertainty?	• Judge overall preclinical study quality (internal, construct, and external validity): the higher the quality, the lower the uncertainty
4. Evaluate whether contextual factors influence the ethical acceptability of initiating early-phase trials	Do contextual factors suggest that it is more or less acceptable to begin early-phase trials?	• Consider whether similar interventions (e.g., similar vectors and/or delivery routes) have demonstrated safety and efficacy: the more therapeutic and better- characterized any similar interventions, the lower the uncertainty surrounding a favorable risk–benefit judgment
5. Judge whether initiating early-phase trials is ethically justified	Is it ethically acceptable to begin early-phase trials in the target population?	• Evaluate relevant contextual factors: the more serious and rapidly progressive the disease, the more limited the alternatives, and the higher the support by the patient community, the more acceptable early-phase trial initiation may be • Make all-things-considered judgment, given the GT's expected risk–benefit profile, its associated uncertainty, and relevant contextual factors

TABLE 15.2
A Regulatory Perspective on Feasibility to Surmount Genome-Editing Safety and Efficacy Challenges

Challenges	Regulatory Feasibility to Overcome	Approaches to Address Challenges
Off-target activity, resulting in insertion or deletion mutations and/or chromosomal translocations	moderate	• assays to predict and identify off-target activity and/or translocations in place • biological assays to evaluate functional consequences of off-target activity still in development
Necessity to maximize efficiency of designer nuclease delivery and to control nuclease expression level and duration	moderate to high	• *in vivo* CRISPR-Cas delivery (mRNA, protein) via lipid nanoparticles may help to fine-tune level and duration of nuclease expression • *ex vivo* delivery of nuclease encoding mRNA by electroporation allows fine-tuning level and duration of nuclease expression • *ex vivo* delivery of nucleases in the form of DNA can be inefficient and induce high cytotoxicity
Inaccurate or random donor DNA (AAV or IDLVs, oligodeoxynucleotide donors) integration in the genome	moderate to high	• assays to detect random integration of AAV and IDLVs in place • randomly integrated oligodeoxynucleotides are difficult to detect
Highly variable tissue distribution of desired *in vivo* genome-editing event	moderate	• collection and assessment of a diverse panel of all major organs and tissues
Potential of immune reaction to nuclease components of current gene-editing systems	moderate	• use of immune suppression may be required

Guideline for investigational ATMPs in terms of quality, non-clinical requirements, and clinical requirements: Trials in clinical settings: National governments approve clinical trials, but the EMA oversees marketing approvals. The distinct development pathway is meeting approval requirements and highlights the distinctive characteristic that ATMPs require extensive non-clinical data before human testing is permitted.

Guidelines for following up on and managing risks associated with ATMP safety and efficacy: The safety and efficacy of ATMPs may not be fully understood at the time of their approval, as they are innovative products. To detect and mitigate against possible risks, this guideline places an increased emphasis on post-authorization oversight of ATMPs, sometimes requiring additional safety and efficacy studies.

Finally, Table 15.2 sourced as it is from Abou-El-Enein M. et al 2017 describes numerous technological issues related to safety and efficacy that are examined and may inform regulatory benefit-risk assessments that would include genome-editing in the clinic.[2]

15.2 THE LEGAL AND REGULATORY LANDSCAPE IN GENE THERAPY

The area of gene therapy holds great potential in treating a variety of diseases. The technology is robust, cost effective, and precise, as evidenced by more than 1,800 currently active gene and cell therapy interventional trials worldwide.[34] On account of constant advancement in the technology, and the inherent complex nature of gene therapy products, the regulatory frameworks for these vary geographically. In fact, due to minimal availability of medical/research infrastructures, most regulatory authorities repurpose policies designed for medicinal products to accommodate gene therapy products. The US, EU, China, and Japan are examples of countries that have made significant advances in designing their regulatory legislature specifically for gene therapy products. Other countries include Korea, Brazil, and Canada, working on similar lines.[34] The following paragraphs give a brief synopsis of the European and American regulatory systems set up for gene therapy products.

In the EU gene therapies are classified as, advanced therapy medicinal product (ATMPs), under the EU legislation, and are governed by the ATMP Regulation (Directive 2001/83/EC, as amended by Regulation [EC] 1394/2007)[35] The European Medical Association's (EMA) Committee for Advanced Therapies (CAT) oversees the regulatory affairs of ATMPs after consultation with the European Commission. The principal EU guideline (under revision) is the basis for development of gene therapies. It provides a thorough strategic plan for gene therapy product development, covering quality, non-clinical, and clinical aspects of the ATMPs. Unlike the regulations for other biotech-based medicinal products, the technical requirements of quality or clinical data required to demonstrate safety of the ATMP product may be highly specific. These guidelines are targeted at mitigating risks to the beneficiaries of the ATMP treatments, once the products are authorized for use. Following points provide details of the EU regulatory guidelines for ATMPs:

15.2.1 EU Guidelines on Quality, Non-Clinical, and Clinical Aspects of Medicinal Products Containing Genetically Modified Cells

Although still in the drafting stage, the main aim of this guideline is to define scientific principles to guide the development and manufacturing of medicinal products containing genetically modified cells intended for humans. Stringent quality requirements for genetic modification of the target cell population (controls, characterization, stability, etc.) and the transduced cell product resulting from the manufacturing process have been discussed keeping in mind a "risk-based approach" when designing the interventions. The non-clinical section states guidelines for designing primary pharmacodynamic studies that demonstrate proof-of-concept, rationale for species and animal models, target selectivity, and biodistribution immunotoxicity of the product, including the pointers on environmental risk assessment. The clinical section defines requirements for pre-authorisation studies required for assessing the safety and efficacy of the gene product, and further includes a section on addressing safety evaluation and pharmacovigilance requirements.

15.2.2 EU (Draft) Guideline on Quality, Non-Clinical, and Clinical Requirements for Investigational Gene Therapy in Clinical Trials

This guideline provides strategies for designing clinical trials for investigational gene therapy products. There is emphasis on taking a risk-based approach while designing the protocols. The dossier defines specific guidelines on the quality of the starting materials, manufacturing process, and the product developed for intended use. The non-clinical guidelines focus on animal models, toxicity studies, minimum supplementary requirements for (first-in human studies) also termed as exploratory studies: a unique feature of the EU regulatory framework and confirmatory studies.

15.2.3 EU (Draft) Guideline on Safety and Efficacy Follow-Up and Risk Management of Gene Therapy Products

The EMA released a draft guideline on safety and efficacy follow-up and risk management of Advanced Therapy Medicinal Products in 2018. The safety and risk management guidelines are designed to identify patient risk pertaining to quality characteristics, storage and distribution of the product, including risks associated with immunogenicity, dosing errors, pharmacokinetics etc. There are measures for risk minimization efficacy and safety follow-up depending on the nature of the different ATMPs.

USFDA via the Center for Biologics Evaluation and Research (CBER) and the National Institute of Health (NIH) via the Recombinant DNA Advisory Committee (RAC) are the apex bodies in the US that oversee regulatory affairs of gene therapy clinical trials and products.[36, 37] Gene therapy products intended for human use are classified and regulated as biological products[38]. The FDA regulatory system is comprised of several statutes, and guidance documents that companies have to comply with before initiating clinical trials for investigational gene therapy products. An approved biologics license (BLA) ensures the market authorization of gene therapy products in the US. Similar to the EU guidelines, gene therapy product trials have to meet strict guidelines in terms of quality, manufacturing process, non-clinical, and clinical aspects of the interventions. Additionally, there is an obligation for gene therapy products to adhere to good manufacturing regulations, good laboratory practice regulations, good clinical practice regulations, and good tissue practice regulations. It is also mandatory to provide proof of concept data and safety profile of the drug in animal models as per the set guidelines[36]. The FDA also oversees clinical trials of these products, unlike the EMA.

The primary aim of any global regulatory framework is to ensure efficiency and safety of gene therapy products with harmonized market availability. Despite their differences, the evaluating regulatory agencies are working toward accelerating the process to facilitate speedy patient access to these therapies. As of 2021, eight gene therapies and a total of five gene therapies were granted market authorization by the EU and the United States respectively[39]. Furthermore, the FDA, EMA, and the Pharmaceuticals and Medical Devices Agency (PMDA) in Japan have provisions to expedite assessments of novel gene therapy products in pipeline. The Priority Medicines (PRIME) scheme is offered by the EU, SAKIGAKE designation by Japan, the breakthrough therapy designation (BTD) or regenerative medicine advanced therapy (RMAT) designation are offered by the FDA to support and accelerate the development and approval of gene therapies[38]. The pace of development of gene therapies has been faster compared to the necessary updates in the regulatory issues. It is imperative that new regulations customized to individual gene therapies be made available to achieve maximum benefit of the technology.

This 'ATMP Regulation' [i.e., Regulation (EC) No. 1394/2007], together with Directive 2009/120, provides definitions of ATMPs, as well as Marketing Authorization Application (MAA) requirements.[35] This regulation described four different product types: GTMPs; somatic cell therapy medicinal products (sCTMPs); tissue-engineered products (TEPs); and combined ATMPs.[35] Table 15.3 (sourced from Carvalho et al 2017) describes the investigational drugs and the outcome in the context of regulatory perspective.[35]

15.2.4 Gene Editing Regulations

Genome editing is a frontier technology that has revolutionized fields of medicine and plant/animal biotechnology. Human, animal, and plant genomes can now be specifically altered using editing tools such as TALENS, ZFNs CRISPER-Cas systems to develop new generation biopharmaceuticals and improved livestock or plant varieties.[40] Regulatory bodies worldwide have already set up regulations for dealing with genetically modified organisms (GMOs). Inclusion or exemption of gene-edited organisms as GMOs is unclear. As a result, there are several inconsistencies in setting up regulatory frameworks for such products. Gene editing can be used to manipulate somatic as well as germline DNA.

15.2.4.1 Human Gene Editing Regulations

Gene editing can be used to manipulate somatic as well as germline DNA. In humans, gene editing can facilitate treatment and prevention of genetic as well as rare diseases. Somatic manipulation of the human genome is the basis for most interventional gene therapies, and as discussed above, currently there are several such clinical trials in the regulatory pipeline. However, application of

TABLE 15.3
Major Objections, Issues, or Concerns Noted in the Assessment of GTMPs at the Level of Quality, and Nonclinical and Clinical Data

	Unsuccessful MAA				Successful MAA		
Commercial name	Advexin	CLG	Cerepro	Cerepro	Glybera	Imlygic	Strimvelis
Year of opinion	2008	2009	2007	2010	2012	2015	2016
Clinical Indication	Li-Fraumeni cancer	Squamous cell carcinoma of head and neck	High-grade glioma	High-grade glioma	Lipoprotein lipase deficiency	Unresectable melanoma	Severe combined immunodeficiency because of adenosine deaminase deficienc
Orphan designation	Orphan	Non-orphan	Orphan	Orphan	Orphan	Non-orphan	Orphan
MAA approval mechanism	Withdrew MAA	Withdrew MAA	Negative opinion	Negative opinion	Approved under exceptional circumstances	Standard approval	Standard approval
QUALITY							
Drug substance							
Manufacture	C	C	B	A	B	A	B
Characterization	C	C	A	A	B	A	A
Specification	C	C	B	B	B	A	B
Stability	F	C	A	A	A	A	A
Drug product							
Pharmaceutical development	A	A	A	A	A	A	A
Manufacture of the product	C	C	B	A	A	A	A
Product specification	C	C	C	A	B	A	A
Stability of the product	C	C	C	A	A	A	A
Adventitious agents	C	A	A	A	B	A	A
NONCLINICAL							
Pharmacology							
Primary pharmacodynamics	A	A	A	A	A	A	A
Secondary pharmacodynamics	C	C	A	A	A	A	A
Safety pharmacology programme	F	F	A	A	F	A	A
Pharmacodynamic drug interactions	A	A	A	A	A	A	A

(continued)

TABLE 15.3 (Continued)
Major Objections, Issues, or Concerns Noted in the Assessment of GTMPs at the Level of Quality, and Nonclinical and Clinical Data

	Unsuccessful MAA		Successful MAA				
Pharmacokinetics							
Biodistribution, persistence, clearance	C	C	A	A	A	A	A
Germline transmission	C	C	A	A	B	A	A
Shedding	F	F	F	F	F	A	A
Toxicology							
Single dose toxicity	A	A	A	A	A	A	B
Repeat dose toxicity with toxicokinetics	C	C	A	A	A	A	A
Genotoxicity	F	F	A	A	B	A	A
Carcinogenicity	F	F	A	A	B	A	A
Reproduction toxicity	F	F	A	A	B	A	A
Local tolerance	F	F	A	A	A	A	A
Other toxicity studies: immunogenicity/ toxicity	F	F	A	A	A	A	F
CLINICAL							
GCP	F	C	C	B	A	A	B
Clinical pharmacology							
Pharmacokinetics	C	C	C	B	B	B	A
Pharmacodynamics	C	C	C	C	B	A	A
Clinical efficacy							
Dose selection and schedule	C	C	A	A	B	A	A
Clinical efficacy data	C	C	C	C	B	B	B
Clinical safety							
Clinical safety data	C	C	C	C	B	B	B
Pharmacovigilance system	C	C	C	A	A	A	A
Risk management plan	C	C	C	C	B	B	B
Environmental risk assessment	C	C	A	A	A	A	A

Note: CLG (Contusugene Ladenovec Gendux).

CRISPER Cas techniques for human germline gene editing are not well conceived worldwide, due to obvious moral and ethical concerns. Regulatory laws for germline editing related to reproductive purposes have been classified as "restrictive," "legally prohibited," ambiguous," or "prohibited by guidelines "by most countries. The US National Academy of Science, engineering and Medicine (NASEM) report 2017, has implicit guidelines for human gene editing related issues.[40]

15.2.4.2 Animal Gene Editing Regulations

As compared to plants, the regulatory landscape for gene-edited animals is rather complicated. Despite adopting a flexible regulatory approach for gene-edited plant products (See below), the regulatory framework for animals is constricted. This is evidenced by the 2017 FDA indication that a mandatory, multigenerational, premarket new animal drug evaluation would be applicable to all "intentional genomic alterations" induced by site-directed nucleases such as CRISPR in food animal genomes irrespective of the novelty of the alteration or the existence of any hazards in the resulting product was proposed. Some have suggested that this stance to be counterproductive in terms of cost and time involved in development of bioengineered animals[41]. The AquAdvantage GE salmon is an example of the only genetically engineered animal to be ever approved as food, after having gone through the new animal drug provisions of Federal Food, Drug, and Cosmetic Act (FD&C Act) where the genetically engineered animal was treated as a drug candidate. The current regulatory asynchronism in the US has warranted a more uniform framework for regulating both plants and animals, wherein genetic alterations using gene editing technology was introduced during the breeding process.[41]

15.2.4.3 Plant Gene Editing Regulations

Plant genome editing has the potential to offer advantageous traits to global agriculture. The CRISPR-Cas system is the most preferred gene editing approach that enables precise genome manipulation to achieve improved plant growth, nutrition, disease control, or sustainability. Nevertheless, there are inconsistencies in the regulatory framework covering gene-edited crops world-wide, thus impacting the public acceptance and marketability of these new plant varieties. Extension of currently established regulations covering conventional GMOs to the products generated through gene editing is a matter of debate needing further elucidation.

Currently most countries have adopted one of the two major regulatory models for gene editing; the process- based approach or the product-oriented approach. Some countries, such as New Zealand and those of the European Union, regulate the end points of gene editing technology based on the process used. As a result of this the end product gets labeled as a "GM product," thus falling under the GMO legislation. This approach overlooks the key benefits of this technology and is ineffective in terms of cost and time. Argentina and Brazil are two countries with precedents where policy makers have clearly defined guidelines to manage gene-edited crops.

The Canadian regulatory framework, on the other hand, is an example of the product-based approach, where the "end product with novel traits," even though obtained by the application of gene editing, conventional breeding, or mutagenesis, is subjected to a standard regulatory scheme managed by the Canadian Food Inspection Agency (CFIA) and Health Canada[40]. A prominent example of this product based approach is the canola variety Falco™ that was declared to be similar to the unmodified canola cultivar despite being a product of gene editing technology by the CFIA in 2013. This product-oriented approach has been suggested to be central to innovations in agricultural biotechnology by some[42]. The United States and Australia have adopted a mixed approach for regulating gene edited plants[43]. Under this hybrid approach, the end products are reviewed on a case-to-case basis, where absence of foreign DNA (genetic material) in the gene-edited variety ensures no regulatory oversite normally applicable for GMOs. In the US the USDA's Animal and Plant Health Inspection Agency (APHIS) is responsible for regulating gene-edited crop varieties. If the novel plant is intended for human consumption, then the regulation for the same is overseen by the FDA. Under this legislation the USDA has, to date waived the GMO regulations for at least two gene-edited crop varieties[44]. The commercial future of gene-edited agricultural produce may be overshadowed by confusion because of the above-described inconsistencies thus demanding a streamlined global regulatory approach that can benefit commercialization as well as innovation.

15.3 CONCLUSION AND FUTURE TRENDS

It is necessary to assess highly complex information about the manufacturing and quality of gene therapies as part of the scientific review. Additionally, premarket studies of reasonable size and duration often fail to fully answer important questions that these products often pose to regulators compared to what traditional drug trials do, such as questions about the durability of response. The FDA and EMA have issued these guidelines that operate against a backdrop of tools, exemptions, and the accelerated pathways for increasing flexibility, facilitating the development of gene editing therapies. Even though the EU and US regulatory framework is evolving to provide some flexibility for these new treatments, it may not be enough. In addition to the exemptions that are currently in place, the use of these medicinal products at scale may prove insufficient to allow their use. There are several existing safety and efficacy issues that can be improved upon if existing regulations are to provide an adequate regulatory framework. Building an ethical and regulatory framework will be the biggest challenge achieving quality and safety standards, while using the products at scale and complying with relevant regulatory requirements societal boundaries deemed appropriate for ethical behavior.

REFERENCES

1. Cornu, T. I., Mussolino, C. and Cathomen, T. Refining strategies to translate genome editing to the clinic. *Nat Med* **23**, 415–423, doi:10.1038/nm.4313 (2017).
2. Abou-El-Enein, M. et al. Human genome editing in the clinic: New challenges in regulatory benefit-risk assessment. *Cell Stem Cell* **21**, 427–430, doi:10.1016/j.stem.2017.09.007 (2017).
3. Abou-El-Enein, M., Elsanhoury, A. and Reinke, P. Overcoming challenges facing advanced therapies in the EU market. *Cell Stem Cell* **19**, 293–297, doi:10.1016/j.stem.2016.08.012 (2016).
4. Naldini, L. Gene therapy returns to centre stage. *Nature* **526**, 351–360, doi:10.1038/nature15818 (2015).
5. Carroll, D. Genome engineering with targetable nucleases. *Annu Rev Biochem* **83**, 409–439, doi:10.1146/annurev-biochem-060713-035418 (2014).
6. Maude, S. L. et al. Chimeric antigen receptor T cells for sustained remissions in leukemia. *N Engl J Med* **371**, 1507–1517, doi:10.1056/NEJMoa1407222 (2014).
7. Marcucci, K. T. et al. Retroviral and lentiviral safety analysis of gene-modified T cell products and infused HIV and oncology patients. *Mol Ther* **26**, 269–279, doi:10.1016/j.ymthe.2017.10.012 (2018).
8. Muul, L. M. et al. Persistence and expression of the adenosine deaminase gene for 12 years and immune reaction to gene transfer components: Long-term results of the first clinical gene therapy trial. *Blood* **101**, 2563–2569, doi:10.1182/blood-2002-09-2800 (2003).
9. Porter, D. L. et al. Chimeric antigen receptor T cells persist and induce sustained remissions in relapsed refractory chronic lymphocytic leukemia. *Sci Transl Med* **7**, 303ra139, doi:10.1126/scitranslmed.aac5415 (2015).
10. Tebas, P. et al. Antiviral effects of autologous CD4 T cells genetically modified with a conditionally replicating lentiviral vector expressing long antisense to HIV. *Blood* **121**, 1524–1533, doi:10.1182/blood-2012-07-447250 (2013).
11. Lundstrom, K. and Boulikas, T. Viral and non-viral vectors in gene therapy: technology development and clinical trials. *Technol Cancer Res Treat* **2**, 471–486, doi:10.1177/153303460300200513 (2003).
12. Krisky, D. M. et al. Deletion of multiple immediate-early genes from herpes simplex virus reduces cytotoxicity and permits long-term gene expression in neurons. *Gene Ther* **5**, 1593–1603, doi:10.1038/sj.gt.3300766 (1998).
13. Zeisig, R., Ress, A., Fichtner, I. and Walther, W. Lipoplexes with alkylphospholipid as new helper lipid for efficient in vitro and in vivo gene transfer in tumor therapy. *Cancer Gene Ther* **10**, 302–311, doi:10.1038/sj.cgt.7700572 (2003).

14. Roder, G. et al. Novel cGMP liposomal vectors mediate efficient gene transfer. *Cancer Gene Ther* **10**, 312–317, doi:10.1038/sj.cgt.7700573 (2003).
15. Iseki, K., Sugawara, M., Saitoh, N. and Miyazaki, K. The transport mechanisms of organic cations and their zwitterionic derivatives across rat intestinal brush-border membrane. II. Comparison of the membrane potential effect on the uptake by membrane vesicles. *Biochim Biophys Acta* **1152**, 9–14, doi:10.1016/0005-2736(93)90225-o (1993).
16. Alexandridou, S., Kiparissides, C., Mange, F. and Foissy, A. Surface characterization of oil-containing polyterephthalamide microcapsules prepared by interfacial polymerization. *J Microencapsul* **18**, 767–781, doi:10.1080/02652040110055243 (2001).
17. Rittner, K. et al. New basic membrane-destabilizing peptides for plasmid-based gene delivery in vitro and in vivo. *Mol Ther* **5**, 104–114, doi:10.1006/mthe.2002.0523 (2002).
18. Abrahams, J. M. et al. Endovascular microcoil gene delivery using immobilized anti-adenovirus antibody for vector tethering. *Stroke* **33**, 1376–1382, doi:10.1161/01.str.0000014327.03964.c0 (2002).
19. Kuntz, E. *HEPATOLOGY : Textbook and atlas*. (SPRINGER, 2016).
20. Ginn, S. L., Christina, S. & Alexander, I. E. Genome editing in the human liver: Progress and translational considerations. *Prog Mol Biol Transl Sci* **182**, 257–288, doi:10.1016/bs.pmbts.2021.01.030 (2021).
21. Nathwani, A. C. et al. Long-term safety and efficacy of factor IX gene therapy in hemophilia B. *N Engl J Med* **371**, 1994–2004, doi:10.1056/NEJMoa1407309 (2014).
22. Rangarajan, S. et al. AAV5-Factor VIII Gene Transfer in Severe Hemophilia A. *N Engl J Med* **377**, 2519–2530, doi:10.1056/NEJMoa1708483 (2017).
23. Pasi, K. J. et al. Multiyear Follow-up of AAV5-hFVIII-SQ Gene Therapy for Hemophilia A. *N Engl J Med* **382**, 29–40, doi:10.1056/NEJMoa1908490 (2020).
24. *Gene Therapy Clinical Trials Worldwide Database*. (John Wiley and Sons Ltd, 2015).
25. Baruteau, J., Waddington, S. N., Alexander, I. E. and Gissen, P. Gene therapy for monogenic liver diseases: clinical successes, current challenges and future prospects. *J Inherit Metab Dis* **40**, 497–517, doi:10.1007/s10545-017-0053-3 (2017).
26. Batty, P. and Lillicrap, D. Advances and challenges for hemophilia gene therapy. *Hum Mol Genet* **28**, R95–R101, doi:10.1093/hmg/ddz157 (2019).
27. Cozmescu, A. C., Counsell, J. and Gissen, P. Gene therapies targeting the liver. *J Hepatol* **74**, 235–236, doi:10.1016/j.jhep.2020.08.003 (2021).
28. Medspace. Gene Therapies in rare diseases from R&D to regulatory approval {White Paper]
29. Yla-Herttuala, S. Bumps in the Road for Commercial Gene Therapy for Rare Diseases. *Mol Ther* **25**, 2225, doi:10.1016/j.ymthe.2017.09.012 (2017).
30. Iyer, A. A. et al. Ethical challenges for a new generation of early-phase pediatric gene therapy trials. *Genet Med*, doi:10.1038/s41436-021-01245-3 (2021).
31. Administration., U. F. a. D. (2020).
32. Kimmelman, J. and London, A. J. Predicting harms and benefits in translational trials: ethics, evidence, and uncertainty. *PLoS Med* **8**, e1001010, doi:10.1371/journal.pmed.1001010 (2011).
33. Food and Drug Administration, U. S. in *Guidances issued today provide regulatory clarity for product developers* (US FDA, https://www.fda.gov/news-events/press-announcements/fda-continues-strong-support-innovation-development-gene-therapy-products, 2020).
34. Drago, D., Foss-Campbell, B., Wonnacott, K., Barrett, D. and Ndu, A. Global regulatory progress in delivering on the promise of gene therapies for unmet medical needs. *Mol Ther – Methods Clin Dev* **21**, 524–529 (American Society of Gene & Cell Therapy, 2021).
35. Carvalho, M., Sepodes, B. and Martins, A. P. Regulatory and Scientific Advancements in Gene Therapy: State-of-the-Art of Clinical Applications and of the Supporting European Regulatory Framework. *Front Med* 4, 182 (Frontiers Media S.A., 2017).
36. Mendicino, M., Fan, Y., Griffin, D., Gunter, K. C. and Nichols, K. *Cytotherapy* **21**, 699–724 (Elsevier, 2019).
37. Iglesias-Lopez, C., Obach, M., Vallano, A. and Agustí, A. Comparison of regulatory pathways for the approval of advanced therapies in the European Union and the United States. *Cytotherapy* **23**, 261–274 (Elsevier, 2021).
38. Lapteva, L., Purohit-Sheth, T., Serabian, M. and Puri, R. K. Global regulatory progress in delivering on the promise of gene therapies for unmet medical needs. *Mol Ther – Methods Clin Dev* **19**, 387–397 (American Society of Gene & Cell Therapy, 2020).

39. Iglesias-López, C., Agustí, A., Obach, M. and Vallano, A. *Front Pharmacol* 10, 921 (Frontiers Media S.A., 2019).
40. Zhang, D. et al. Genome editing with the CRISPR-Cas system: an art, ethics and global regulatory perspective. *Plant Biotechnol J* 18, 1651–1669 (John Wiley and Sons Inc., 2020).
41. Van Eenennaam, A. L., Wells, K. D. and Murray, J. D. Proposed U.S. regulation of gene-edited food animals is not fit for purpose. *NPJ Sci Food* 3, 3 (Nature Publishing Group UK, 2019).
42. Turnbull, C., Lillemo, M. and Hvoslef-Eide, T. A. K. *Frontiers in Plant Science* 12, 630396 (Frontiers Media S.A., 2021).
43. Zhang, Y., Restall, J., Crisp, P., Godwin, I. and Liu, G. *In vitro cellular & developmental biology. Plant: journal of the Tissue Culture Association* 1–10 (Springer US, 2021).
44. Jones, H. D. *GM Crops & Food* 6, 223–232 (Taylor & Francis, 2015).

Index

Page numbers in *Italics* refers to figures and **bold** refers to tables.

C

D

E

F

G

H

M

N

Q

R

S

T

U